Organic Structure Determination Using 2-D NMR Spectroscopy

Organic Structure Determination Using 2-D NMR Spectroscopy

A Problem-Based Approach

Jeffrey H. Simpson

Department of Chemistry Instrumentation Facility
Massachusetts Institute of Technology
Cambridge, Massachusetts

AMSTERDAM • BOSTON • HEIDELBERG • LONDON • OXFORD • NEW YORK
PARIS • SAN DIEGO • SAN FRANCISCO • SINGAPORE • SYDNEY • TOKYO
Academic Press is an imprint of Elsevier

Academic Press is an imprint of Elsevier
30 Corporate Drive, Suite 400, Burlington, MA 01803, USA
525 B Street, Suite 1900, San Diego, California 92101-4495, USA
84 Theobald's Road, London WC1X 8RR, UK

(∞) **This book is printed on acid-free paper.**

Library of Congress Cataloging-in-Publication Data
Simpson, Jeffrey H.
   Organic structure determination using 2-D NMR spectroscopy / Jeffrey
     H. Simpson,
       p. cm.
     Includes bibliographical references and index.
     ISBN 978-0-12-088522-0 (pbk. : alk. paper) 1. Molecular structure. 2. Organic
compounds—Analysis. 3. Nuclear magnetic resonance spectroscopy. I. Title.
       QD461.S468 2008
       541'.22—dc22

                                                                    2008010004

British Library Cataloguing-in-Publication Data
A catalogue record for this book is available from the British Library

ISBN: 978-0-12-088522-0

> For information on all Academic Press publications
> visit our Web site at www.books.elsevier.com

Printed in Canada

09 10 11   9 8 7 6 5 4 3 2

Dedicated to

Alan Jones
mentor, friend, and tragic hero

Contents

Preface

I wrote this book because nothing like it existed when I began to learn about the application of nuclear magnetic resonance spectroscopy to the elucidation of organic molecular structure. This book started as 40 two-dimensional (2-D) nuclear magnetic resonance (NMR) spectroscopy problem sets, but with a little cajoling from my original editor (Jeremy Hayhurst), I agreed to include problem-solving methodology in Chapters 9 and 10, and after that concession was made, the commitment to generate the first 8 chapters was a relatively small one.

Two distinct features set this book apart from other books available on the practice of NMR spectroscopy as applied to organic structure determination. The first feature is that the material is presented with a level of detail great enough to allow the development of useful 'NMR intuition' skills, and yet is given at a level that can be understood by a junior-level chemistry major, or a more advanced organic chemist with a limited background in mathematics and physical chemistry. The second distinguishing feature of this book is that it reflects my contention that the best vehicle for learning is to give the reader an abundance of real 2-D NMR spectroscopy problem sets. These two features should allow the reader to develop problem-solving skills essential in the practice of modern NMR spectroscopy.

Beyond the lofty goal of making the reader more skilled at NMR spectra interpretation, the book has other passages that may provide utility. The inclusion of a number of practical tips for successfully conducting NMR experiments should also allow this book to serve as a useful resource.

I would like to thank D.C. Lea, my first teacher of chemistry, Dana Mayo, who inspired me to study NMR spectroscopy, Ronald Christensen, who took me under his wing for a whole year, Bernard Shapiro, who taught the best organic structure determination course I ever took, David Rice, who taught me how to write a paper, Paul Inglefield and Alan Jones, who had more faith in me than I had in myself, Dan Reger who was the best boss a new NMR lab manager could have and who let me go without recriminations, and of course Tim Swager, who inspired me to amass the data sets that are the heart of this book. I thank Jeremy Hayhurst, Jason Malley, Derek Coleman, and Phil Bugeau of Elsevier, and Jodi Simpson, who graciously agreed to come out of retirement to copyedit the manuscript.

I also wish to thank those that reviewed the book and provided helpful suggestions. Finally, I have to thank my wife, Elizabeth Worcester, and my children, Grant, Maxwell, and Eva, for putting up with me during manuscript preparation.

Any errors in this book are solely the fault of the author. If you find an error or have any constructive suggestions, please tell me about it so that I can improve any possible future editions. As of this writing, e-mail can be sent to me at jsimpson@mit.edu.

Jeff Simpson
Epping, NH, USA
January 2008

Introduction

1.1 WHAT IS NUCLEAR MAGNETIC RESONANCE?

Nuclear magnetic resonance (NMR) spectroscopy is arguably the most important analytical technique available to chemists. From its humble beginnings in 1945, the area of NMR spectroscopy has evolved into many overlapping subdisciplines. Luminaries have been awarded several recent Nobel prizes, including Richard Ernst in 1991 and Kurt Wüthrich in 2002.

Nuclear magnetic resonance spectroscopy is a technique wherein a sample is placed in a homogeneous (constant) magnetic field, irradiated, and a magnetic signal is detected. Photon bombardment of the sample causes nuclei in the sample to undergo transitions (resonance) between states. Perturbing the equilibrium distribution of state populations is called excitation. The excited nuclei emit a magnetic signal called a free induction decay (FID) that we detect with electronics and capture digitally. The digitized FID(s) is(are) processed by using computational methods to (we hope) reveal meaningful things about our sample.

Although excitation and detection may sound very complicated and esoteric, we are really just tweaking the nuclei of atoms in our sample and getting information back. How the nuclei behave once tweaked conveys information about the chemistry of the atoms in the molecules of our sample.

The acronym NMR simply means that the nuclear portions of atoms are affected by magnetic fields and undergo resonance as a result.

Homogeneous. Constant throughout.

Signal. An electrical current containing information.

Excitation. The perturbation of spins from their equilibrium distribution of spin state populations.

Free induction decay, FID. The analog signal induced in the receiver coil of an NMR instrument caused by the xy component of the net magnetization. Sometimes the FID is also assumed to be the digital array of numbers corresponding to the FID's amplitude as a function of time.

1.2 CONSEQUENCES OF NUCLEAR SPIN

Observation of the NMR signal requires a sample containing atoms of a specific atomic number and isotope, i.e., a specific nuclide such as

Table 1.1 NMR-active nuclides.

Nuclide	Element-isotope	Spin	Natural abundance (%)	Frequency relative to ^{1}H
^{1}H	Hydrogen-1	½	99.985	1.00000
^{13}C	Carbon-13	½	1.108	0.25145
^{15}N	Nitrogen-15	½	0.37	0.10137
^{19}F	Fluorine-19	½	100.	0.94094
^{31}P	Phosphorus-31	½	100.	0.40481
^{2}H (or ^{2}D)	Deuterium-2	1	0.015	0.15351

Spin state. Syn. spin angular momentum quantum number. The projection of the magnetic moment of a spin onto the z-axis. The orientation of a component of the magnetic moment of a spin relative to the applied field axis (for a spin-½ nucleus, this can be +½ or −½).

Magnetic moment. A vector quantity expressed in units of angular momentum that relates the torque felt by the particle to the magnitude and direction of an externally applied magnetic field. The magnetic field associated with a circulating charge.

Nuclear spin. The circular motion of the positive charge of a nucleus.

protium, the lightest isotope of the element hydrogen. A magnetically active nuclide will have two or more allowed nuclear spin states. Magnetically active nuclides are also said to be NMR-active. Table 1.1 lists several NMR-active nuclides in approximate order of their importance.

An isotope's NMR activity is caused by the presence of a magnetic moment in its nucleus. The nuclear magnetic moment arises because the positive charge prefers not to be well located, as described by the Heisenberg uncertainty principle. Instead, the nuclear charge circulates; because the charge and mass are both inherent to the particle, the movement of the charge imparts movement to the mass of the nucleus. The motion of all rotating masses comes in units of angular momentum; in a nucleus this motion is called nuclear spin. Imagine the motion of the nucleus as being like that of a wild animal pacing in circles in a cage. Nuclear spin (see column three of Table 1.1) is an example of the motion associated with zero-point energy in quantum mechanics, whose most well known example is perhaps the harmonic oscillator.

The small size of the nucleus dictates that the spinning of the nucleus is quantized. That is, the quantum mechanical nature of small particles forces the spin of the NMR-active nucleus to be quantized into only a few discreet states. Nuclear spin states are differentiated from one another based on how much the axis of nuclear spin aligns with a reference axis (the axis of the applied magnetic field).

We can determine how many allowed spin states there are for a given nuclide by multiplying the nuclear spin number by 2 and adding 1. For a spin-½ nuclide, there are therefore 2 (½) + 1 = 2 allowed spin states.

In the absence of an externally applied magnetic field, the energies of the two spin states of a spin-½ nuclide are degenerate (the same).

The circulation of the nuclear charge, as is expected of any circulating charge, gives rise to a tiny magnetic field called the nuclear magnetic moment—also commonly referred to as a spin for short (recall that the mass puts everything into a world of angular momentum). Magnetically active nuclei are rotating masses, each with a tiny magnet, and these nuclear magnets interact with other magnetic fields according to Maxwell's equations.

1.3 APPLICATION OF A MAGNETIC FIELD TO A NUCLEAR SPIN

Placing a sample inside the NMR magnet puts the sample into a very high strength magnetic field. Application of a magnetic field to this sample will cause the nuclear magnetic moments of the NMR-active nuclei of the sample to become aligned either partially parallel (α spin state) or antiparallel (β spin state) with the direction of the magnetic field.

Alignment of the two allowed spin states for a spin-½ nucleus is analogous to the alignment of a compass needle with the Earth's magnetic field. A point of departure from this analogy comes when we consider that nearly half of the nuclear magnetic moments in our sample line up opposed to the directions of the magnetic field lines we apply (applied field). A second point of departure from this analogy is due to the small size of the nucleus and the Heisenberg uncertainty principle (again!). The nuclear magnetic moment cannot align itself exactly with the applied field. Instead, only part of the nuclear magnetic moment (half of it) can align with the field. If the nuclear magnetic moment were to align exactly with the applied field axis, then we would essentially know too much, which nature does not allow. The Heisenberg uncertainty principle forbids mathematically the attainment of this level of knowledge.

The energies of the parallel and antiparallel spin states of a spin-½ nucleus diverge linearly with increasing magnetic field. This is the Zeeman effect (see Figure 1.1). At a given magnetic field strength, each NMR-active nuclide exhibits a unique energy difference between its spin states. Hydrogen has the second greatest slope for the energy

Degenerate. Two spin states are said to be degenerate when their energies are the same.

Applied field, B_0. Syn. applied magnetic field. The area of nearly constant magnetic flux in which the sample resides when it is inside the probe, which is in turn inside the bore tube of the magnet.

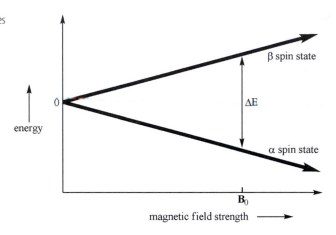

■ **FIGURE 1.1**   Zeeman energy diagram showing how the energies of the two allowed spin states for the spin-½ nucleus diverge with increasing applied magnetic field strength.

Zeeman effect. The linear divergence of the energies of the allowed spin states of an NMR-active nucleus as a function of applied magnetic field strength.

Gyromagnetic ratio, γ. Syn. magnetogyric ratio. A nuclide-specific proportionality constant relating how fast spins will precess (in radians .sec^{-1}) per unit of applied magnetic field (in T).

divergence (second only to its rare isotopic cousin, tritium, ^{3}H or ^{3}T). This slope is expressed through the gyromagnetic ratio, γ, which is a unique constant for each NMR-active nuclide. The gyromagnetic ratio tells how many rotations per second (gyrations) we get per unit of applied magnetic field. Equation 1.1 shows how the energy gap between states (ΔE) of a spin-½ nucleus varies with the strength of the applied magnetic field $\mathbf{B_0}$ (in tesla). By necessity, the units of γ are joules per tesla.

$$\Delta E = \gamma B_0 \tag{1.1}$$

To induce transitions between the allowed spin states of an NMR-active nucleus, photons with their energy tuned to the gap between the two spin states must be applied (Equation 1.2).

$$\Delta E = h\upsilon = \hbar\omega \tag{1.2}$$

where h is Planck's constant in joule · seconds ν is the frequency in events per second, $\hbar$ ("h bar") is Planck's constant divided by 2π, and ω is the angular frequency in radians per second.

From Equations 1.1 and 1.2 we can calculate the NMR frequency of any NMR-active nuclide on the basis of the strength of the applied magnetic field alone (Equations 1.3a and 1.3b). In practice, the gyromagnetic ratio we look up will already have the factor of Planck's constant included; thus the units of γ will be in radians per tesla per second. For hydrogen, γ is 2.675×10^8 radians/tesla/second (radians are used because the radian is a "natural" unit for oscillations and rotations), so the frequency is:

$$\nu = \gamma B_0/h \tag{1.3a}$$

or,

$$\omega = \gamma B_0/\hbar \tag{1.3b}$$

To calculate NMR frequency correctly, it is important we make sure our units are consistent. For a magnetic field strength of 11.74 tesla (117,400 gauss), the NMR frequency for hydrogen is:

$$\nu = 2.675 \times 10^8 \,\text{radians/tesla/second} \times 11.74 \,\text{tesla}/2\pi \,\text{radians/cycle}$$
$$= 4.998 \times 10^8 \,\text{cycles/second} = 500 \,\text{MHz} \tag{1.4}$$

Thus, an NMR instrument operating at a frequency of 500 MHz requires an 11.74 tesla magnet. Each spin experiences a torque from the applied magnetic field. The torque applied to an individual nuclear magnetic moment can be calculated by using the right hand rule because it involves the mathematical operation called the cross product. Because a spin cannot align itself exactly parallel to the applied field, it will always feel the torque from the applied field. Hence, the rotational axis of the spin will precess around the applied field axis just as a top's rotational axis precesses in the Earth's gravitational field. The amazing fact about the precession of the spin's axis is that its frequency is the same as that of a photon that can induce transitions between its spin states. That is, the precession frequency for protons in an 11.74 Tesla magnetic field is also 500 MHz! This nuclear precession frequency is called the Larmor (or NMR) frequency; the Larmor frequency will become an important concept to remember when we discuss the rotating frame of reference.

1.4 APPLICATION OF A MAGNETIC FIELD TO AN ENSEMBLE OF NUCLEAR SPINS

Only half of the nuclear spins align with a component of their magnetic moment parallel to an applied magnetic field because the energy difference between the parallel and antiparallel spin states is extremely small relative to the available thermal energy, kT. The omnipresent thermal energy kT randomizes spin populations over time.

NMR instrument. A host computer, console, preamplifier, probe, cryomagnet, pneumatic plumbing, and cabling that together allow the collection of NMR data.

Cross product. A geometrical operation wherein two vectors will generate a third vector orthogonal (perpendicular) to both vectors. The cross product also has a particular handedness (we use the right-hand rule), so the order of how the vectors are introduced into the operation is often important.

Precession frequency. Syn. Larmor frequency, NMR frequency. The frequency at which a nuclear magnetic moment rotates about the axis of the applied magnetic field.

Larmor frequency. Syn. precession frequency, nuclear precession frequency, NMR frequency, rotating frame frequency. The rate at which the xy component of a spin precesses about the axis of the applied magnetic field. The frequency of the photons capable of inducing transitions between allowed spin states for a given NMR-active nucleus.

Thermal energy, kT. The random energy present in all systems which varies in proportion to temperature.

This nearly complete randomization is described by using the following variant of the Boltzmann equation:

$$N_\alpha/N_\beta = \exp(\Delta E/kT) \qquad (1.5)$$

In Equation 1.5, $N\alpha$ is the number of spins in the α (lower energy) spin state, N_β is the number of spins in the β (higher energy) spin state, ΔE is the difference in energy between the α and β spin states, k is the Boltzmann constant, and T is the temperature in degrees kelvin. Because $\Delta E/kT$ is very nearly zero, both spin states are almost equally populated. That is, because the spin state energy difference is much less than kT, thermal energy equalizes the populations of the spin states. Mathematically, this equal distribution is borne out by Equation 1.5, because raising e (2.718 . . .) to the power of almost 0 is very nearly 1, thus showing that the ratio of the populations of the two spin states is almost 1:1.

An analogy here will serve to illustrate what may seem to be a rather dry point. Suppose we have an empty paper box that normally holds ten reams of paper. If we put 20 ping pong balls in it and then shake up the box with the cover on, we expect the balls will become distributed evenly over the bottom of the box (barring tilting of the box). If we add the thickness of one sheet of paper to one half of the bottom of the box and repeat the shaking exercise, we will still expect the balls to be evenly distributed. If, however, we put a ream of paper (500 sheets) inside the box (thus covering half of the area of the box's bottom) and shake, not too vigorously, we will find upon the removal of the top of the box that most of the balls will not be on top of the ream of paper but rather next to the ream, resting in the lower energy state. On the other hand, with vigorous shaking of the box, we may be able to get half of the balls up on top of the ream of paper.

Most of the time when doing NMR, we are in the realm wherein the thickness of the step inside the box (ΔE) is much smaller than the amplitude of the shaking (kT). Only by cooling the sample (making T smaller) or by applying a greater magnetic field (or by choosing an NMR-active nuclide with a larger gyromagnetic ratio) are we able to significantly perturb the grim statistics of the Boltzmann distribution.

Let's say we have a sample containing 10 mM chloroform (the solute concentration) in deuterated acetone (acetone-d_6). If we have

0.70 mL of the sample in a 5 mm diameter NMR tube, the number of hydrogens atoms from the solute (chloroform) would be

$$\begin{aligned} \text{Number of hydrogens atoms} &= 0.010 \text{ moles/liter} \times 0.00070 \text{ liters} \\ &\quad \times 6.0 \times 10^{23} \text{ units/mol} \\ &= 4.2 \times 10^{18} \text{ hydrogen atoms} \end{aligned}$$

The number of hydrogen atoms needed to give us an observable NMR signal is significantly less than 4.2 quintillion. If we were able to get all 4.2 quintillion spins to adopt just one spin state, we would, with a modern NMR instrument, see a booming signal. But the actual signal we see is not that due to summing the magnetic moments of 4.2 quintillion hydrogen nuclei because a great deal of cancellation occurs.

The cancellation takes place in two ways. The first form of cancellation take place because nuclear spins in any spin state will (at equilibrium) have their **xy** components (those components perpendicular to the applied magnetic field axis, **z**) distributed randomly along a cone (see Figure 1.2). Recall that only a component of the nuclear magnetic moment can line up with the applied magnetic field axis. Because of the random distribution of the nuclear magnetic moments along the cone, the **xy** components will cancel each other out, leaving only the **z** components of the spins to be additive. To better understand this, imagine dropping a bunch of pins point down into an empty pointed ice-cream cone. If we shake the cone a little while holding the cone so the cone tip is pointing straight down, then all the pin heads will become evenly distributed along the inside surface of the cone. This example illustrates how the nuclear magnetic moments will be distributed for one spin state at equilibrium, and thus how the pins will not point in any direction except for straight down. That is, the **xy** (horizontal) components of the spins (or pins) will cancel each other, leaving only half of the nuclear magnetic moments lined up along the **z**-axis.

The second form of cancellation takes place because, for a spin-½ nucleus, the two cones corresponding to the two allowed spin states (α and β) oppose each other (the orientations of the two cones is opposite—don't try this with pins and an actual ice cream cone or we will have pins everywhere on the floor!). The Boltzmann equation dictates that the number of spins (or pins) in the two cones is very nearly equal under normal experimental conditions. At 20°C (293 K), perhaps only 1 in about 25,000 hydrogen nuclei will

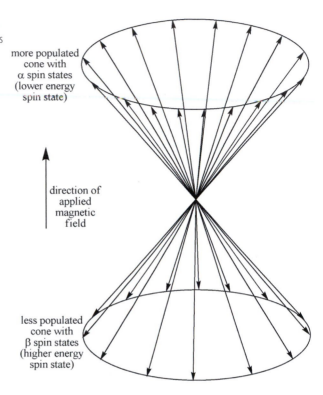

■ **FIGURE 1.2** The two cones made up by the more-populated α spin state (top cone) and the less-populated β spin state; each arrow represents the magnetic moment μ of an individual nuclear spin.

more populated
cone with
α spin states
(lower energy
spin state)

direction of
applied
magnetic
field

less populated
cone with
β spin states
(higher energy
spin state)

reside in the lower energy spin state in a typical NMR magnetic field (11.74 tesla).

The small difference in the number of spins occupying the two spin states can be calculated by plugging our hydrogen ΔE at 11.74 Tesla ($h\nu$ or $h \times 500\,\text{MHz}$, see Equation 1.4) and the absolute temperature (293 K) into Equation 1.5:

$$
\begin{aligned}
N_\alpha/N_\beta &= \exp(\Delta E/kT) \\
&= \exp(6.63 \times 10^{-34}\text{J's} \times 5.00 \times 10^8 \text{ s}^{-1}/1.38 \times 10^{-23}\text{J/K}/293 \text{ K}) \\
&= \exp(0.0000820) \\
&= 1 + 0.0000820
\end{aligned}
\tag{1.6}
$$

Note that e (or any number except 0) raised to a power near 0 is equal to 1 plus the number to which e is raised, in this case

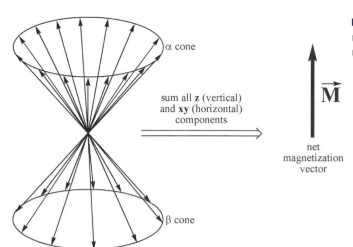

α cone

sum all **z** (vertical)
and **xy** (horizontal)
components

$\overrightarrow{\textbf{M}}$

net
magnetization
vector

β cone

■ **FIGURE 1.3** Summation of all the vectors of the magnetic moments that make up the α and β spin state cones yields the net magnetization vector **M**.

0.0000820 (only the first two terms of the Maclaurin power series expansion are significant). Because $1/0.0000820 = 12{,}200$, we can see that only one more spin out of every 24,400 spins will be in the lower energy (α) spin state.

The simple result is this: Cancellation of the nuclear magnetic moments has the unfortunate result of causing approximately all but 2 of every (roughly) 50,000 spins to cancel each other (24,999 spins in one spin state will cancel out the net effect of 24,999 spins in the other spin state), leaving only 2 spins out of our ensemble of 50,000 spins to contribute the z-axis components to the net magnetization vector **M** (see Figure 1.3).

Thus, for our ensemble of 4.2 quintillion spins, the number of nuclear magnetic moments that we can imagine being lined up end to end is reduced by a factor of 50,000 (25,000 for the excess number in the lower energy or α spin state, and 2 for the fact that only part of each nuclear magnetic moment is along the z-axis) to give a final number of 1.7×10^{14} spins or 170 trillion (in the UK, a 170 billion) spins. Even though 170 trillion is still a big number, nonetheless it is more than four orders of magnitude less than what we might have first expected on the basis of looking at one spin.

Performing vector addition of the 170 trillion excess α spins gives us the net magnetization vector for our 5 mm sample containing 0.70 mL of 10 mM chloroform solution at 20°C in a 500 MHz NMR.

Ensemble. A large number of NMR-active spins.

Net magnetization vector, M. Syn. magnetization. The vector sum of the magnetic moments of an ensemble of spins.

It is common to refer to this and comparable numbers of spins as an ensemble.

The net magnetization vector **M** is the entity we detect, but only **M**'s component in the **xy** plane is detectable. Sometimes we refer to a component of **M** simply as magnetization or polarization.

The gyromagnetic ratio γ affects the strength of the signal we observe with an NMR spectrometer in three ways. One, the larger the γ, the more spins will reside in the lower energy spin state (a Boltzmann effect). Two, for each additional spin we get to drop into the lower energy state, we add the magnitude of that spin's nuclear magnetic moment μ (which depends on γ) to our net magnetization vector **M** (a length-of-μ effect). Three, the precession frequency of **M** depends on γ, so our detector will have less noise interfering with it. This last point is the most difficult to understand, but it basically works as follows: The higher the frequency of a signal, the easier it is to detect. DC (direct current) signals are notoriously hard to make stable in electronic circuitry, but AC (alternating current) signals are much easier to generate stably. These three factors mean that the signal-to-noise ratio we obtain depends on the gyromagnetic ratio γ raised to a power greater than two!

Once we have summed the behavior of individual spins into the net magnetization vector **M**, we no longer have to worry about some of the restrictions discussed earlier. In particular, the length of the vector or whether it is allowed to point in a particular direction is no longer restricted. **M** can be manipulated with electromagnetic radiation in the radio-frequency range, often simply referred to as RF. **M** can be tilted away from its equilibrium position along the **z**-axis to point in any direction. The ability to visualize **M**'s movement will become important later when we discuss RF pulses and pulse sequences. For now, though, just try to accept that **M** can be tilted from equilibrium and can grow or shrink depending on its interactions with other things, be they other spins, RF, or the lattice (the rest of the world).

In other ways, however, the net magnetization vector **M** behaves in a manner similar to the individual spins that it comprises. One very important similarity has to do with how **M** will behave once it is perturbed from its equilibrium position along the z-axis. **M** will itself precess at the Larmor frequency if it has a component in the **xy** plane (i.e., if it is no longer pointing in its equilibrium direction). Detection of signal requires magnetization in the **xy** plane, because

Polarization. The unequal population of two or more spin states.

Signal-to-noise ratio, S/N. The height of a real peak (measure from the top of the peak to the middle of the range of baseline noise) divided by the amplitude of the baseline noise over a statistically reasonable range.

Radio frequency, RF. Electromagnetic radiation with a frequency range from 3 kHz to 300 GHz.

Lattice. The rest of the world. The environment outside the immediate vicinity of a spin.

only a precessing magnetization generates a changing magnetic flux in the receiver coil—what we detect!

1.5 TIPPING THE NET MAGNETIZATION VECTOR FROM EQUILIBRIUM

The nuclear precession (Larmor) frequency is the same frequency as that of photons that can make the spins of the ensemble undergo transitions between spin states.

The precession of the net magnetization vector **M** at the Larmor frequency (500 MHz in the preceding example) gives a clue as to how RF can be used to tip the vector from its equilibrium position.

Electromagnetic radiation consists of a stream of photons. Each photon is made up of an electric field component and a magnetic field component, and these two components are mutually perpendicular. The frequency of a photon determines how fast the electric field component and magnetic field component will pulse, or beat. Radio-frequency electromagnetic radiation at 500 MHz will thus have a magnetic field component that beats 500 million times a second, by definition.

Radio-frequency electromagnetic radiation is a type of light, even though its frequency is too low for us to see or (normally) feel. Polarized RF therefore is polarized light, and it has all its magnetic field components lined up along the same axis. Polarized light is something most of us are familiar with: Light reflecting off of the surface of a road tends to be mostly plane-polarized, and wearing polarized sunglasses reduces glare with microscopic lines in the sunglass lenses (actually individual molecules lined up in parallel). The lines selectively filter out those photons reflected off the surface of a road or water, most of whose electric field vectors are oriented horizontally.

If polarized 500 MHz RF is applied to our 10 mM chloroform sample in the 11.74 tesla magnetic field, the magnetic field component of the RF will, with every beat, tip the net magnetization vector of the ensemble of the hydrogen atoms in the chloroform a little bit more from its equilibrium position. A good analogy is pushing somebody on a swing set. If we push at just the right time, we will increase the amplitude of the swinging motion. If our pushes are not well timed, however, they will not increase the swinging amplitude. The same timing restrictions are relevant when we apply RF to our spins.

Pulse. Syn. RF pulse. The abrupt turning on of a sinusoidal waveform with a specific phase for a specific duration, followed by the abrupt turning off of the sinusoidal waveform.

Beat. The maximum of one wavelength of a sinusoidal wave.

If we do not have a well-timed application of the magnetic field component from our RF, then the net magnetization vector will not be effective in tipping the net magnetization vector. In particular, if the RF frequency is not just randomly mistimed but is consistently higher or lower than the Larmor frequency, the errors between when the push should and does occur will accumulate; before too long our pushes will actually serve to decrease the amplitude of the net magnetization vector **M**'s departure from equilibrium. The accumulated error caused by poorly synchronized beats of RF with respect to the Larmor frequency of the spins is well known to NMR spectroscopists and is called pulse roll-off.

The reason why pulse roll-off sometimes occurs is that not all spins of a particular nuclide (e.g., not all ^{1}H's) in a sample will resonate at exactly the same Larmor frequency; consequently, the frequency of the applied RF cannot always be tuned optimally for every chemically distinct set of spins in a sample.

1.6 **SIGNAL DETECTION**

Scan. A single execution of a pulse sequence ending in the digitization of a FID.

Receiver coil. An inductor in a resistor-inductor-capacitor (RLC) circuit that is tuned to the Larmor frequency of the observed nuclide and is positioned in the probe so that it surrounds a portion of the sample.

If the frequency of the applied RF is well tuned to the Larmor frequency (or if the pulse is sufficiently short and powerful), the net magnetization vector **M** can be tipped to any desired angle relative to its starting position along the z-axis. To maximize observed signal for a single event (one scan), the best tip angle is 90°. Putting **M** fully into the **xy** plane causes **M** to precess in the **xy** plane, thereby inducing a current in the receiver coil which is really nothing more than an inductor in a resistor-inductor-capacitor (RLC) circuit tuned to the Larmor frequency. Putting **M** fully into the **xy** plane maximizes the amplitude of the signal generated in the receiver and gives the best signal-to-noise ratio if **M** has sufficient time to fully return to equilibrium between scans. **M** can be broken down into components, each of which may correspond to a chemically unique magnetization (e.g., $\mathbf{M_a}$, $\mathbf{M_b}$, $\mathbf{M_c}$ . . .) with its own unique amplitude, frequency, and phase.

Following excitation, the net magnetization vector **M** will almost always have a component precessing in the **xy** plane; this component returns to its equilibrium position through a process called relaxation. Relaxation occurs when an ensemble of spins are distributed among their available allowed spin states contrary to the Boltzmann equation (Equation 1.5). Relaxation occurs through a number of different relaxation pathways and is itself a very demanding and rich subdiscipline of NMR. The two basic types of relaxation of which we need be aware at this point are spin-spin relaxation and spin-lattice relaxation. As their names imply, spin-spin relaxation involves one spin interacting with another spin so that one or both sets of spins can return to equilibrium, whereas spin-lattice relaxation involves spins relaxing through their interaction with the rest of the world (the lattice).

1.7 THE CHEMICAL SHIFT

The inability to tune RF to the exact Larmor frequency of all spins of one particular NMR-active nuclide in a sample is often caused by a phenomenon known as the chemical shift. The term chemical shift was originally coined disparagingly by physicists intent on measuring the gyromagnetic ratio γ of various NMR-active nuclei to a high degree of precision and accuracy. These physicists found that for the ^{1}H nuclide, the γ they measured depended on what hydrogen-containing material they used for their experiments, thus casting into serious doubt their ability to ever accurately measure the true value of γ for ^{1}H. Over the years, the attribute known as the chemical shift has come to be reasonably well understood, and many chemists and biochemists are comfortable discussing chemical shifts.

The chemical shift arises from the resistance of the electron cloud of a molecule to the applied magnetic field. Because the electron itself is a spin-½ particle, it too is affected by the applied field, and its response to the applied field is to shield the nucleus from feeling the full effect of the applied field. The greater the electron density in the immediate vicinity of the nucleus, the greater the amount to which the nucleus will be protected from feeling the full effect of the applied field. Increasing the strength of the applied field in turn increases how much the electrons resist allowing the magnetic field to penetrate to the nucleus. Therefore, the nuclear shielding is directly proportional to the strength of the applied field, thus making the chemical shift a unitless quantity.

1.8 THE 1-D NMR SPECTRUM

The one-dimensional NMR spectrum shows amplitude as a function of frequency. To generate this spectrum, an ensemble of a particular NMR-active nuclide is excited. The excited nuclei generate a signal that is detected in the time domain and then converted mathematically to the frequency domain by using a Fourier transform.

Relaxation. The return of an ensemble of spins to the equilibrium distribution of spin state populations.

Spin-lattice relaxation. Syn. T_1 relaxation. Relaxation involving the interaction of spins with the rest of the world (the lattice).

Spin-spin relaxation. Syn. T_2 relaxation. Relaxation involving the interaction of two spins.

Chemical shift (δ). The alteration of the resonant frequency of chemically distinct NMR-active nuclei due to the resistance of the electron cloud to the applied magnetic field. The point at which the integral line of a resonance rises to 50% of its total value.

1-D NMR spectrum. A linear array showing amplitude as a function of frequency, obtained by the Fourier transformation of an array with amplitude as a function of time.

Time domain. The range of time delays spanned by a variable delay (t_1 or t_2) in a pulse sequence.

Fourier transform, FT. A mathematical operation that converts the amplitude as a function of time to amplitude as a function of frequency.

Resonance. An NMR signal consisting of one or more relatively closely spaced peaks in the frequency spectrum that are all attributable to a unique atomic species in a molecule.

Chemical shift axis. The scale used to calibrate the abscissa (x-axis) of an NMR spectrum. In a one-dimensional spectrum, the chemical shift axis typically appears underneath an NMR frequency spectrum when the units are given in parts-per-millions (as opposed to Hz, in which case the axis would be termed the frequency axis).

Pulse sequence. A series of timed delays, RF pulses, and gradient pulses that culminates in the detection of the NMR signal.

90° RF pulse. Syn. 90° pulse. An RF pulse applied to the spins in a sample to tip the net magnetization vector of those spins by 90°.

Proton decoupling. The irradiation of ^1H's in a molecule for the purpose of collapsing the multiplets one would otherwise observe in a ^{13}C (or other nuclide's) NMR spectrum. Proton decoupling will also likely alter the signal intensities of the observed spins of other nuclides through the NOE. For ^{13}C, proton decoupling enhances the ^{13}C signal intensity.

Older instruments called continuous wave (CW) instruments do not simultaneously excite all the spins of a particular nuclide. Instead, the magnetic field is varied while RF of a fixed frequency is generated. As various spin populations come into resonance, the complex impedance of the NMR coil changes in proportion to the number of spins at a particular field and RF frequency. Thus, we can speak of observing a resonance at a particular point in a spectrum we collect. This process of scanning the magnetic field is slow and inefficient compared to how today's instruments work, although there is a certain aesthetic appeal in the intuitively more obvious nature of the CW method.

All 1-D NMR time domain data sets must undergo one Fourier transformation to become an NMR spectrum. The Fourier transformation converts amplitude as a function of time to amplitude as a function of frequency. Therefore the spectrum shows amplitude along a frequency axis that is normally the chemical shift axis.

The signal we detect to ultimately generate a 1-D NMR spectrum is generated using a pulse sequence. A pulse sequence is a series of timed delays and RF pulses that culminates in the detection of the NMR signal. Sometimes more than one RF channel is used to perturb the NMR-active spins in the sample. For example, the effect of the spin state of ^1H's on nearby ^{13}C's is typically suppressed using ^1H decoupling (proton decoupling) while we acquire the signal from the ^{13}C nuclei.

Figure 1.4 shows a simple 1-D NMR pulse sequence called the one-pulse experiment. The pulse sequence consists of three parts: relaxation, preparation, and detection. A relaxation delay is often required because obtaining a spectrum with a reasonable signal-to-noise ratio often requires repeating the pulse sequence (scanning) many times to accumulate sufficient signal, and following preparation (putting

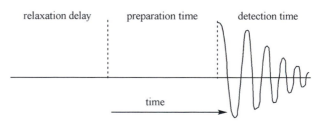

■ FIGURE 1.4 The three distinct time periods of a generic 1-D NMR pulse sequence.

magnetization into the **xy** plane), the NMR spins will often not return to equilibrium as quickly as we might like, so we must wait for this return to equilibrium before starting the next scan. Some relaxation will take place during detection, but often not enough to suit our particular needs.

1.9 THE 2-D NMR SPECTRUM

A 2-D NMR spectrum is obtained after carrying out two Fourier transformations on a matrix of data (as opposed to one Fourier transform on an array of data for a 1-D NMR spectrum). A 2-D NMR spectrum will generate cross peaks that correlate information on one axis with information on the other; usually, both axes are chemical shift axes, but this is not always the case.

The pulse sequence used to collect a 2-D NMR data set differs only slightly (at this level of abstraction) from the 1-D NMR pulse sequence. Figure 1.5 shows a generic 2-D NMR pulse sequence. The 2-D pulse sequence contains four parts instead of three. The four parts of the 2-D pulse sequence are relaxation, evolution, mixing, and detection. The careful reader will note that preparation has been split into two parts: evolution and mixing.

Evolution involves imparting phase character to the spins in the sample. Mixing involves having the phase-encoded spins pass their phase information to other spins. Evolution usually occurs prior to mixing and is termed t_1 (not to be confused with T_1 the relaxation time!), but in some 2-D NMR pulse sequences the distinction is blurred, for example in the correlation spectroscopy (COSY) experiment. Evolution often starts with a pulse to put some magnetization

1-D NMR pulse sequence. A series of delays and RF pulses culminating in the detection, amplification, mixing down, and digitization of the FID.

One-pulse experiment. The simplest 1-D NMR experiment consisting of only a relaxation delay, a single RF pulse, and detection of the FID.

Preparation. The placement of magnetization into the **xy** plane for subsequent detection.

Relaxation delay. The initial period of time in a pulse sequence devoted to allowing spins to return to equilibrium.

Cross peak. The spectral feature in a multidimensional NMR spectrum that indicates a correlation between a frequency position on one axis with a frequency position on another axis. Most frequently, the presence of a cross peak in a 2-D spectrum shows that a resonance on one chemical shift axis somehow interacts with a different resonance on the other chemical shift axis. In a homonuclear 2-D spectrum, a cross peak is a peak that occurs off of the diagonal. In a heteronuclear 2-D spectrum, any observed peak is, by definition, a cross peak.

Mixing. The time interval in a 2-D NMR pulse sequence wherein t_1-encoded phase information is passed from spin to spin.

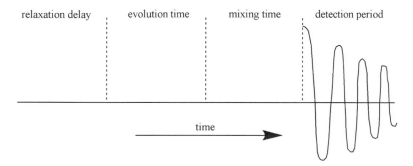

relaxation delay   evolution time   mixing time   detection period

time

■ **FIGURE 1.5**   The four distinct time periods of a generic 2-D NMR pulse sequence.

Phase character. The absorptive or dispersive nature of a spectral peak. The angle by which magnetization precesses in the xy plane over a given time interval.

Evolution time, t_1. The time period(s) in a 2-D pulse sequence during which a net magnetization is allowed to precess in the xy plane prior to (separate mixing and) detection. In the case of the COSY experiment, the evolution and mixing times occur simultaneously. Variation of the t_1 delay in a 2-D pulse sequence generates the t_1 time domain.

Detection period. The time period in the pulse sequence during which the FID is digitized. For a 1-D pulse sequence, this time period is denoted t_1. For a 2-D pulse sequence, this time period is denoted t_2.

t_1 time. The first time delay in a pulse sequence used to establish a time domain that will subsequently be converted to the frequency domain f_1.

Frequency domain. The range of frequencies covered by the spectral window. The frequency domain is located in the continuum of all possible frequencies by the frequency of the instrument transmitter's RF (this frequency is also that of the rotating frame) and by the rate at which the analog signal (the FID) is digitized.

f_1 frequency domain. The frequency domain generated following the Fourier transformation of the t_1 time domain. The f_1 frequency domain most often used for ^{1}H or ^{13}C chemical shifts.

into the **xy** plane. Once in the **xy** plane, the magnetization will precess or evolve (hence the name "evolution") and, depending on the t_1 evolution time, will precess a certain number of degrees from its starting point. How far each set of chemically distinct spins evolves is a function of the t_1 evolution time and each spin set's precession frequency which in turn depends on its chemical environment. Thus, a series of passes through the pulse sequence using different t_1's will encode each chemically distinct set of spins with a unique array of phases in the **xy** plane. During the mixing time, the phase-encoded spins are allowed to mix with each other or with other spins. The nature of the mixing that takes place during a 2-D pulse sequence varies widely and includes mechanisms involving through-space relaxation, through-bond perturbations (scalar coupling), and other interactions.

During the detection period denoted t_2 (not the relaxation time T_2!) the NMR signal is captured electronically and stored in a computer for subsequent workup. Although detection occurs after evolution, the first Fourier transformation is applied to the time domain data detected during the t_2 detection period to generate the f_2 frequency axis. That is, the t_2 time domain is converted using the Fourier transformation into the f_2 frequency domain before the t_1 time domain is converted to the f_1 frequency domain. This ordering may seem counterintuitive, but recall that t_1 and t_2 get their names from the order in which they occur in the pulse sequence, and not from the order in which the data set is processed.

Following conversion of t_2 to f_2, we have a half-processed NMR data matrix called an interferogram. The interferogram is not a particularly useful thing in and of itself, but performing a Fourier transformation to convert the t_1 time domain to the f_1 frequency domain renders a data matrix with two frequency axes (f_1 and f_2) that will (hopefully) allow the extraction of meaningful data pertaining to our sample.

1.10 INFORMATION CONTENT AVAILABLE USING NMR

NMR spectroscopy can provide a wealth of information about the nature of solute molecules and solute-solvent interactions. At this point, it is best to highlight the simplest and most basic features present in typical solution state NMR spectra, especially those of proton NMR spectra.

- NMR provides chemical shifts (denoted δ) for atoms in differing chemical environments. For example, an aldehyde proton will show a different chemical shift than a methyl proton.

- NMR also can give the relative population of spins in each chemical environment through peak integration.

- NMR shows how one spin may be near (in terms of number of bonds distant) another spin through scalar coupling or J-coupling.

- NMR shows through the J-coupling of spins only a few bonds distant how a molecule may be folded or bent.

- NMR also can show how one atom in a molecular may be nearby (in space) to another atom in the same or even a different molecule through the nuclear Overhauser effect.

- NMR can reveal how molecular dynamics and chemical exchange may be taking place over a wide range of time scales.

If we acquire a reasonable grasp of the first five bulleted items above as a result of reading this book and working its problems, then we will have done well. Attaining a limited awareness of the sixth bulleted item is also hoped for. As with many disciplines (perhaps all except particle physics), we have to accept limits to understanding, accept the notion of the black box wherein some behavior goes in and something happens as a result that is unfathomed (but not unfathomable), and relegate the particulars to others more well-versed in the particular field in question. Being simply aware of the realm of molecular dynamics and knowing whom we might ask is probably a good start. In general, this quest begins with us consulting our local NMR authority. If we are lucky, that person will be a distinguished faculty member, senior scientist, or the manager of the NMR facility in our institution. The author can personally attest to the helpfulness of the Association of Managers of Magnetic Resonance Laboratories (AMMRL), and while membership is limited, there are ways to query the group (perhaps through someone we may know in the group) and obtain possible suggestions and answers to delicate NMR problems. The NMR vendors monitor AMMRL e-mail traffic and often make it a point to address issues raised relating to their own products in a timely manner.

f_2 frequency domain. The frequency domain generated following the Fourier transformation of the t_2 time domain. The f_2 frequency domain is almost exclusively used for ^{1}H chemical shifts.

Interferogram. A 2-D data matrix that has only undergone Fourier transformation along one axis to convert the t_2 time domain to the f_2 frequency domain. An interferogram will therefore show the f_2 frequency domain on one axis and the t_1 time domain on the other axis.

F_1 axis, f_1 axis. Syn. f_1 frequency axis. The reference scale applied to the f_1 frequency domain. The f_1 axis may be labeled with either ppm or Hz.

F_2 axis, f_2 axis. Syn. f_2 frequency axis. The reference scale applied to the f_2 frequency domain. The f_2 axis may be labeled with either ppm or Hz.

Integration. The measurement of the area of one or more resonances in a 1-D spectrum, or the measurement of the volume of a cross peak in a 2-D spectrum.

Instrumental Considerations

The modern NMR instrument is a complex combination of equipment that can reveal simple and profound truths when conditions permit. Unfortunately, a large number of factors must be controlled precisely to find such wondrous answers. The evolution of the NMR instrument from its first manifestation in the 1940s is a fascinating tale of technological development. Suffice it to say that this chapter cannot describe in detail every nuance and pitfall associated with the practice of NMR spectroscopy, but some attempt is made to provide a reasonable overview and thus put at least some of the NMR dogma in its place.

2.1 SAMPLE PREPARATION

As discussed in Chapter 1, the temperature, the frequency of the nuclide being observed, and the number of spins in the sample all affect the strength of the signal observed.

Efforts aimed at improving sensitivity start with maximizing sample concentration and lowering sample temperature. But gains employing these two signal enhancement approaches are not always realized because increasing solution viscosity from both increased solute concentration and lowered temperature often degrades spectral resolution and hence lowers the signal-to-noise ratio through viscosity-induced resonance broadening.

Preparation of high-quality samples is a prerequisite for obtaining high-quality NMR data. The following sample attributes are recommended.

Sensitivity. The ability to generate meaningful data per unit time.

Resonance broadening. The spreading out, in the frequency spectrum, of one or more peaks. Resonance broadening can either be homogeneous or inhomogeneous. An example of homogeneous resonance broadening is the broadening caused by a short T_2^*. An example of inhomogeneous resonance broadening is the broadening caused by the experiencing of an ensemble of molecular environments (that are not averaged on the NMR time scale).

Viscosity-induced resonance broadening. Syn. viscosity broadening. The increase in the line width of peaks in a spectrum caused by the decrease in the T_2 relaxation time that results from a slowing of the molecular tumbling rate. Saturated solutions and solutions at a temperature just above their freezing point often show this broadening behavior.

NMR probe. Syn. probe. A non-ferrous metal housing consisting of a cylindrical upper portion that fits inside the lower portion of the magnet bore tube. The probe contains electrical conductors, capacitors, and inductors, as well as a Dewared air channel with a heater coil and a thermocouple. It may also contain one or more coils of wire wound with a geometrical configuration such that passing current through these coils will induce a magnetic field gradient across the volume occupied by the sample when it is in place.

2.1.1 NMR Tube Selection

We use the highest quality NMR tube we can afford. We match the diameter of the sample tube to the coil diameter of the NMR probe in the magnet. We do not put a 5 mm tube in a 10 mm probe unless we have no choice, and we NEVER use an NMR tube with a diameter larger than that the probe is designed to accommodate! For most organic samples comparable to those whose spectra are found in this book, a Wilmad 528-pp or similar tube suffices. Cheap tubes contain regions where the tube wall thickness varies, and this variation makes our sample not just difficult, but nearly impossible, to shim well. Variations in concentricity, camber, and diameter all limit data quality. Those interested in saving a little on tubes should examine Equation 2.1 where t is time and $ is money and do the math for themselves—we can spend an extra $10 on our tube or we can shim for an hour. Consider that tubes are reusable and that the extra cost associated with the purchase of a quality NMR tube can be amortized easily over the course of several years.

$$t = \$ \qquad\qquad (2.1)$$

2.1.2 Sample Purity

We make our sample as pure as possible. While a high solute concentration is good, a high sample purity is better; it is better to have a 5 mM sample of pure product than a 20 mM sample containing other isomers and reaction by-products. To maximize our ^{13}C signal, however, we may elect to collect the ^{13}C one-dimensional (1-D) spectrum before we purify for ^{1}H detection. The ^{1}H-detected HMQC/HSQC and HMBC data will help rule out spurious peaks from the ^{13}C 1-D spectrum we obtain from our crude mixture. We may fail to collect a ^{13}C spectrum with sufficiently high enough signal-to-noise ratio from our purified sample given competition for instrument time in our research environment. That is, collecting a ^{13}C 1-D spectrum from our purified-but-less-concentrated sample for 24 hours may give us a spectrum showing only noise in the chemical shift ranges where we expect to observe the resonances of our nonprotonated ^{13}C molecular sites. Scanning over and over for four days to double the signal-to-noise ratio may be discouraged in a multiuser environment. If we only have a small amount of product and wish to avoid repeating the synthesis, isolation, purification, and sample preparation, we may still be able to fully assign the ^{1}H and ^{13}C resonances of our molecule without observing all our ^{13}C resonances directly in the ^{13}C 1-D spectrum.

2.1.3 **Solvent Selection**

We use a high-quality deuterated solvent. For precious samples, we try to use individual ampoules rather taking solvent out of a bottle that originally contained 50 or 100 g of solvent. Deuterated chloroform more than six months old may be acidic enough to exchange away labile protons from our solute molecule; we must take particular care if our molecule contains hydrogen atoms with low pK_a's or is particularly susceptible to acid-catalyzed degradation.

2.1.4 **Cleaning NMR Tubes Prior to Use or Reuse**

Whenever we use a tube—even for the first time—we may wish to wash it out thoroughly. If rinsing with appropriate solvents fails to properly clean an NMR tube, the tube may not be visually free of residue. That is, it will appear cloudy or translucent instead of transparent. Immersion of the tube for 30 seconds in a saturated base and alcohol bath may suffice. Caution: when we perform this step, we wear gloves, a laboratory coat or an apron, and safety glasses or a face shield—we are only born with, after all, one perfect suit of skin and one set of eyes, hands, and feet.

Sometimes physical abrasion is needed to properly clean a tube. In this case, we GENTLY scrub the inside and outside of the tube. We can use pipe cleaners to clean 5 mm NMR tubes effectively, but we must take care not to scratch the inside of our tube with the exposed wire at the end.

We may even have access to an NMR tube washer, a device available from vendors of chemical laboratory equipment. If possible, we use high-performance liquid chromatography (HPLC) or spectroscopy ("Spec") grade water or acetone for the final rinse. We never use dimethyl sulfoxide (DMSO) for the final rinse unless we are immediately going to reconstitute our sample in DMSO, because the low vapor pressure of DMSO prevents its evaporation. Caution: we always wear gloves when working with DMSO. If our solute/DMSO solution comes in contact with our skin, the DMSO will transport our solute directly through our skin into our bloodstream.

2.1.5 **Drying NMR Tubes**

We dry expensive (and, most properly, all) tubes by laying them flat on a paper towel or clean cloth. We never store NMR tubes upright inside a beaker or an Erlenmeyer flask, and never put tubes in a drying oven for more than a minute or two. The most expensive conventional NMR tubes have the highest degree of concentricity, camber, and the most uniform wall thickness and glass composition. The thinner the wall, the faster the glass making up the wall will flow. If we lean our tubes in a beaker in the drying oven, gravity will bend the tube and make it out of camber. If we lay our tubes flat for too long, though, they will develop an oval cross section and thus will no longer be concentric. Thin-walled tubes are easier to destroy.

NMR tubes can be tested for camber and concentricity by using an NMR tube checker. These tube checkers are available from Wilmad and other vendors.

2.1.6 **Sample Mixing**

If we prepare a sample with a limited quantity of readily soluble solute and have just added the solute to the solvent, we must make sure the solution is well mixed. However, we must be careful how we mix our sample, because the standard-issue NMR tube caps of high-density polyethylene dissolve (or at least release pigment) in commonly used NMR solvents. A vortexer will afford effective mixing, but may not be readily available. Old salts in the NMR community can sometime be observed holding the tube gently in one hand and using deft whacks of the finger (be extra careful with thin-walled tubes, e.g., Wilmad 535-pp or higher) to induce mixing. Repeated withdrawal and reintroduction of a portion of the sample with a long necked Pasteur pipette will also facilitate mixing. Some samples are prone to foaming during the dissolution process, so we must take care not to mix too vigorously at first.

2.1.7 **Sample Volume**

How much solution we dispense into our NMR tube will affect our ability to quickly adjust the applied field to make it of constant strength in the detected region. In a 5 mm diameter NMR tube, a volume of between 0.6 and 0.7 mL is normally optimal.

Each NMR instrument has a depth gauge to allow us to position the NMR tube correctly with respect to the spinner. Figure 2.1 shows the correct spatial relationships between the tube, the spinner that holds it, and the region of the NMR tube that will occupy the probe's detector (a coil of wire that is an inductor in a resistor-inductor-capacitor circuit) when the spinner-tube assembly is in the instrument. Note that not all of the sample volume occupies the detected region.

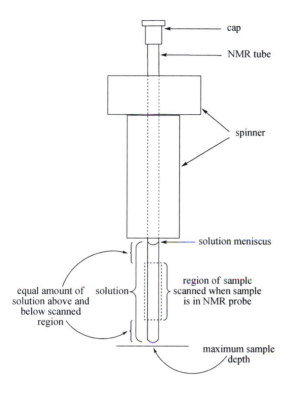

cap

NMR tube

spinner

solution meniscus

equal amount of solution above and below scanned region

solution

region of sample scanned when sample is in NMR probe

maximum sample depth

■ **FIGURE 2.1**   Schematic diagram of an NMR spinner containing a capped and solution-filled NMR tube. The region of the tube from which the signal is detected when the spinner/tube combination is placed in the probe inside the magnet is indicated. A depth gauge will normally indicate the detected region and the maximum allowed sample depth.

Prior to introducing our sample into the spinner, we wipe off the NMR tube scrupulously (fingerprints give a signal and also hinder smooth rotation). We align the NMR tube in the spinner with the aid of the depth gauge so that the solution's top interface (solution meniscus) and bottom interface (tube bottom) will be equidistant from the center of the detected region once the tube-spinner assembly is lowered pneumatically into the NMR magnet.

We must NEVER allow our tube to exceed the maximum allowed sample depth in the spinner because this error may cause damage upon sample insertion. If we have an excess of solution in our tube, we cannot center our solution volume about the midpoint of the detected region because this will exceed the maximum allowable depth; instead we put the tube into the spinner only down to the maximum depth. The result will be that the distance from the meniscus to the center of the detected region will be greater than the distance from the tube bottom to the center of the detected region.

Magnetic susceptibility. The ability of a material to accommodate within its physical being magnetic field lines (magnetic flux).

Line broadening. Syn. Apodization (not strictly correct). Any process that increases the measured width of peaks in a spectrum. This can either be a natural process we observe with our instrument, or the post-acquisition processing technique of selectively weighting different portions of a digitized FID to improve the signal-to-noise ratio of the spectrum obtained following conversion of the time domain to the frequency domain with the Fourier transformation.

Field heterogeneity. The variation in the strength of the applied magnetic field within the detected or scanned region of the sample. The more heterogeneous the field, the broader the observed NMR resonances. Field heterogeneity is reduced through adjustment of shims and, in some cases, through sample spinning.

Field homogeneity. The evenness of the strength of the applied magnetic field over the volume of the sample from which signal is detected. The more homogeneous the field, the narrower the observed NMR resonances. Field homogeneity is achieved through adjustment of shims and, in some cases, through sample spinning.

2.1.8 **Solute Concentration**

Ideally, we try to strike a balance between having our sample too concentrated and having it too dilute. If our sample is too dilute, we will find that a simple 1-D spectrum may take hours to acquire. If our sample is too concentrated, we will observe only broad resonances because a high solution viscosity slows molecular tumbling. Slow molecular tumbling only partially averages the dipolar and chemical shift tensors, depriving us of the full orientational averaging that occurs with rapid molecular tumbling; only complete orientational averaging allows us to observe narrow resonances.

Case 1. Excess solute. When we have the luxury of copious amounts of solute, our prepared solution should (still) be homogeneous. We must avoid having solids present in the tube. The one exception to this ban on solids is the presence of one dry Molecular Sieve™ (or comparable drying agent) in the very bottom of the NMR tube—well out of the detected region. If we want to use a saturated solution and are not worried about viscosity broadening of the NMR resonances, we can filter the solution after adding excess solute.

Unfiltered solutions can still be run—even those that are obviously heterogeneous—but this practice is discouraged because we may miss fine detail due to the broadness of the NMR resonances we will observe. The magnetic susceptibilities (the ability of a material to have magnetic field lines pass through it) of solids and solutions almost always differ, so we try to avoid the condition of having avoidable line broadening mechanisms. Solution heterogeneity causes field heterogeneity for which we cannot compensate effectively. A layman might describe the bits of solid in a solution as floaties (solids at the top of the solution), sinkies (solids at the bottom), and swimmies (solids with neutral buoyancy). Of the three, the swimmies will cause the most problems because they will drift in and out of the detected region. The passage of each undissolved solute particle through the detected region of the sample will bring with it an accompanying field homogeneity distortion. If we only have a few solid particles traversing our detected region, we will observe their deleterious effects either at random or periodically as a result of convection.

We filter a heterogeneous solution before putting it in our NMR tube. Adding a tiny splash of extra solvent to a saturated solution to get below the precipitation threshold may also help minimize the line broadening caused by the microscopic nucleation of colloidal or crystalline particles present in saturated solutions. Alternatively, we may raise the sample temperature 5° above the temperature at which our solution was prepared. When using a conventional NMR tube, we always keep our sample temperature well below the solvent's boiling point, especially when working with corrosive solvents such as trifluoroethanol (TFE) and trifluoroacetic acid (TFA). If we create excessive pressure in our NMR tube from heating our sample, the tube cap may come off and the contents of the tube will then spray up into the magnet's upper bore tube and then drip back down into the NMR probe, thereby creating a huge mess. Wrapping vinyl tape or Parafilm™ on top of the cap of a conventional NMR tube to keep the cap from popping off during sample heating is one measure we can take, but a more prudent approach is for us to resort to the use of a special NMR tube such as the J. Young™ NMR tube.

Case 2. Limited solute. When our amount of solute is limited and its solubility is high, we may be tempted to increase concentration at the expense of the total volume of solution. In most cases, we resist this temptation because lower than optimal solution volumes decrease the observed signal-to-noise ratio as the result of resonance broadening. Broadening a resonance with a fixed area decreases its amplitude, and the amplitude (height) of a resonance is the measure of how strong the signal is when we calculate the signal-to-noise ratio. The unwanted resonance broadening we observe with low volume samples is caused by the field heterogeneity associated with the detected region's close proximity to either or both the solution-vapor interface at the top of the solution and the solution-tube-air (or nitrogen) interface at the bottom of the tube. That is, skimping on solution volume to pump up the concentration often does more harm than good.

Upper bore tube. Syn. upper magnet bore assembly. A second metal tube (plus air lines, and possibly spin sensing components and PFG wiring), residing inside the upper portion of the magnet bore tube, through which the spinner/tube assembly passes via pneumatics en route between the top of the magnet and its operating position just above the probe inside the magnet.

If the number of moles of material we have (mass divided by molecular weight, MW) is small enough to make the solution we would prepare to dilute to carry out our desired NMR experiment(s) in the instrument time available to us, we may wish to resort to the use of susceptibility plugs, a Shigemi™ tube, or even a special probe (with special sample tubes available at especially high prices) such as the Bruker microprobe with a 80 μL sample volume or the Varian nanoprobe with a 40 μL sample volume. On a 400 or 500 MHz NMR instrument, sample concentrations below about 2 mM often prove problematic for ^{1}H-detected two-dimensional (2-D) NMR work. Sample concentrations below 5 mM for ^{13}C 1-D spectrum collection are similarly problematic. On a 200 or 300 MHz NMR instrument, the required sample concentrations approximately double.

2.1.9 **Optimal Solute Concentration**

For most of the spectra featured in this book, the sample concentrations were in the 20–50 mM concentration range. The point at which the onset of viscosity-induced dipolar and/or chemical shift anisotropy (CSA) broadening occurs (giving unacceptable results) will vary as a function of solute amount, solvent, temperature, and the judgment of the NMR operator (or the operator's research advisor, client, supervisor, or other superior). Viscosity broadening arises from incomplete averaging of the chemical shift and/or dipolar coupling tensor.

Chemical shift anisotropy, CSA. The variation of the chemical shift as a function of molecular orientation with respect to the direction of the applied magnetic field.

To obtain the proper solution height in a 5 mm diameter NMR tube with a 20 mM solution, use 14 μmol of solute dissolved in enough solvent to yield 0.70 mL of solution. If the MW is 300 g mol^{-1}, the amount of pure material required will be 4.2 mg. If the MW is 600 g mol^{-1}, the mass required is 8.4 mg.

To appreciate the importance of concentration, suppose that we find a four-scan ^{1}H 1-D NMR spectrum obtained from a 20 mM sample of one compound (compound 1) requires 30 s of instrument time to obtain a signal-to-noise ratio of 100:1 for an uncoupled methyl resonance. We may then wonder what signal-to-noise ratio we will obtain for a similarly uncoupled methyl resonance arising from a minor component in the same sample (compound 2) that is present with a concentration of 1.4 mM.

Signal S_i accumulates in direct proportion to the number of scans (n) and the sample concentrations c_i (Equation 2.2) but the noise (N) partially cancels and only accumulates as the square root of the number of scans (Equation 2.3). Thus, the signal-to-noise ratio (S/N) for each component will grow as the square root of the number of scans (Equation 2.4).

$$S_i \propto n \cdot c_i \tag{2.2}$$

$$N \propto n^{1/2} \tag{2.3}$$

$$S_i/N \propto n^{1/2} c_i \tag{2.4}$$

If compound 1 gives an S/N of 100:1 with a concentration of 20 mM in 30 s, then compound 2 at a concentration of 1.4 mM will give an S/N of 1.4/20 × 100:1 or 7:1.

To get the same S/N for compound 2 (1.4 mM) as we got for compound 1 (20 mM) in 30 s, however, we must increase the original experiment time (30 seconds) by the square of the ratio of the two concentrations! That is, if compound 2 is 14 times less concentrated than compound 1, we will require 14^2 or 98 times more NMR instrument time (49 minutes) to obtain the same S/N for a given resonance of comparable line width.

2.1.10 Minimizing Sample Degradation

As mentioned under Section 2.1.3, we use fresh solvent from an individual ampoule for our more precious samples if possible to prevent our solute molecule from undergoing further chemical reactions including labile proton exchange.

For cases in which an organic solvent is being used and the sample is to be kept as water-free as possible, we can introduce one molecular sieve™ (go quickly from the drying oven to the tube) in the bottom of our 5 mm NMR tube so long as the bead is well outside the detected region of the sample.

To minimize sample exposure to oxygen and water vapor, we prepare our sample in a glove box or glove bag. If we cannot ensure an inert atmosphere around the tube, we can use a latex septum to seal our NMR tube instead of a polyethylene cap and then cannulate our solution into the tube through the septum. Unfortunately, latex septa are permeable to oxygen and water vapor over time. Putting Parafilm™ or another comparable barrier to gaseous diffusion over the NMR cap (whatever type it may be) will also reduce the degradation caused by the entry of water vapor and oxygen. Also be aware that some NMR tube vendors do not wash their products prior to sale, they may only rinse the tube with potable tap water.

2.2 LOCKING

Solvent deuteration serves two important functions in the sample. First, while observing the ^1H nuclide, solvent deuteration prevents a large solvent signal from overwhelming the weaker solute signal (imagine trying to listen to one person whisper something important while a second person is yelling something we already know), and second, it provides a separate NMR signal whose frequency can be recorded to compensate for the inexorable downward drift of a superconducting cryomagnet (in the absence of a feedback mechanism, even an electromagnet will drift over time).

Locking. The act of establishing the condition of a stable deuterium lock.

Lock frequency. The Larmor frequency of the ^2H's in the solvent.

Deuterium lock channel. Syn. lock channel. The RF channel in the NMR console devoted to maintaining a constant applied magnetic field strength through the monitoring of the Larmor frequency of the ^{2}H's in the solvent and adjusting the field with the z_0 (Varian) or FIELD (Bruker) shim to keep the ^{2}H Larmor frequency constant.

Lock. Syn. field lock. The maintenance of a constant applied field strength through the use of an active feedback mechanism.

Field lock. Syn. deuterium field lock, ^{2}H lock, lock. The holding constant of the strength of the applied magnetic field through the monitoring of the Larmor frequency of one nuclide (normally ^{2}H, but possibly ^{19}F) in the solution and making small field strength adjustments.

Shim (n). One of a number of coils of wire surrounding the sample and probe wrapped so that a current passing through this coil induces a change in the strength of the applied magnetic field with a prescribed geometry.

Shim (v). The variation of current in a number of coils of wire, each wrapped in such a way as to produce a different geometrical variation in the strength of the applied magnetic field, in order to make the magnetic field experienced by the portion of the sample residing in the detected region of the NMR probe as homogeneous as possible.

The deuterium lock channel is the part of the NMR instrument that monitors the frequency of the ^{2}H's in the sample and adjusts the strength of the applied field so that the frequency of whatever nuclide is being observed is known. The frequency of the ^{2}H NMR signal is monitored every 500 ns and applied field strength is adjusted to maintain a constant value.

The field lock is established by slightly varying the strength of the applied field $\mathbf{B_0}$ until the ^{2}H frequency being generated in the NMR console is the same as the Larmor frequency of the ^{2}H's of the solvent in the sample. At this point, a phase-locked loop circuit is used to lock onto the ^{2}H frequency. From then on, magnet drift is compensated for, unless the limit of the ability of the instrument to adjust the applied field strength is reached.

2.3 SHIMMING

Shimming is a process in which we adjust a number of magnetic field gradients parallel and perpendicular to the applied field axis $\mathbf{B_0}$. Shims are adjusted by varying the amount of current traveling through the shim coils that make up the shim set, which lies between the probe and the magnet bore tube.

Shimming is normally carried out by maximizing the amplitude of the detected ^{2}H signal in the lock channel. Because the number of ^{2}H's in the detected region is constant, the area of the ^{2}H signal from the solvent is also constant as long we do not apply too much RF power that is tuned to the ^{2}H Larmor frequency.

By maximizing the amplitude of the signal in the lock channel through adjustment of the various shims, the width of the range of the ^{2}H solvent resonance is minimized. Imagine that the peak from the solvent is a triangle (it is probably a Lorentzian line, but a triangle is close enough for right now). If the area of a triangle is constant, reducing the width of the triangle must necessarily increase the height of the triangle. The width of the roughly triangular solvent resonance at half-height represents the range of frequencies being detected from the ^{2}H's in the solvent of the sample. The narrower the range of frequencies, the more homogeneous (even) the field. That is, if we have a range of $\mathbf{B_0}$'s being felt by the solvent molecules, the energy gap ΔE between the allowed spin states will also exhibit a range of values.

In most cases we adjust the shims along the **z**-axis, but sometimes we will also shim in the **xy** plane without sample spinning. Sample spinning serves to partially average field inhomogeneities present in the detected region of the sample. Shim sets with 40 or more adjustable shims are available.

Shimming by hand can be tedious. Fortunately, there are a number of automated applications that can expedite our arrival at a good shim set for a given sample. In practice, we often load a starting shim set (set of shim currents) and make improvements from there. Because every sample is unique, the optimal shims for each sample will similarly be unique.

2.4 TEMPERATURE REGULATION

Whenever possible, we regulate the sample temperature by using the variable temperature (VT) regulation hardware found on modern NMR instruments. Even regulating the sample at 25°C can improve the quality of the NMR data sets we generate, especially if the NMR experiment (the total amount of time we scan the sample) lasts more than a few minutes. If possible, we regulate the sample temperature when conducting long-term (overnight) 2-D NMR runs. However, we may be subject to limitations imposed by hardware, compressed gas reliability, our ability to operate the VT hardware, and by financial constraints. We generally use nitrogen gas at temperatures below 5°C and above 50°C.

2.5 MODERN NMR INSTRUMENT ARCHITECTURE

The modern NMR instrument contains three principal components: the magnet, the computer, and the electronic hardware between the two.

An NMR experiment is carried out as follows: we select an experiment to run. We tell the NMR software on the host computer (the computer attached directly to the NMR instrument) what experiment we want to run, and the computer sends the appropriate set of instructions to the computer built into the NMR console. The two computers most often communicate via an Ethernet connection. The host computer will have a second Ethernet card dedicated to talking to the NMR console computer. These instructions will include:

- how fast to spin the sample (if at all),
- the temperature at which to maintain the sample,
- whether or not to insert a new sample (if an automated sample changer is present),

Shimming. The act of varying the currents in the shims to achieve a more homogeneous applied magnetic field. Shimming most often entails maximizing the level of the signal of the lock channel, as rendering the field more homogeneous reduces the solvent's ^{2}H line width, which, given that the area of the ^{2}H peak is constant, must necessarily increase the height of the ^{2}H solvent peak, i.e., the lock level.

Shim set. A group of shims.

Magnet bore tube. Syn. bore, bore tube. The hollow, cylindrical tube that runs vertically (for an NMR magnet, horizontally for an MRI magnet) through the interior of a cryomagnet. The magnetic field maximum occurs within the interior of the bore tube. The room temperature (RT) shims are a hollow cylinder that is inserted inside the bore tube, and the probe is inserted inside the RT shims. Samples are lowered pneumatically down the upper bore tube, which is a smaller tube that rests on top of the RT shims and probe assembly.

Sample spinning. The rotation, using an air bearing, of the NMR tube/spinner assembly, used to average, on the NMR time scale, the strength of the applied magnetic field experienced by molecules in the sample solution. Sample spinning narrows the line widths of the peaks we observe and is almost exclusively employed in the collection of 1-D spectra.

Host computer. The computer attached directly to the NMR instrument.

Ethernet connection. The link between two or more digital devices through their respective Ethernet cards.

Ethernet card. A printed circuit board that resides in a digital device (a host computer or NMR console) that allows communication between the digital device and one or more other digital devices.

Console computer. The computer built into the NMR console (usually one or two 19" wide electronics racks). The console computer normally communicates via an Ethernet connection with the host computer, which is the computer the NMR operator uses to initiate experiments, etc.

Phase. The point along one wavelength of a sine wave where the waveform starts. The phase of an RF pulse also determines the direction in the rotating frame of reference that the net magnetization vector will tip relative to its initial orientation. The phase of an RF pulse is denoted with a subscript to indicate the axis of the rotating frame axis about which rotation occurs.

Oversampling. The collection of data points at a rate faster than that called for by the sweep width being used, thus allowing the subsequent averaging of the extra points to yield more accurate amplitude values spaced at the correct dwell time.

- whether or not to establish the deuterium field lock (only likely if a sample changer is being used),

- whether or not to shim the sample by using an automated shimming routine,

- whether or not and how to disrupt the magnetic field at the start of the pulse sequence,

- how long to wait before the first pass through the pulse sequence,

- the timing, amplitude, and phase of the one or more RF pulses tuned to the NMR frequency of the NMR-active nuclide being observed,

- the timing, amplitude, and phase of the one or more RF pulses tuned to the NMR frequencies of other NMR-active nuclides in the sample (besides those being observed),

- at what point or points in the pulse sequence to collect data,

- how many data points to collect per repetition of the pulse sequence, and how often to collect data points,

- how many repetitions of the pulse sequence to carry out,

- how many repetitions of the pulse sequence to periodically discard for the purposes of achieving the NMR steady state,

- whether or not to average the collection of extra points (oversampling),

- what to do in the event of an error,

- whether or not to periodically test the data for some criterion that indicates whether or not the experiment is done, e.g., has the solute signal exceeded a given signal-to-noise ratio?

If we are familiar with modern NMR instrumentation, then we will likely recall specifics concerning most if not all of the preceding items. If we lack modern NMR knowledge, then we may only recognize a handful.

The core of the NMR experiment can be addressed here without delving too deeply into many of the features that have been added to NMR instruments to make them easier and more convenient to use.

2.5.1 **Generation of RF and Its Delivery to the NMR Probe**

Radio-frequency electromagnetic radiation (RF) is generated by using a frequency synthesizer, which is just a box that generates a sinusoidal wave with a particular frequency. Because a single sinusoid has a constant phase and amplitude, a number of circuit boards within the NMR console are dedicated to controlling the phase and amplitude of the RF coming from the frequency synthesizer and eventually going to the sample.

Once the RF has been delayed (to control its phase) and chopped up into discreet bursts with constant or varying amplitudes, these RF pulses are sent to an RF amplifier. The RF amplifier takes in a low-power signal and amplifies it (this is one of the two points at which the RF pulse amplitudes are controlled). The high power RF pulse is then routed through cables from the console to a box that sits next to the magnet and is often called the preamplifier or preamp; this box may also contain other components. The RF passes straight through the preamp on its way to the sample via cables to the NMR probe. The NMR probe resides partly inside a tube running vertically through the NMR magnet (called the magnet bore tube or bore for short). The RF is then transmitted along a tuned electrical network that runs inside the probe and to the sample which is also inside the magnet.

Frequency synthesizer. A component of the NMR instrument that generates a sinusoidal signal at a specific frequency.

Sinusoid. A sine wave.

Preamplifier. Syn. preamp. An electronic device housed inside a metal box very close to the magnet containing circuitry to amplify the low-level NMR signal coming from the probe.

2.5.2 **Probe Tuning**

To maximize the efficiency of the delivery of the RF to the sample via the probe, one or more electrical components may need to be adjusted. This process is called probe tuning or just tuning, and is often a prerequisite for obtaining good data.

Probe tuning is required because the efficiency of the delivery of the RF power to the sample depends on the complex impedance of the transmitter coil in the NMR probe. NMR probes are normally tuned to a complex impedance (resistance as a function of frequency) of 50 ohms at the NMR frequency of interest. A few basic physics equations, show why probe tuning is important.

Power (P) is current (I) times voltage drop (V) across a circuit element (the probe's transmitter coil), or more simply,

Probe tuning. The adjustment of the complex impedance of the probe to maximize the delivery of RF power to the sample (forward power), to minimize reflected RF power, and to maximize the sensitivity of the instrument receiver to the NMR signal emanating from the sample following the application of the pulse sequence.

Complex impedance. Electrical resistance as a function of frequency.

$$P = IV \qquad (2.5)$$

If the resistance of the transmitter coil is infinite, then no current will flow through the coil and the power P will be zero.

The voltage drop (V) across an element in a circuit is current (I) times resistance (R) or

$$V = IR \qquad (2.6)$$

If the resistance of the transmitter coil is zero, then the voltage drop across the coil will also be zero. A zero voltage drop across the transmitter coil will again cause the power P (Equation 2.5) to be zero.

Somewhere between zero resistance and infinite resistance, the power dissipated by the transmitter coil (and delivered to the sample) will have a maximum. Most NMR probes are designed to deliver maximum power at 50 ohms.

Poor probe tuning degrades RF both coming and going. Failure to properly tune the NMR probe will make delivery of the RF to the sample inefficient. Even more significantly, poor probe tuning prevents efficient delivery of signal from the sample to the preamplifier. Because the NMR transmitter coil and receiver coil are normally the same (except in cryoprobes), poor probe tuning results in both inefficient delivery of RF to the sample and inefficient delivery of the NMR signal from the sample to the preamplifier.

Not only can a poorly tuned probe cause ineffective sample excitation and inefficient signal detection, but it can also cause damage. If only a portion of the total power generated makes it into the sample (forward power), the remaining power (reflected power) must be dissipated somewhere else. In some cases, a poorly tuned probe can reflect enough power to damage the NMR hardware, especially in high power applications such as solid-state NMR (not covered in this book).

2.5.3 When to Tune the NMR Probe and Calibrate RF Pulses

Many NMR probes now available are designed so that probe tuning response is relatively insensitive to sample changes. Consequently, we do not need to tune our probe or calibrate our RF pulses before carrying out various 1-D and 2-D NMR experiments. A probe of this nature is an essential feature of a high-throughput instrument.

Not all NMR data sets should be collected using a high-throughput NMR probe, because damping the probe's tuning response sacrifices

Forward power. Power delivered by the NMR instrument to the sample.

Reflected power. The portion of the power of an applied RF pulse that fails to be dissipated in the sample and instead returns through the cable connecting the probe to the rest of the instrument.

probe performance. Certain NMR experiments require (or at least benefit greatly from) probe tuning for every sample as well as meticulous pulse calibration. These experiments include the heteronuclear single quantum correlation (HSQC) [1] experiment, the heteronuclear multiple bond correlation (HMBC) [2] experiment, and the nuclear Overhauser effect spectroscopy (NOESY) [3] experiment, among others. Experiments relatively insensitive to moderately miscalibrated pulses (from marginal probe tuning or other instrument misadjustments) include most 1-D ^{1}H and ^{13}C experiments, the correlation spectroscopy (COSY) [4] experiment, the total correlation spectroscopy (TOCSY) [5] experiment, and even the heteronuclear multiple quantum correlation (HMQC) [6] [7] experiment.

With practice, we will be able to tune two RF channels of an NMR probe in under a minute, but we must take care not to damage the relatively delicate glass capacitors found in NMR probes (some newer NMR probes no longer contain these electrical elements—a point for rejoicing for those charged with maintaining NMR instrumentation in the multiuser environment).

2.5.4 **RF Filtering**

Probe tuning is not the final topic in the delivery of RF to our sample. Highpass, lowpass, bandstop, and bandpass analog filters are used to select which portion(s) of the RF frequency spectrum will get from one side of the filter to the other. Different filters are used, depending on need, and there are many variations from instrument to instrument. Nonetheless, many NMR instruments have similar filtering and RF configurations. Many NMR instruments normally operate with three active RF channels.

The first RF channel we consider is the proton or highband channel. This channel is dedicated either to ^{1}H exclusively or to ^{1}H and ^{19}F (the frequency of ^{19}F is 94% that of ^{1}H at a given field strength, see Table 1.1). This channel contains a bandpass filter that allows only the ^{1}H (or ^{1}H and ^{19}F) frequencies through, or it may contain a highpass filter to allow frequencies from ^{19}F and higher (up to infinite frequency) through because no NMR-active nuclide except ^{3}H (or ^{3}T, tritium) resonates at a frequency higher than that of ^{1}H. When not being used as the observation channel, the proton/highband channel may be used to decouple ^{1}H or ^{19}F nuclei in the sample so their persistence in the α and β spin states will not broaden other NMR-active nuclides we wish to observe. There is more discussion of decoupling in Section 6.7.

RF channel. The portion of the instrument devoted to generating RF with a specific frequency. There are four types of RF channels that may be found in an NMR instrument: a high-band channel (for ^{1}H and ^{19}F, and maybe ^{3}H), a broadband channel, for all nuclides with Larmor frequencies at that of ^{31}P and lower, a lock channel (devoted exclusively ^{2}H), and a fullband channel (any nucleus). Most instruments have one of the first three channels listed above.

Filter. Syn. RF filter. An electronic device used to limit the passage of RF of specific frequencies from one side of the filter to the other side. There are four types of filters. A high-pass filter only allows RF with frequencies above a given value to traverse the filter. A low-pass filter only allows RF with frequencies below a given value to traverse the filter. A band-pass filter only allows RF with frequencies that fall within a certain frequency range to traverse the filter. A band-stop filter only prevents RF with a certain range of frequencies from traversing the filter.

Filtering. The limiting of the frequencies of RF that may pass from one side of a filter to the other.

Proton channel. Syn. highband channel. The RF channel of an NMR instrument devoted to the generation and detection of the highest frequencies of which the instrument is capable. The highband channel can also normally generate the RF suitable for carrying out ^{19}F NMR experiments. Although ^3H (tritium) has a higher Larmor frequency than $_1H$, in practice this frequency is rarely called for.

Decoupling. The practice of irradiating one set of spins to simplify or otherwise perturb the appearance of other sets of spins through the suppression of one or more spin-spin couplings. Decoupling can be homonuclear or heteronuclear. Decoupling can be applied either continuously or in discreet bursts.

Broadband channel. The portion of a spectrometer capable of generating RF to excite nuclei with Larmor frequencies less than or equal to that of ^{31}P (at 40% of the ^1H Larmor frequency). In some cases, the broadband channel may be capable of exciting ^{205}Tl (at 57% of the ^1H Larmor frequency). The preamplifier in the broadband channel's receiver will normally feature a low-pass filter and a ^2H bandstop (notch reject) filter.

Stray field. The magnetic field lines that extend beyond the physical dimensions of the NMR magnet's cryostat.

Pulse calibration. The correlation of RF pulse duration (at a given transmitter power) to net magnetization tip angle.

The second RF channel we consider is the broadband channel. The broadband channel covers a wide frequency range, from that of ^{31}P at 40% of the ^1H frequency down past that of ^{13}C at 25% of the ^1H frequency on to the frequency of ^{15}N at 10% of the ^1H frequency. The broadband channel may operate at frequencies even lower than that of ^{15}N, and nuclides with NMR frequencies below ^{15}N are typically called low-γ nuclides because their gyromagnetic ratios (their γ's) are less than 10% of the magnitude of the γ of ^1H. This channel typically sees less use than the proton/highband channel. The broadband channel normally contains two filters, a lowpass filter to prevent the highband RF being used to irradiate ^1H's or ^{19}F's in the sample from interfering with the receiver when broadband nuclei such as ^{13}C, ^{31}P, and ^{15}N are observed, and a bandstop filter to reject the oft-forgotten ^2H lock RF that falls into the broadband frequency range (^2H resonates at 15% of the ^1H frequency).

The third RF channel we consider is the lock channel. This channel contains a bandpass filter that only allows RF at the NMR frequency of ^2H's to pass from console to probe and back. The purpose of the lock channel is to calibrate the frequency of the other NMR channels of the instrument. The lock channel compensates for fluctuations in the strength of the applied magnetic field by slightly exciting deuterons in the solvent many times per second and measuring the precession frequency of the net magnetization vector for the deuterons in the sample. Because the ratio of the frequency of any other NMR-active nuclide relative to that of ^2H is constant, a slight change in the observed frequency of the ^2H in the sample requires that the other NMR frequencies be adjusted accordingly. By keeping the observed ^2H frequency constant through a shim that varies B_0, the ^2H lock channel compensates for both the slow run down of the strength of a superconducting magnet over time (in the range of <1–$10\,Hz/hr$ for ^1H) and the transient perturbations of the field strength due to the movement of ferromagnetic and other objects in the relatively weak stray field surrounding the magnet. This compensation helps keep the NMR lines we observe over the course of a long NMR run to a minimum.

2.6 PULSE CALIBRATION

To tip the net magnetization of a spin ensemble by 90°, we apply a timed pulse of RF at the correct frequency and amplitude to our sample. To ensure our tip angle is 90°, we first determine the amplitude and duration of the pulse of the RF pulse to use in a process called pulse calibration. Pulse calibration may be carried out only periodically or with each sample.

Figure 2.2 shows the time periods of the one-pulse NMR experiment. The area (amplitude times duration) of the RF pulse will determine how much the net magnetization vector will be tipped from its equilibrium position along the z-axis (the applied field axis). In practice, we place a standard sample in the NMR instrument, set a reasonable transmitter power level, and adjust the duration of the RF pulse to determine the 90° pulse width, usually 8–10 μs for ^{1}H using a 5 mm probe. To minimize the amount of time we must wait between pulses, we look for a null by applying pulses four times longer than the expected 90° pulse, as returning the net magnetization vector to a place at least close to its starting point (following 360° of rotation) minimizes the wait time needed between pulses to allow for relaxation.

For conducting 2-D NMR experiments such as the HSQC, HMBC, and NOESY experiments, we may elect to tune and calibrate the ^{1}H 90° pulse for each sample.

Calibration of ^{13}C pulse widths is often done with a standard sample containing a copious amount of solute (or one that is ^{13}C enriched). Calibration of ^{15}N pulse widths also benefits from the use of an isotopically enriched standard sample. Once we put in our real world sample (with a concentration of perhaps less than 5 mM), often the best we can do is tune the probe and hope the calibration arrived at while using a different sample will be sufficiently accurate. We must assume that pulse calibrations determined using a standard are valid for our sample as well.

Figure 2.3 shows how the amplitude of an NMR resonance varies sinusoidally as a function of the duration of the applied RF pulse (following a lengthy relaxation delay in between each scan). This figure shows how the signal generated by the net magnetization vector M does NOT increase without bound as the pulse width is increased; instead, there is an optimal RF pulse width that generates maximum NMR signal.

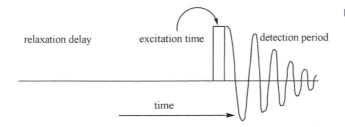

■ **FIGURE 2.2**   Time periods in the one-pulse NMR experiment.

■ **FIGURE 2.3**   A single ^{1}H NMR resonance traces a sinusoidal pattern as a function of increasing pulse width; the first pulse is 1 μs, and each subsequent pulse is 1 μs longer.

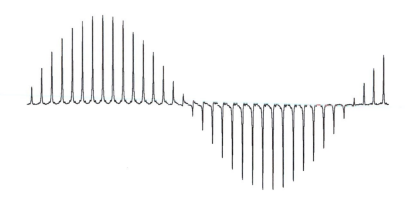

2.7 SAMPLE EXCITATION AND THE ROTATING FRAME OF REFERENCE

Once the RF pulse travels from the transmitter coil of the NMR probe to the sample, the net magnetization vector summed from the ensemble of NMR-active spins in the sample will be perturbed from its equilibrium position. Recall from Chapter 1 that the net magnetization vector **M** of our spin ensemble will feel a little push from each beat of the magnetic field component of the RF being applied to the sample. The magnetic field component of the applied RF pulse is called the **B$_1$** field (recall that the static applied magnetic field is **B$_0$**). If a 10 μs pulse at 125 MHz is applied to our sample, then there will be 1×10^{-5} s $\times$ 1.25×10^{8} cycles s^{-1} or 1250 little pushes (cycles), with each push occurring every 8 ns. Each push will tilt the net magnetization vector by 90°/1250 or 0.075°.

Figure 2.4 illustrates the concept of the rotating frame. The rotating frame is distinct from the laboratory (static) frame. The rotating frame is a frame of reference we can use to view the net magnetization vector without having to worry about how it precesses at its Larmor (NMR) frequency (in this example 125 MHz). It is a second Cartesian (**xyz**) coordinate system in which the **z**-axis is stationary and parallel to the **z**-axis of the laboratory frame of reference, but in which the **x**- and **y**-axes remain perpendicular to each other (and to the **z**-axis) and rotate at the Larmor frequency in the laboratory frame's **xy** plane. The axes of the rotating frame are denoted **x'**, **y'**, and **z'**.

Recall from Chapter 1 that, once the net magnetization vector is tipped away from equilibrium along the **z**-axis and has a component in the **xy** plane, it will feel a torque from the applied field **B$_0$** and

Rotating frame. An alternate Cartesian coordinate system (**x'**, **y'**, **z'**) sharing its z-axis with that of the laboratory (stationary) frame of reference. The rotating frame of reference rotates at the Larmor frequency of the nuclide being observed.

Static frame. Syn. laboratory frame. The frame of reference corresponding to the physical world in which the experiment is carried out.

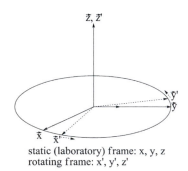

static (laboratory) frame: x, y, z
rotating frame: x', y', z'

■ **FIGURE 2.4**   The relationship between the static (laboratory) frame of reference and the rotating frame of reference.

precess at the NMR frequency as a result of this torque. When the net magnetization vector has felt the first of the 1250 B_1 field pushes and is viewed from the perspective of the rotating frame, the vector does not precess because the frame of reference (the rotating frame) is rotating in synchrony with the precession. That is, the net magnetization vector will just be tilted a wee bit (0.075°) away from its initial alignment along the z′-axis (Figure 2.5a).

As subsequent B_1 pushes occur, the net magnetization vector **M** will continue to tilt further from the z′-axis (Figure 2.5b). After 1250 pushes, the net magnetization vector will reside completely in the **xy** plane (Figure 2.5c).

2.8 **PULSE ROLL-OFF**

Each group of chemically distinct spins in a molecule may precess at slightly different frequencies due to chemical shift. Figure 2.6a shows a perspective of the rotating frame as viewed from above. Two chemically distinct sets of spins (A and B) allow the net magnetization vector **M** to be split into two components: M_A and M_B. When tilted from their equilibrium positions along the z′-axis, M_A and M_B precess at different rates, so only one can be stationary with respect to the rotating frame. A single rotating frame of reference will not suffice to keep all components of the net magnetization vector stationary once M_A and M_B have been tipped from their equilibrium positions along the z′-axis. If we apply a single frequency of RF to our sample, then we can have only one chemical shift exactly on-resonance (stationary) with respect to the rotating frame. All spins that are off resonance will accumulate a small phase difference once we start to apply our 90° pulse comprising the 1250 B_1 pushes (Figure 2.6b). That is, once M_B has a component in the **xy** plane, this component will precess during the course of the RF pulse. The result is the accumulation of a phase difference between the **xy** component of M_B and the **xy** component of M_A (recall that M_A is stationary with respect to the rotating frame).

If a pulse is long enough, the phase difference between an off-resonance spin and the phase of the RF pulse will eventually reach 90° (M_B in Figure 2.6c), after which point the off-resonance spin will no longer continue to tip away from its equilibrium position but instead will begin to return to where it started (the +z′-axis, see M_B in Figure 2.6d). The dashed arrow in Figure 2.6d represents the **xy** component of M_B, and is shorter than the **xy** component of M_A. The dashed ellipse in Figure 2.6 is the measure of the **xy** component of M_A.

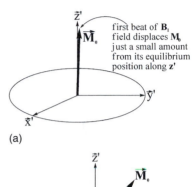

first beat of B_1 field displaces **M**, just a small amount from its equilibrium position along **z′**

(a)

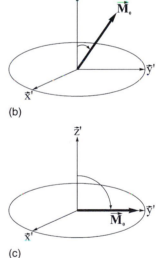

(b)

(c)

■ **FIGURE 2.5** Tipping of the net magnetization vector $\mathbf{M_0}$ from its equilibrium position along **z′** through a series of B_1 field pushes.

Off resonance. A spin is off resonance when the spin's resonant frequency is not at the center of the spectral window (the center of the spectral window corresponds to the frequency of the rotation of the rotating frame of reference).

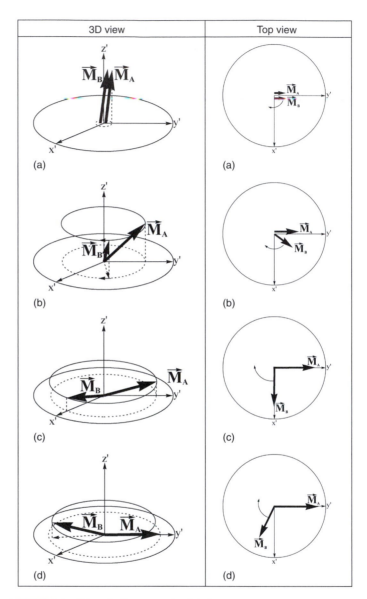

■ **FIGURE 2.6**   Pulse roll-off attenuates signals far from on-resonance.

On resonance. A spin is said to be on resonance if the spin's resonant frequency lies at the center of the spectral window.

After a 10 μs pulse of RF at 125 MHz is applied to our sample, those chemical shifts 25 kHz from being on resonance (those at 125.025 MHz and at 124.975 MHz) will just begin to turn around and head toward the +z′-axis.

The longer the RF pulse, the narrower the excited range of spins with precession frequencies in the vicinity of the NMR frequency (also referred to as the Larmor frequency, the spectrometer frequency, the carrier frequency or the transmitter frequency, or as on-resonance). Because the relative intensities of signals lessen gradually with increasing distance from being on resonance, this phenomenon is referred to as pulse roll-off—a term implying the gradual lessening of the breadth of excitation effectiveness of applied RF pulses. Figure 2.7 shows a series of ^{13}C 1-D spectra obtained at 125 MHz; these spectra show how increasing pulse width (the power was adjusted to achieve a constant tip angle for on resonance spins) decreases the excitation range, gives nodes (points where there is no excitation), and even negative peaks. Notice how the nodes (Figures 2.7b-d panels only) move closer to the left side of the spectra (where the transmitter is) as the pulse width increases.

We can calculate the frequency of the precession of the net magnetization vector in the rotating frame as a result of the action of the $\mathbf{B_1}$ field. This $\mathbf{B_1}$ frequency represents the edge of the excitation range for a 90° pulse. To preserve true intensities, it is best to limit resonances to within only 10% of this range.

In the example above, if 90° of rotation takes 10 μs, then 360° of rotation must take 40 μs. Therefore the $\mathbf{B_1}$ frequency excitation limit is plus or minus $1/4 \times 10^{-5}$s or ± 25 kHz.

2.9 PROBE VARIATIONS

NMR probes come in many different varieties. The three main features of conventional liquid-state NMR probes (for a narrow bore magnet with an inside diameter of 51–54 mm) are (1) the diameter of the NMR tube they are designed to accommodate; (2) the configuration of the coils (normal versus inverse), and (3) the presence or absence of gradients coils to allow the employment of pulsed field gradients (PFGs). Most probes sold currently are pulsed field gradient probes.

Beyond variations of these three features, there are other commercially available probes that can be used for liquids NMR work, but these probes will not be discussed in detail in this text. We should for now simply be aware that other types of liquids NMR probes exist.

Most NMR magnets have a magnet bore tube (the vertical tube through the center of the magnet) with an inside diameter of

Spectrometer frequency, sfrq. The frequency of the RF applied to the sample for the observe channel, also the frequency of the rotating frame for the observe nuclide.

Carrier frequency. Syn. NMR frequency, Larmor frequency, on-resonance frequency, transmitter frequency. The frequency of the RF being generated by a particular channel of the spectrometer. The carrier frequency is located at the center of the observed spectral window for the observed (detected) nuclide.

Transmitter frequency. Syn. carrier frequency, NMR frequency, on-resonance frequency. The rate at which the maxima of the sinusoidal wave of the RF generated by the observe nuclide's RF channel occur.

Pulse roll-off. The diminution of tip angle that results from the accumulated error caused by the difference between the frequency of the applied RF pulse and the frequency of a given resonance.

Node. A point where a function— e.g., the excitation profile of an RF pulse—has zero amplitude.

Pulsed field gradient, PFG. The transient application of an electric current through a coil of wire wound to induce a change in magnetic field that varies linearly with position along the x-, y-, or z-axis of the probe.

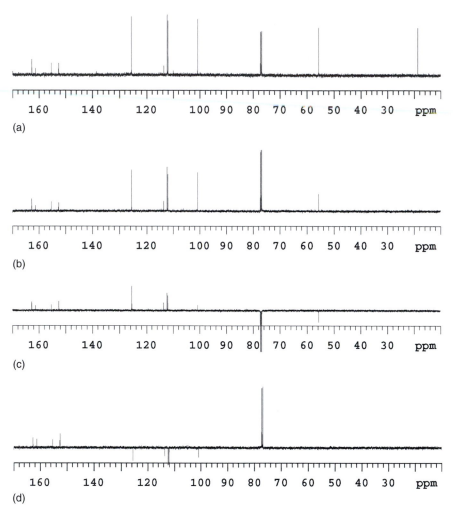

■ FIGURE 2.7   Four 125 MHz ^{13}C NMR spectra of the same sample collected with the transmitter at the left edge of the displayed spectra (170 ppm) and a pulse width of (a) 26 μs; (b) 52 μs; (c) 104 μs; and (d) 208 μs.

51–54 mm. Magnets used for solids and other high-power NMR applications often have larger bore tubes with inside diameters of 89 mm or larger; these magnets are called widebore magnets. A larger bore diameter is required to accommodate NMR probes whose components are bulky (e.g., magic-angle spinning stator assemblies) and must be spread apart to avoid electrical arcing associated with the high voltages. The extra space in widebore magnets also allows probe components to be physically larger, thereby making them better able to withstand the larger amounts of power dissipated when large amounts of RF power are applied.

A portion of the magnet bore tube space is occupied by a sleeve of insulated wires called a room temperature (RT) shim set. The RT shim set should never be removed from a magnet's bore tube unless it is defective. On the other hand, the NMR probe fits inside the RT shim set in the bore and can be more or less freely exchanged with other NMR probes, depending on need. Both the RT shim set and the NMR probe are inserted into the NMR magnet's bore tube from the bottom. In most cases there will be an upper magnet bore assembly (also called the upper bore tube) through which the spinner/tube assembly will travel as the spinner/tube assembly is raised and lowered pneumatically.

The outside diameter of an NMR probe is less than the inside diameter of the magnet bore tube because the NMR probe must fit inside the RT shim set, which is also inside the magnet bore tube.

2.9.1 Small Volume NMR Probes

There are probes specifically designed for small volumes of sample (e.g., the Bruker microprobe, which handles 80 µL volumes and the Varian nanoprobe, which handles 40 µL volumes), but these probes suffer from problems with either low spectral resolution and/or low throughput, not to mention the order-of-magnitude-plus higher cost associated with sample vessels versus conventional NMR tubes. Nonetheless, if our application requires and benefits from the use of one of these small-sample-volume probes, we will be glad that these probes exist.

2.9.2 Flow-Through NMR Probes

The pharmaceutical industry and others carrying out combinatorial chemical methods have embraced the advent of the flow-through NMR probe, because this probe can be used to examine fractions eluted from liquid chromatographs. This probe uses no NMR tubes, instead an autosampler pumps the solute-containing solution through the probe and into the detected region of the probe. These probes require rinsing between samples and thus consume significant quantities of solvent. Pulsed-field gradients are often used to suppress the signal from nondeuterated solvents, because the use of deuterated solvents for the preceding chromatographic work is costly—in some cases prohibitively so, depending on the choice of solvent.

Pulsed field gradient probe, PFG Probe. Syn. gradient probe. An NMR probe equipped with one or more gradient coils capable of altering linearly the strength of the applied field as a function of position.

Field gradient pulse. Syn. gradient pulse. The short application (10 ms is typical) of an electrical current in a coil of wire in the probe that surrounds the detected region of the sample and causes the strength of the applied magnetic field to vary as a function of displacement along one or more axes.

2.9.3 **Cryogenically Cooled Probes**

As of this writing (2007), cryogenically cooled probes are commercially available but not yet financially viable for all but those institutions and firms with the deepest pockets.

The principle of cooling the probe is a sound one: using a superconducting (or nearly so) receiver coil to allow the NMR signal coming from the sample to travel with minimal resistance through cooled conductors (to reduce thermal or Johnson noise) to the preamplifier where it undergoes amplification to increase the signal current to a level less susceptible to corruption by shot noise. The signal-to-noise ratio obtained using a probe with a cooled receiver coil rather than a conventional NMR probe with its more-resistive receiver coil makes the cooled probe a truly attractive proposition. Unfortunately, at the time of this writing, commercially available versions generally lack both reliability and affordability.

The high cost to play the cooled probe game arises from the myriad problems associated with the probe's technical requirements. Requirements and weaknesses include:

- Power intended for the sample may heat and possibly damage the cooled coil (especially if probe tuning is poor) and may increase the resistance of the receiver coil, thus diminishing sensitivity. Repair is not trivial.

- Keeping the sample at ambient temperature and having a cryogenically cooled receiver coil ($\sim$10 K) perhaps 1 mm distant creates huge thermal gradients, and the thermal stresses associated with cooling and rewarming the receiver coil cause mechanical fracture. Again, repair is not trivial.

- The delicate and expensive coil assembly in the probe can also be damaged by improper sample insertion and/or imperfect NMR tubes.

- The heat exchanger must operate with its moving metal parts in a strong magnetic field (although shielded/screened magnets reduce this problem). Failure due to wear caused by the force of the NMR magnet's stray or fringe field on the nearby moving parts of the helium heat exchanger is possible.

- The acquisition of high-quality NMR data requires vibration isolation of the probe from the helium compressor, which is only a short distance away.

Johnson noise. Syn. thermal noise. The electrical noise caused by the Brownian motion of ions in a conducting material, e.g., a wire. This type of electrical noise varies with temperature.

Shot noise. Electrical noise resulting from the movement of individual charge quanta, like raindrops on a tin roof. With a low flux, individual drops are heard in a random pattern, but as the flux increases, the impact of the individual drop is lost in the continuum of many drop impacts per unit time.

The high costs of frequently repairing cryogenically cooled probes have prompted vendors to require or at least strongly suggest that their cooled probe customers purchase an annual probe plus heat exchanger maintenance contract for sums in excess of $25,000.

When we consider that the price of this accessory starts around $125,000 (if we drive a hard bargain; $200,000 if we do not) we may conclude that liquids NMR experimentation employing cooled receiver coil probes has to evolve more before it becomes accessible to those with more limited finances.

It is expected that high maintenance costs will lessen and poor reliability will diminish in the next five to ten years. As a result of the extensive engineering required, the cost of the cooled receiver coil probe is expected to remain more than double the cost of a conventional probe for longer than that.

2.9.4 Probe Sizes (Diameter of Recommended NMR Tube)

Probes for liquids NMR work typically are designed to accommodate NMR tubes with outside diameters of 3, 4, 5, 6.5, 7.5, 8, and 10 mm (and larger). Probes designed for 5 mm NMR tubes are most common. Probes designed to accommodate 15 and 20 mm NMR tubes, while once common, are no longer offered by NMR vendors for normal applications (for the right amount of money, they will probably agree to build us one, but delivery may take years).

The advantage to having an NMR probe that can accommodate a large diameter NMR tube (8 or 10 mm) is that more solute molecules at a particular concentration can be put into the detected region of the sample if solute is not limited. The disadvantages of a large diameter NMR tube is that (1) larger amounts of solute will be required to attain the desired sample concentration, and (2) the region of the magnetic field to be made homogeneous through shimming is larger. If we are given the choice between running 2.8 mL of a 2.5 mM solution in a 10 mm NMR probe and 0.70 mL of a 10 mM solution (the number of moles of solute is the same), we choose the latter unless the viscosity of the 10 mM unacceptably broadens the resonances.

2.5 and 3 mm NMR probes have become popular with scientists carrying out bioNMR work, because this discipline often deals with small molar amounts of material. Having a more concentrated sample with less solvent is generally preferred as long as viscosity broadening, precipitation, and aggregation are not significant.

2.9.5 Normal Versus Inverse Coil Configurations in NMR Probes

Not counting the lock channel, an NMR probe will typically have two or possibly three RF channels. One of the channels is almost always a highband (^{1}H and ^{19}F) or ^{1}H-only channel, whereas the remaining channel or channels will be tunable to the NMR frequency of one or more of the nuclides that fall in to the broadband frequency range (^{31}P to ^{15}N or lower—this range includes ^{13}C).

Single-channel NMR probes were initially built to allow the observation of the ^{1}H NMR signal, and these probes produced fine spectra. A single-channel NMR probe for observing ^{13}C cannot suppress the ^{13}C resonance broadenings caused by J-coupling to nearby ^{1}H's. The solution is addition of a second coil to allow application of a continuous stream of RF tuned to the ^{1}H NMR frequency to rapidly scramble (decouple) the ^{1}H's.

When two-channel NMR probes were first developed, the coil closest to the sample was the coil used for the detection of ^{13}C. A second coil, positioned outside of the first coil and tunable in the ^{1}H frequency range, was there to allow the decoupling of the protons near ^{13}C nuclei (decoupling is discussed in Chapter 6), thereby simplifying the observed ^{13}C multiplets and enhancing the ^{13}C signals through the nuclear Overhauser effect (NOE, discussed in Chapter 7). Because the broadband coil was closer to the sample, this coil was more sensitive to the broadband NMR signals received from the sample, whereas the ^{1}H coil (for decoupling) was on the outside. Just as putting your ear next to somebody's mouth will allow you to hear their whisper more clearly, so too does the close proximity of the receiver coil to the sample affect the magnitude of the detected signal. This type of dual coil configuration is known as the normal coil configuration: the broadband coil is inside the proton coil. Normally configured two-channel NMR probes are best suited to the observation of the broadband nucleus or nuclei to whose frequency the broadband coil can be tuned.

If, on the other hand, we wish to observe a weak proton signal from a sample, we prefer to use a probe with a receiver coil configuration in which the ^{1}H (or highband) coil is the coil closest to the sample.

Multiplet. A resonance showing multiple maxima; the amplitude distribution, often showing a high degree of symmetry, in a frequency spectrum arising from a single NMR-active atomic site in a molecule that is divided (split) into multiple peaks, lines, or legs.

Coil configuration. The relative orientation of the highband and broadband coils with respect to their placement around the sample-containing NMR tube. A normal coil configuration locates the broadband coil closer to the sample, while an inverse configuration locates the highband (normally tuned to ^{1}H, but possibly ^{19}F) closer to the sample.

When modern NMR was in its infancy, few if any scientists realized that they would ever require a two-coil NMR probe in which the ^1H coil lies inside the broadband coil. It was not until the demonstration that experiment time can be drastically reduced for the acquisition of 2-D heteronuclear correlation NMR data sets by employing ^1H detected methods and a two-coil NMR probe with the coil configurations inverted (relative to the normal coil configuration described above) that the inverse probe really caught on. Now the inverse probe (a probe with the ^1H coil closer to the sample than the broadband coil) is a staple in nearly every modern NMR laboratory.

2.10 ANALOG SIGNAL DETECTION

When a component of the net magnetization vector **M** precesses in the **xy** plane at the NMR frequency, it induces an analog (continuous) signal, in this case a current in the receiver coil. The current induced in the receiver coil in the NMR probe passes out of the NMR probe, through a coaxial (insulated) cable, and into the preamplifier. Because the signal coming from the sample is so weak, it must be amplified as soon as possible to minimize its corruption by noise. Therefore, a preamplifier (or preamp) is located outside of the console and closer to the probe. A separate preamp box is located near the magnet, or maybe even inside one of the legs of the magnet.

After the analog signal has been amplified to a level that will not allow it to become easily corrupted, its frequency is reduced from ten or hundreds of MHz to the kHz (audio) range. Frequency reduction (mixing down) is accomplished with a mixer that subtracts the NMR transmitter (or carrier) frequency from the signal, leaving only low frequency (audio) signals due the movement of the net magnetization vectors in the rotating frame of reference. That is, mixing down converts precession frequencies relative to the laboratory frame to precession frequencies relative to the rotating frame.

2.11 SIGNAL DIGITIZATION

We use a computer to process the data we observe. However, before we can process with a computer, we convert the mixed-down-but-still-analog signal to a digital signal using an analog-to-digital converter (A/D, vocalized as "a to d"). Reduction of the analog signal from the MHz range to the kHz range is required to allow digitization to take

Analog-to-digital converter, A/D. An electronic device that converts an analog voltage into a binary number composed of discreet digits (a series of 1's and 0's).

Digitization. The conversion of an analog voltage to a digital, binary number amenable to subsequent computational manipulation.

Spectral window, SW. The range of frequencies spanned by a spectrum, whose location in the frequency spectrum is determined by both the dwell time and the frequency subtracted from the time domain analog signal prior to digitization.

Sweep width, SW. The amount of the frequency spectrum spanned, which is controlled by the dwell time. (Note that spectral window is not the same thing but is also denoted SW—a great source of confusion).

Dwell time. The time interval between sampling events for the digitization of the analog signal arising from the FID; equal to the reciprocal of the sampling rate.

t_2 time increment. The second time delay in a pulse sequence used to establish a time domain that will subsequently be converted to the frequency domain f_2.

place properly. The signal is mixed down so that the A/D will not be overwhelmed with a signal that oscillates too rapidly.

In the present state of technology, A/D conversion cannot adequately characterize a sinusoidal signal with a frequency at the NMR frequency. At the time of this writing (2007), the fastest 16-bit A/D available on www.analog.com operates at 100 MHz and the fastest 12-bit A/D operates at 400 MHz. The state of A/D conversion technology continues to improve, but the A/D capable of sampling at 1–2 GHz is not likely to be forthcoming in the near future. A 12-bit A/D produces 12 ones and zeros to characterize an analog signal. A 16-bit A/D similarly generates 16 ones and zeroes.

Most modern liquid-state NMR instruments have a maximum receiver sampling rate of 100–500 kHz (instruments for solid-state applications have faster A/D's). The Nyquist sampling theorem states that, to properly characterize a sinusoidal signal of frequency v, the sampling rate must be greater than or equal to 2v. Therefore, an A/D sampling at 100 kHz can only characterize a 50 kHz signal. If we add a second receiver channel with a 100 kHz A/D perpendicular to the first channel, we will be able to distinguish positive from negative precession of NMR signals in the rotating frame of reference, and therefore our spectral window will span +50 kHz to −50 kHz for a total sweep width of 100 kHz. The width of the window we observe in the frequency domain (i.e., in frequency space) is controlled by the sampling rate of our A/D('s)—that is, by how rapidly the A/D('s) is(are) operating in the time domain.

The time interval between A/D sampling events is the reciprocal of the sampling rate and is called the dwell time. For a 100 kHz sampling rate, the dwell time is $1/(1 \times 10^5$ Hz) or 10 µs. For a 10 kHz sampling rate the dwell time is therefore 100 µs. The dwell time is the t_2 time increment for a 2-D experiment.

Any signal (or noise) outside the spectral window will alias (or fold back) into our spectral window. We can prevent this from happening by using either conventional analog filters (different filters from those we put between the probe and the rest of the NMR) in our receiver or by using digital filters. Digital filters have the advantage of cutting off signals outside the spectral window more abruptly than conventional filters.

Bruker and Varian NMR instruments differ in their method of sampling the NMR signal to produce a phase-sensitive NMR spectrum.

A phase-sensitive NMR spectrum allows us to adjust the proportions of the signals given us by the two orthogonal receiver channels. This adjustment is called phasing and can be used to generate fully absorptive (as opposed to dispersive) resonances in most cases. A phase-sensitive NMR spectrum is preferred over an absolute-value or power NMR spectrum because spurious signals can be easily identified if they do not become fully absorptive when the authentic NMR resonances are all properly phased.

Some older Bruker NMR instruments sample at a frequency that is two times the sweep width and use one A/D, whereas Varian NMR instruments sample at the frequency of sweep width and use two orthogonal receiver channels. The difference in the two sampling methods results in how aliasing (a.k.a. folding) takes place. On a Bruker NMR instrument, a NMR signal outside the spectral window will reflect from the same side it is on, whereas on a Varian NMR instrument the folded signal will appear on the opposite side of the spectrum. This difference in sampling also carries over into the 2-D realm insofar as most Bruker pulse sequences use the time-proportional phase incrementation (TPPI) [7] method for making the second dimension phase sensitive, whereas most Varian pulse sequences use two perpendicularly phased pulses (Ruben-States-Haberkorn [8] or simply the States method) to generate phase sensitivity in the second dimension of 2-D NMR spectra.

An A/D is really just a mapping device. It takes an analog signal in and puts out a number. If we have a sine wave that varies between +1.0 volts and −1.0 volts (in electronics, most sine waves vary about zero volts) and we want to be able to convert the voltage of that sine wave at any moment into a digital number, we will need to keep one of the bits of our A/D to use as the sign bit. That is, if the signal is negative, then set bit #1 to a one; otherwise, set bit #1 to a zero. The other 15 bits in our 16-bit A/D word are used to describe the amplitude of the voltage being converted. Thus, our 16-bit A/D can go from −32,767 to +32,768 ($2^{15} = 32,768$, but we have to devote one of the possible combinations of ones and zeroes to a zero voltage).

Whether or not we worry how computer scientists represent a negative number by using a word of memory composed of ones and zeroes is largely irrelevant. It is important, however, that we understand that each unique voltage in the allowed range will in turn be mapped into some unique combination of ones and zeroes in the output.

Phasing. The manipulation of a frequency spectrum through the weighting of points from two orthogonal data arrays (or matrices) to generate spectral features that are most often purely absorptive and positive.

Time-proportional phase incrementation method, TPPI method. A method for imparting phase sensitivity into either the indirectly (f_1) detected dimension of a 2-D experiment or the directly detected dimension of a 1-D experiment (or the f_2 dimension of a 2-D experiment). The TPPI method involves sampling points at half the dwell time prescribed for a given sweep width (for the directly detected dimension), or using a t_1 time increment that is half of that prescribed for the f_1 sweep width.

Phase sensitive. The collection of an NMR data set involving the use of a 90° phase shift in the receiver and also possibly in the phase of one of the RF pulses of the pulse sequence, thus allowing the storage of the digitized data points into two separate memory locations to allow phase correction during processing.

Ruben-States-Haberkorn method. Syn. States method. A phase-cycling method for making the indirectly detected dimension (f_1) in a 2-D spectrum phase sensitive. Phase sensitivity is realized by varying the phase of one of the RF pulses in the pulse sequence by 90° for pairs of digitized FIDs obtained using the same t_1 time increment.

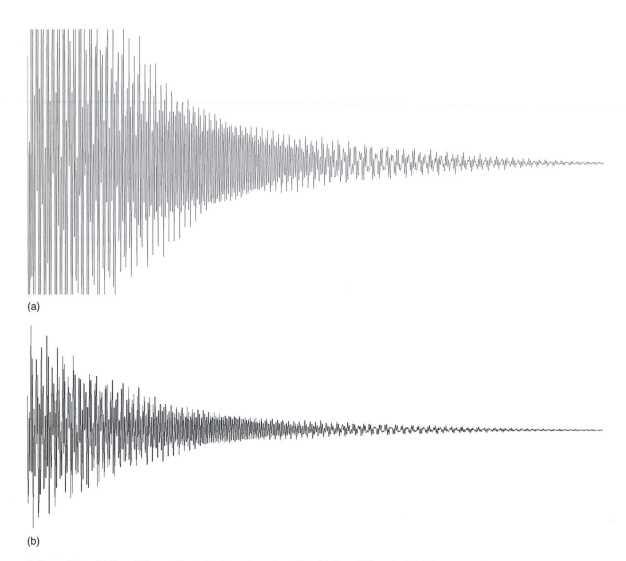

(a)

(b)

■ **Figure 2.8**   (a) Clipped FID caused by setting the receiver gain too high; (b) Normal FID acquired with the proper receiver gain.

Word. A portion of computer memory devoted to the storage of one number. A word will normally consist of four or eight bytes (one byte is eight bits, or eight 1s or 0s).

The A/D is actually a very primitive device. It works by guessing a number in the middle of the range, generating the voltage that the guess corresponds to, and comparing it to the input voltage. If the guessed voltage is too high, the A/D guesses a lower number, converts it to a voltage, compares the two, and so on until it gets it right. This obviously takes time, so guessing all 16 bits two hundred thousand times every second is actually a remarkable achievement.

Problems arise if the range of input voltages is too small, that is, if the variation of the sine wave is so small that the A/D always produces the same number. This difficulty arises when there is not enough signal or when the receiver gain is too low.

Too much signal also creates difficulties. If the receiver gain is too high or if the analog signal is too strong, the A/D will generate a number like +32,678 or −32,767 over and over again. When this happens, we see a clipped sine wave with a flat top and bottom instead of a normal sine wave. Figure 2.8a shows a clipped FID caused by having the receiver gain set too high. Notice the flat top and bottom of the wave. Figure 2.8b shows a normal FID without the flat top and bottom.

■ REFERENCES

[1] G. Bodenhausen, D. J. Ruben, *Chem. Phys. Lett.,* **69**, 185–189, (1980).

[2] A. Bax, M. F. Summers, *J. Am. Chem. Soc.,* **108**, 2093–2094 (1986).

[3] J. Jeener, B. H. Meier, P. Bachmann, R. R. Ernst, *J. Chem. Phys.,* **71**, 4546–4553 (1979).

[4] W. P. Aue, E. Bartholdi, R. R. Ernst, *J. Chem. Phys.,* **64**, 2229–2246 (1975).

[5] L. Braunschweiler, R. R. Ernst, *J. Magn. Reson.,* **53**, 521–528 (1983).

[6] L. Müller *J. Am. Chem. Soc.,* **101**, 4481–4484 (1979).

[7] A. Bax, R. H. Griffey, B. L. Hawkins *J. Magn. Reson.,* **55**, 301–315 (1983).

[8] A. G. Redfield, S. D. Kunz, *J. Magn. Reson.,* **19**, 250–254 (1975).

[9] D. J. States, R. A. Haberkorn, D. J. Ruben, *J. Magn. Reson.,* **48**, 286–292 (1982).

Chapter 3

Data Collection, Processing, and Plotting

When an NMR signal is generated by tipping the net magnetization vector of a sample into the **xy** plane by using the magnetic field component of applied RF radiation, the electrical (analog) signal induced in the receiver coil of the NMR probe (the FID) is amplified, the frequency of the signal is reduced from the MHz range to the kHz range through mixing, and finally the mixed-down (reduced frequency) signal is digitized to yield a data array with signal amplitude as a function of time. How many points we have in the data array that is our digitized NMR signal depends on a number of factors.

Mixing down. Syn. mixdown. The reduction of an analog signal from a high frequency (typically tens or hundreds of MHz) to a lower frequency range (typically below 100 kHz).

3.1 SETTING THE SPECTRAL WINDOW

Normally we set our spectral window (SW) to ensure that every frequency component of the net magnetization vector (from a particular nuclide) is observed. Sometimes the term sweep width is used when spectral window is meant. These two terms are commonly used interchangeably, but this is not strictly correct. The term spectral window denotes not only how wide a range of frequency is covered, but also where this range is centered relative to the frequency of a standard.

We can control both the width and the location of the SW. The width is controlled by varying the sampling rate of the analog-to-digital converter (A/D). We can translate (move side to side) the center of the spectral window by varying the transmitter frequency. The transmitter frequency is also known as the transmitter offset or the carrier frequency, or sometimes simply the transmitter (this is an imprecise term and should be avoided unless the context is well understood). On a Bruker instrument, the transmitter frequency is determined by a coarse value listed in MHz (sfrq, for spectrometer frequency)

Offset. The small amount a coarse frequency value (typically tens or hundreds of MHz) may be adjusted up or down.

and a fine value listed in Hz called o1 (for offset, channel 1). On a Varian instrument the transmitter frequency is determined similarly by the spectrometer frequency listed in MHz (again called sfrq) and is adjusted up or down by the contribution of a finer frequency variable in units of Hz called the tof (transmitter offset).

If we have a sample that we expect to yield a spectrum with an excellent signal-to-noise ratio from the collection of only a few scans, we initially set the SW to cover a very wide range of frequencies. We try to ensure that every resonance (of the nuclide being observed) falls within the frequency range spanned by the SW. Just to be safe, we may estimate we need a SW that is 10 kHz wide (and appropriately centered) and then use a SW that is 15 kHz wide. After collecting data while using this wide SW, we then reduce the size of the SW to leave only a small amount of baseline outside the observed NMR resonances. If we have the edges of the SW too close to some of our outlying resonances, we may find that integrals and intensities of the outlying resonances are attenuated because of filters in the receiver that reduce degradation of the signal-to-noise ratio of the spectrum caused by noise with frequencies just outside the range covered by the SW. We also avoid having the exact center of the SW on a peak of interest, because the center of the SW often contains an artifact known as the transmitter glitch, which is caused when a small amount of the RF synthesized by the transmitter channel gets through to the receiver.

Transmitter glitch. A small spectral artifact often observed in the very center of the spectral window that is caused by a small amount of the RF generated in the console getting through to the receiver.

If, however, we have a sample that requires many scans before the signal can be observed, we cannot determine the optimal SW to use with one or only a few initial scans. In this case, we position the SW to leave extra room on either side of where we think we may observe our most outlying resonances. If the SW we use is not wide enough (or if we do not center it appropriately), we may find ourselves unable to simultaneously phase all the resonances in our spectrum so that they are fully absorptive (with no dispersive character). The inability to simultaneously phase all peaks is (in this case) caused by folding or aliasing. Figure 3.1 shows two 1-D NMR spectra of the same organic compound. The top spectrum (Figure 3.1a) contains no folded resonances, whereas the bottom spectrum (Figure 3.1b) has the right-most resonance from top spectrum resonance folded over to the left side of the spectrum; also notice that the folded resonance is not phased properly relative to the unfolded resonances.

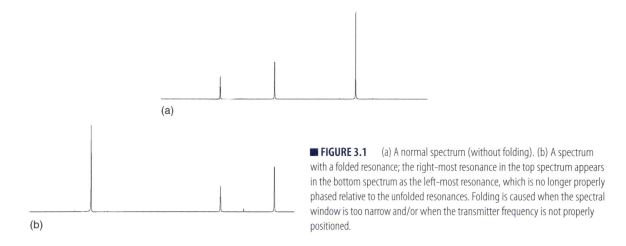

(a)

(b)

■ **FIGURE 3.1**   (a) A normal spectrum (without folding). (b) A spectrum with a folded resonance; the right-most resonance in the top spectrum appears in the bottom spectrum as the left-most resonance, which is no longer properly phased relative to the unfolded resonances. Folding is caused when the spectral window is too narrow and/or when the transmitter frequency is not properly positioned.

3.2 **DETERMINING THE OPTIMAL WAIT BETWEEN SCANS**

The net magnetization vector **M**, once tipped from its equilibrium position along the z-axis (**M₀**) and made to precess in the **xy** plane, will relax back toward its equilibrium position with a characteristic time constant. In the absence of any additional perturbations and starting with no **z** component of net magnetization—i.e., after applying a 90° pulse and completely tipping the **M** into the **xy** plane— the z component of the net magnetization vector (**M**$_z$) will return to its equilibrium value as follows:

$$\mathbf{M_z}(t) = \mathbf{M_0}[1 - \exp(-t/T_1)] \qquad (3.1)$$

where t is the time allowed for relaxation and T_1 is the time constant for the return (relaxation) of the net magnetization vector to its equilibrium position along the **z** axis. T_1 is also called the spin-lattice relaxation time because it describes the time constant associated with the decay of magnetization through the communication of the NMR spins to the outside world (the lattice).

Because $1 - e^{-5}$ is 0.993, a delay of five T_1's is normally sufficient between the last RF pulse and the application of the next RF pulse (getting 99.3% of the original signal back is generally thought to be sufficient). We sometimes estimate the T_1 value for the slowest relaxing resonance of interest and multiply this value by five to arrive at an appropriate relaxation delay for our sample. When waiting five

Integral. The numerical value generated by integration.

Ernst angle. The optimal tip angle for repeated application of a 1-D pulse sequence based on the relaxation time of the spin being observed and the time required to execute the pulse sequence a single time.

Paramagnetic relaxation agent. A nonreactive chemical additive (often containing europium) introduced into a sample containing unpaired electrons that has the effect of reducing the spin-lattice relaxation times for the spins in the solute molecules.

times the longest T_1 value (measured or estimated) will make our experiment inordinately lengthy, we may have to acknowledge that some of the resonances we observe will or may not be relaxed fully at the start of each scan. Failure to wait long enough between scans, coupled with different T_1's for the various resonances in a molecule, will introduce small but readily detectable errors in the NMR resonance areas (integrals) we measure in a process called integration (see Section 3.14).

When we conduct a 2-D experiment or a 1-D experiment that requires a large number of scans, we rarely wait five times the longest T_1 relaxation time between the pulse preceding NMR signal acquisition and start of the first pulse in the pulse sequence.

If the collection of 1-D NMR data sets requires many thousands of repetitions of the pulse sequence (scans), the use of less than a 90° pulse is sometimes found to yield a higher S/N using the same amount of instrument time. For every relaxation delay and T_1 value, there is an optimal tip angle (referred to as the Ernst angle) whose value can be determined by using relatively simple calculus. In practice, we rarely if ever know the T_1's of the resonances to which we have yet to devote many hours of instrument time. We are often left to guess at the T_1; or we may simply elect to use a 30° or a 45° ^{13}C pulse along with a 2–3 s relaxation delay.

However, not waiting for the net magnetization vector to return fully to equilibrium can generate undesirable artifacts when we conduct certain experiments. A good example of this artifact generation occurs when we conduct a 2-D NOESY experiment where the number of passes through the pulse sequence for a given set of delays is smaller than the size of the complete phase cycle. Phase cycling serves to cancel out various artifacts such as a direct current (DC) offset in the receiver.

In some cases, the sample may be spiked (treated) with a small amount of a paramagnetic relaxation agent to shorten the T_1's. These agents, often containing an element such as europium, facilitate rapid relaxation of nuclear spins due to the presence of unpaired electrons. There are two principal disadvantages accompanying the addition of paramagnetic agents to our sample: (1) the sample purity goes down (if the sample is precious, we have just added something that makes it less pure), and (2) faster T_1 relaxation may contribute to broadening the resonances we observe.

To keep the total amount of time required for a particular experiment to a minimum, methods are sometimes employed to eliminate residual magnetization, thereby allowing the relaxation delay to be shorter than five times the longest T_1. These residual magnetization elimination methods are generally referred to as homospoil methods because they involve spoiling (disrupting) the homogeneity of the applied magnetic field for a short period of time prior to the relaxation delay. During the relaxation delay that follows a homospoil pulse, all spin ensembles generate net magnetization vectors with only a **z** component.

Homospoil methods work either by changing the electrical current passing through the room-temperature (RT) z^1 shim for perhaps 10 ms prior to the relaxation delay or by passing a current through a dedicated magnetic field gradient coil within the probe. Both changing the current in the RT z^1 shim and passing current through a dedicated gradient coil in the probe vary the strength of the applied field as a function of z-axis position. Probes that contain this dedicated gradient coil are called pulsed field gradient (PFG) probes or gradient probes for short.

Figure 3.2 shows how the **xy** components of a number of spins spaced evenly along the z-axis in an NMR tube will behave upon application of a magnetic field gradient pulse along the z-axis. A gradient pulse of some reasonable duration—perhaps 10 ms—applied to the sample causes any residual magnetization in the **xy** plane to dephase (all **xy** components cancel each other). Dephasing (depolarization) can be readily understood if we visualize breaking up the net magnetization vector into many tiny vectors along the z-axis in the detected region of the sample. If a magnetic field gradient along the z-axis is introduced, the magnetic field will vary as a function of its position along the z-axis, and therefore the precession frequency of each portion of the net magnetization vector will vary. Turning on the field gradient causes the individual portions of the net magnetization vector to wind up into a helical array. Summing the helical array of the vectors gives a zero net result—thus showing that the gradient pulse destroys any residual **xy** (transverse) magnetization present. Only the **z** component of the net magnetization vector (longitudinal magnetization) will survive a gradient pulse along the z-axis, as briefly varying the strength of the applied field will not significantly affect magnetization aligned parallel to the z-axis.

Homospoil methods. A method of eliminating residual net magnetization achieved through temporary disruption of the applied magnetic field's homogeneity so that the spins precess at different frequencies and become dephased.

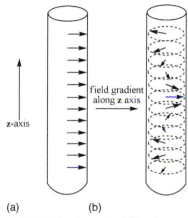

(a)   (b)

■ **Figure 3.2** A magnetic field gradient pulse along the **z**-axis randomizes the **xy** components of the spins in the sample: (a) **xy** components are aligned; (b) **xy** components cancel each other after a magnetic field gradient changes the spin precession frequencies as a function of **z**-axis position.

Dephasing. The spreading out of the individual components that comprise a net magnetization vector so the summation is zero.

Depolarization. The equalization of the populations of two or more spin states.

Splat-90-splat. A method for destroying residual magnetization involving the applying a z-gradient pulse, applying a 90° pulse, and then another z-gradient pulse.

Full cannon homospoil. Application of an RF pulse at or near full transmitter power with a duration 200 times greater than that required to tip the net magnetization by 90° for the express purpose of dephasing the net magnetization in **xz** or **yz** plane. Small phase errors accumulate with this absurdly large tip angle because of the slight inhomogeneities of the applied RF's **B₁** field across the volume of the sample excited by the pulse.

If we follow the first gradient pulse with a 90° pulse, then the surviving **z** magnetization gets tipped into the **xy** plane (see Section 1.6). A second gradient pulse can then be applied to completely wipe out whatever magnetization survived the first gradient pulse. In this way, we can ensure that absolutely no net magnetization is present at the start of the relaxation delay in the pulse sequence. This procedure is sometimes referred to as the "splat-90-splat" method.

Another method used to destroy residual magnetization involves applying to the sample an RF pulse that is on the order of 200 times the 90° pulse width. The same transmitter power as that used for the shortest attainable 90° pulse (called a hard 90° pulse) is used for this long pulse: if the 90° pulse is 10 μs, then 200 times the pulse is 2 ms! The combination of RF inhomogeneity in the detected region of the sample coupled with the effects of pulse roll-off serve to effectively destroy any residual net magnetization. This "full cannon" method may test the limits of the transmitter amplifiers and the power handling capability of the probe, and may also reflect a large amount of power if the probe is not properly tuned. Furthermore, a high ionic strength (salty or buffered) sample may undergo significant RF heating. If our sample contains a fragile protein and undergoes RF heating, denaturation may occur.

3.3 SETTING THE ACQUISITION TIME

For a typical 1-D spectrum, we want to collect data until the signal has faded into the background noise. After this signal decay has taken place we may wait and then pulse and detect again and again. The signal fades away in the **xy** plane with a time constant that is shorter than or perhaps equal to the time constant for the return of the net magnetization vector $\mathbf{M}$ to its equilibrium length and position. A different relaxation time, called the T_2 or spin-spin relaxation time, governs how quickly the **xy** component of the net magnetization vector ($\mathbf{M_{xy}}$) will decay in the **xy** plane. If we start by tipping the equilibrium net magnetization vector ($\mathbf{M_0}$) into the **xy** plane, then, barring any subsequent perturbation (no more RF pulses), the T_2 relaxation of $\mathbf{M_{xy}}$ occurs according to Equation 3.2:

$$\mathbf{M_{xy}}(t) = \mathbf{M_0}\exp(-t/T_2)]$$ (3.2)

Thus, after five T_2's (plug in t=5T_2) have elapsed, the amount of the net magnetization vector left in the **xy** plane is only 0.7% of its starting value. The failure to shim the sample perfectly also contributes to the decay of the signal in the **xy** plane. To include this influence, we often speak of T_2^*; but this point is a subtle one and can be ignored most of the time. In many cases, T_1 is approximately equal to T_2 (and T_2^*), so distinctions are not often made between the two.

For most ^{1}H's in organic compounds in solution, the T_1 will be on the order of 1 to 2 seconds; therefore an acquisition time of 10 seconds is normally sufficient to completely capture the decaying NMR signal. For the collection of 1-D NMR spectra, we use a relaxation delay that is five times the longest T_1 whenever possible. In practice, this condition cannot (or may not) always be met because our sample may contain resonances with one or more lengthy T_1's. Collecting only a few scans on a fully-relaxed sample may not give us a reasonable signal-to-noise ratio for any of the solute resonances in our sample. If we shorten the relaxation delay, we give up quantitation but improve the signal-to-noise ratio for the resonances that relax more quickly. We also may shorten the relaxation delay because obtaining nearly perfect integrals is of a lower priority than expediency.

The NMR signal we observe by using pulsed NMR methods is called the free induction decay (FID). This name arises because the net magnetization vector is tipped and then allowed to freely precess, thereby inducing a current in the receiver coil of the NMR; and this signal decays as a result of T_2 (really T_2^*) relaxation. We digitize the FID during the acquisition period of the pulse sequence to generate a data array that we will subsequently process with our computer software.

3.4 HOW MANY POINTS TO ACQUIRE IN A 1-D SPECTRUM

When we are collecting a ^{1}H NMR spectrum using a 500 MHz instrument, we normally want to observe the frequency range of ^{1}H resonances with chemical shifts from -1 ppm to $+11$ ppm. Therefore, we set our spectral window to be 12 ppm wide and locate the transmitter frequency at 5 ppm in the spectrum. Each ppm for a ^{1}H spectrum on a 500 MHz instrument is, by definition, 500 Hz wide, so a sweep width of 12 ppm corresponds to 6 kHz. Consequently, we must either sample one channel with a single A/D sampling at a rate of 12 kHz or sample with two orthogonal channels, each of which has an A/D sampling at a rate of 6 kHz. In 1 s of acquisition we collect 12,000 data points, because a sampling rate of 12 kHz means we are collecting 12,000 points per second. The time interval between successive points is called the dwell time and is 1 divided by the sampling rate. In the single A/D example the dwell is therefore $12,000\,Hz^{-1}$, or 83 µs. If we collect data for 10 s, then the number of data points we collect is 120,000. It is a good thing we do not record these values by hand.

When we want to collect ^{13}C data on a 500 MHz instrument, our ^{13}C NMR frequency will be one quarter of the ^1H frequency, or 125 MHz. Consequently, each ppm of a ^{13}C spectrum collected on a 500 MHz instrument spans only 125 Hz. That is the good news. The bad news is that the ^{13}C chemical shift range is far greater than that for ^1H. For ^{13}C, the chemical shift range may be in excess of 250 ppm. If we scan from -10 ppm to 240 ppm, we locate our transmitter frequency at the average of the edges of the spectrum, or 115 ppm ($(-10 + 240)/2 = 115$). We use a sweep width of 250 ppm $\times$ 125 Hz ppm^{-1}, or 31.25 kHz. Therefore, for proper sampling, we employ one A/D sampling at 62.5 kHz or we employ two A/D's sampling at 31.25 kHz. With a single A/D, we sample with a dwell time of 16 µs, and with a pair of A/D's we sample with a dwell time of 32 µs.

More good news is that we rarely collect ^{13}C NMR data for extremely long periods of time (per scan). That is, we normally will collect the ^{13}C FID for only 2–3 s; seldom will we collect data for longer (of course, we may end up scanning over and over all night in pursuit of a sufficient signal-to-noise ratio). Three seconds of ^{13}C FID digitization will require that we collect 62,500 points per second times 3 s or 187,500 data points. Again, this number of points is big, but not so big that a computer cannot handle the task.

Zero filling. Addition of 0's at the tail end of time domain data for the purpose of improving the digital resolution in the frequency spectrum following Fourier transformation. Zero filling increases the number of points per unit frequency (a good thing).

Digital resolution. The number of Hz per data point in a spectrum. The digital resolution is the sweep width divided by the number of data points in one channel of the spectrum.

3.5 ZERO FILLING AND DIGITAL RESOLUTION

Although the intensity of the NMR signal may have decayed to the background noise level by the time the data collection of the FID ceases, we can garner an additional benefit by increasing the size of the Fourier-transformed data set beyond the next power of 2. This practice is called zero filling, and its name arises from the fact that, before carrying out the Fourier transform, we add a long string of zeroes to the end of the array of numbers that is our data set (our digitized FID). This increase in the size of the data set improves the digital resolution of the frequency spectrum once we carry out the Fourier transform. Digital resolution is the number of Hz per data point. The smaller the digital resolution, the smoother the curves we

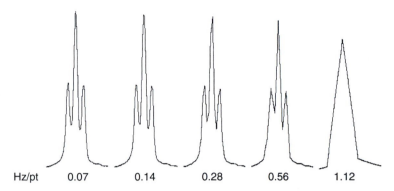

| Hz/pt | 0.07 | 0.14 | 0.28 | 0.56 | 1.12 |

■ **FIGURE 3.3**   The effect of digital resolution on the appearance of NMR spectral features. On the left is a portion of a spectrum with sufficient digital resolution. Successive moves to the right show the effect of reducing the digital resolution.

observe when we view spectral features whose fineness (spacing in Hz) is comparable to the inverse of the FID's acquisition time.

Figure 3.3 shows an expansion of a ^{1}H NMR spectrum that contains a single resonance weakly coupled ($J = 1.15\,$Hz) to two other spins. The resonance is split into three lines with an intensity ratio of 1:2:1, and is called a 1:2:1 triplet or, if our context is well understood, simply a triplet. From left to right, the digital resolution doubles with each successive trace, starting at 0.07 Hz/pt and ending at 1.12 Hz/pt (less than the 1.15 Hz splitting). Notice that the middle trace (0.28 Hz/pt) already shows a noticeable connect-the-dots quality and that further reductions in the number of points per unit frequency dramatically degrade the symmetry of the multiplet.

Zero filling is extremely common. One-dimensional data sets are often doubled and sometimes even quadrupled or more in size to improve the continuity of the observed lines. Reviewers and others passing judgment on the quality of our NMR spectra will rarely, if ever, object to zero filling. With that said, the improvements to be realized with zero filling are limited. This subject will be discussed at greater length in Section 3.15 in the context of measuring chemical shifts and J-couplings.

3.6 SETTING THE NUMBER OF POINTS TO ACQUIRE IN A 2-D SPECTRUM

A significant problem arises once we make the leap to the second dimension. For a basic 2-D ^{1}H COSY spectrum, we might wish to

Data point. Two numerical values (i.e., an x, y pair) corresponding to intensity as a function of time, or to intensity as a function of frequency.

Triplet. Three evenly spaced peaks in the frequency spectrum caused by the splitting of a single resonance by J-coupling to two identical spin-½ nuclei to give a multiplet with three peaks with relative intensity ratios of 1:2:1, or to one spin-1 nucleus to give a multiplet with three peaks with relative intensity ratios of 1:1:1.

digitize 128 FIDs and store the result for subsequent processing. If each digitized FID has 240,000 points, we have 30,720,000 points in our data matrix. This number of points could be generated in as little as 1280 s or about 21 min.

Double-precision word. A computer memory allocation used to store a single number that contains twice the number of bytes as a normal (single-precision) word.

Furthermore, if we consider that each point of the digitized FID is stored on our computer as a double-precision word (32 bits or 4 bytes, because 8 bits = 1 byte) to allow for signal accumulation, then we have 4 bytes per word times 30,720,000 data points (one point occupies a double precision word). Our simple 2-D ^{1}H COSY data set is now going to occupy a whopping 122.88 Mbytes of memory. If our computer has a 64-bit word, the size is doubled. If we produce only two data sets of this size per hour (given extra time for sample insertion, probe tuning, shimming, experiment setup, etc.), we still will be filling 5.9 Gbytes of memory per day. Storing five 123 Mbyte data sets per 770 Mbyte CD-RW disc requires that we burn ten discs every day just to keep up with the data backed up on a busy NMR instrument. Clearly, the size of the 2-D data sets we generate needs to be pared down.

Although computer disk storage and data archiving capacities now allow files of this size to be handled, a problem arises when we start processing the data set. An array of NMR data with 240,000 data points will normally be Fourier transformed using a data array with a dimension that is the next higher power of 2 ($2^{18} = 262,144$, also known as 256 k, where k denotes 1024). Using the Cooley-Tukey algorithm for the fast Fourier transform, a typical NMR host computer may process a 256 k data set in about a second. If we process 128 FIDs, our processing time goes from a barely noticeable second to over two minutes. In this age of faster and faster computers, nobody likes to wait 2 min for a mathematical operation to take place. And that wait is only for the processing of the data matrix acquired in the directly detected dimension (the t_2 time domain is converted to the f_2 frequency domain)! If we then process the data matrix to transform the t_1 time domain to the f_1 frequency domain, we must carry out 262,144 FTs of at least 128 data points each (and more if we zero fill). Because of the time associated with shuttling the data along the data bus of the computer, the time required for the processing of each 128-point data array will be more than we might expect if we assume that it takes one second to FT a 256 k data array. Even if our computer can carry out a thousand FTs every second (likely it will be slower than this), completion of the processing of the data matrix will take an additional 262 s, or 4.4 minutes.

Waiting 6+ min to process a basic 2-D ^{1}H COSY NMR data set is never going to be something we want to do.

In practice, for the acquisition of 2-D NMR data sets, we limit the collection of the time domain data to 2 k (2048) or 4 k (4096) data points. In the case of the collection of 2-D ^{1}H-^{13}C HMQC and HSQC NMR data sets, we may want to further restrict the number of data points we collect to keep the acquisition time as short as possible (perhaps 150 ms), because the extended use of decoupling with the ^{13}C RF channel of the NMR may result in excessive RF heating of the sample and/or probe damage.

3.7 TRUNCATION ERROR AND APODIZATION

There is a way around the problem of cutting short the generation of time domain data before the NMR signal has faded to the level of the background noise (through $T_2{}^*$ relaxation). We do not need to digitize the FID until it relaxes away into nothingness. Instead, we collect our digital data points for only a short period of time and multiply the time-domain data (the digitized FID) by a mathematical function to force the data points to become 0 (or close to it) at the tail end of the digitized FID (the right edge of the time domain data array). Put another way, we damp the NMR signal selectively at the right side of the data array that is the digitized FID.

Multiplying the digitized FID by a mathematical function is called apodization. There are many different functions that we can apply to time-domain data arrays. The most common functions are Lorentzian line broadening (commonly referring to as line broadening), Gaussian line broadening, the shifted sine bell, and the shifted, squared sine bell. Different apodization functions should be used in different situations, because no one function is always appropriate. Another term for an apodization function is a window function. The term window function is used because the function is essentially a window through which the data is viewed.

Because the application of the apodization function can be done after an NMR data set has been collected, we are free to experiment with different functions to evaluate the effect various functions have on the quality of the processed NMR data. We always disclose exactly what post-acquisition processing has taken place to generate our final spectrum, and many researchers, research advisors, supervisors, reviewers, and regulators do not hold excessive and/or fancy NMR data processing manipulations in high regard. It is best to keep our NMR processing

Apodization. The application (multiplication) of a mathematical function to an array of numbers that represent the time domain signal. An apodization function is often used to force the right edge of a digitized FID to zero to eliminate truncation error, but apodization functions can also be used to enhance resolution or to modulate the time domain signal in other ways. Common apodization functions are Lorentzian line broadening (lb), Gaussian line broadening (gf), (phase-)shifted sine bell, and (phase-)shifted squared sine bell.

Apodization function. Syn. window function. The mathematical function multiplied by the time domain signal.

Line broadening function. A mathematical function multiplied by the time domain data to smooth out noise by emphasizing the signal-rich beginning of the digitized FID and deemphasizing the noise-rich tail of the digitized FID. In a 2-D interferogram, application of a line broadening function emphasizes the signal-rich f_2 spectra (usually those with shorter t_1 evolution times) and deemphasizes the noise-rich f_2 spectra (usually with longer t_1 evolution times).

Lorentzian line broadening function. The apodization function most commonly used to emphasize the signal-rich initial portion of a digitized FID.

Gaussing line broadening function. Syn. Gaussian function, Gaussian, Gaussian apodization function. An apodization function commonly used to weight the signal-rich initial portion of a digitized FID relative to the noise-rich tail portion of a digitized FID. The weighted digitized FID is then Fourier transformed to convert the time domain to the frequency domain. In the frequency spectrum, the Gaussian function is more effective than the Lorentzian function at suppressing truncation error, improving the spectrum's signal-to-noise ratio, and minimizing the peak broadening inherent in the use of most apodization functions.

Shifted sine bell function. An apodization function with the amplitude of the sinusoidal pattern starting at a maximum and dropping to zero. The first quarter of a cosine waveform.

Shifted squared sine bell function. An apodization function with the amplitude of a squared sinusoidal pattern starting at a maximum and dropping to zero. The first quarter of a squared cosine waveform.

Fourier ripples. Constantly spaced bumps in the frequency spectrum found on either side of a peak. In a 1-D spectrum or a half-transform 2-D data matrix, these ripples are found when the apodized intensity has not faded to the level of the background noise by the time the digitization of the FID ceases. In a fully transformed 2-D spectrum, Fourier ripples parallel to the f_1 frequency axis are observed when an inappropriate t_1 apodization function is used prior to conversion of the t_1 time domain to the f_1 frequency domain.

methods simple and limit our choice of function to those that are already accepted in the pertinent literature in our area of study.

If we fail to use an apodization function that will force the values of our digitized FID to 0 at the right edge of the time domain window, we find that upon Fourier transformation the narrow peaks in our spectrum show a series of bumps or ripples on the baseline on either side of one or more of the resonances in our spectrum. These bumps—resembling a corduroy pattern and possibly impinging upon nearby peaks in the spectrum—are sometimes called Fourier ripples. They are the result of truncation error, which is caused by the abrupt change in a digitized FID's intensity at its right edge.

Figure 3.4 illustrates truncation error. Figure 3.4a shows a digitized FID that has been truncated (notice that the intensity of the digitized FID has not decayed to 0 at the right edge); Figure 3.4b shows the result of carrying out a Fourier transformation on the digitized FID in Figure 3.4a; the resulting spectrum clearly shows the artifacts resulting from truncation.

The Fourier transform is a mathematical operation that converts amplitude as a function of time to amplitude as a function of frequency. An inverse Fourier transform can be used to go from the frequency domain back to the time domain. To use an inverse Fourier transform to generate a (possibly damped) data array whose amplitude varies sinusoidally and then abruptly drops to 0, we must sum up many different frequencies. The corduroy pattern we observe in the vicinity of a narrow peak in the frequency spectrum reflects the wide range of nearby frequencies required to make our digitized FID's intensity drop abruptly to 0 in the time domain.

Narrow resonances are the most susceptible to truncation error. A broad resonance generates a sinusoidal signal whose maximum amplitude decays rapidly down to the level of the background noise by the time the FID is completely digitized. That is, the T_2^* relaxation time of a broad resonance is shorter than that of a narrow resonance; the broad resonance's **xy** component will cease to be observed in the digitized FID sooner than that of a longer lived (longer T_2^*) resonance.

3.8 THE RELATIONSHIP BETWEEN T_2^* AND OBSERVED LINE WIDTH

The relationship between how long we observe the signal from a particular resonance in the time domain and how narrow the resonance

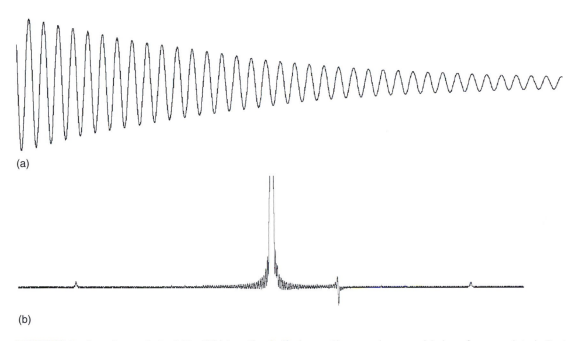

(a)

(b)

■ FIGURE 3.4   Truncation error in the digitized FID (a) manifests itself as bumps, ridges, or corduroy around the base of narrow peaks in the Fourier-transformed spectrum (b).

is in the frequency domain (i.e., in the spectrum) is relatively simple to explain. Consider two ^{1}H's, ^{1}H$_A$ and ^{1}H$_B$, in different chemical environments. Let us suppose that, after the application of a 90° RF pulse, the net magnetization vectors from ^{1}H$_A$ and ^{1}H$_B$ precess at frequencies whose difference is small relative to their T_2* relaxation times. We will therefore observe that as the net magnetization vectors **M$_A$** and **M$_B$** (from ^{1}H$_A$ and ^{1}H$_B$, respectively) precess, only a small phase difference will accumulate between **M$_A$** and **M$_B$** before the **xy** components of **M$_A$** and **M$_B$** decay to 0 as the result of T_2* relaxation. In the case of rapidly decaying net magnetizations, our receiver will not be able to distinguish between the two resonances. The Fourier-transformed spectrum will show poorly resolved resonances from ^{1}H$_A$ and ^{1}H$_B$, as illustrated in Figure 3.5a.

If, on the other hand, **M$_A$** and **M$_B$**'s **xy** components precess in the **xy** plane for an appreciable length of time (due to long T_2*'s), a large phase difference between the signals generated by the two net magnetizations is detected, and our receiver will thus resolve the frequencies of the two chemically distinct net magnetizations; the Fourier-transformed spectrum will show that the resonances from ^{1}H$_A$ and ^{1}H$_B$ are well resolved (Figure 3.5b).

Truncation error. Regularly spaced bumps or ripples observed on either side of the narrow peaks in a frequency spectrum that are caused by the failure of the digitized FID's amplitude to fall to the level of the background noise before the end of the digitization of the FID.

Doublet. A resonance splitting pattern wherein the resonance appears as two peaks, lines, or legs.

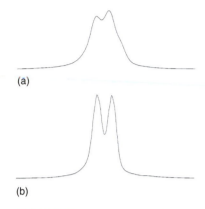

(a)

(b)

■ **FIGURE 3.5** (a) A poorly resolved doublet due to a short T_2 relaxation time; (b) a well-resolved doublet with a longer T_2 relaxation time.

Put another way, if the sinusoidal signal we detect does not persist in the **xy** plane long enough for the detector to accurately measure its frequency, then there will be uncertainty in the measured frequency. A resonance with an indeterminate frequency will possess a large line width in the frequency domain. (The line width of a resonance is measured at half of the peak height). A broad line corresponds to a large uncertainty in frequency.

Therefore narrow resonances, with amplitudes that do not decay fully at the end of the digitization of the FID, will be subject to distortion from truncation error.

Apodization can be used to artificially broaden the narrower resonances in a spectrum and thus to remove the visually disturbing aesthetic caused by truncation error.

When we collect a 2-D spectrum, we limit the number of points we collect in the t_2 time domain because we digitize the FID for only a short time. Even if we know our FID signal will not die away in the **xy** plane for 10 seconds, we can still limit our digitization of the FID to 1 s without suffering the ill effects of truncation error by applying an apodization function to force the digitized FID's intensity to 0 at its end. The disadvantage to apodizing our t_2 time domain data is that the peaks whose line widths at half height we may have shimmed to 0.3 Hz or less become broad, and therefore we may fail to resolve closely spaced resonances in the 2-D spectrum even though these same resonances are clearly resolved in the 1-D spectrum.

3.9 **RESOLUTION ENHANCEMENT**

The use of certain apodization functions improves the frequency resolution we obtain in our Fourier-transformed spectrum, but caution should be exercised when employing this technique. The use of negative line broadening and shifted Gaussian or squared sine bells (with the maximum to the right of the start of the FID) can be used to resolve a small peak that formerly appeared as the shoulder of a larger peak, but supervisors and reviewers frown upon the excessive application of these methods; the starting NMR spectroscopist would do well to exercise restraint in this area.

Although the appropriateness of the application of resolution-enhancing apodization is sometimes in doubt, the method does have a physical basis. By multiplying the digitized FID by a function that enhances the intensity of points that are collected later in the

FID relative to those collected at the start of the FID, the $T_2{}^*$-induced decay of the NMR signal is modified in such a way that the apparent $T_2{}^*$ is longer. The result is a narrower line width in the Fourier-transformed spectrum. That is, the portion of the digitized FID that is collected at later times contains more information about the accumulated phase difference of the various populations of precessing spins. More heavily weighting these difference-rich points emphasizes the differences in the frequency spectrum.

Resolution enhancement also has the unfortunate consequence of decreasing the signal-to-noise ratio in our spectrum. This effect arises from deemphasizing the points corresponding to the signal-rich initial portion of the digitized FID relative to the points from the more noise-rich later portion of the digitized FID. In those cases where severe resolution enhancement methods are used, we will typically observe a dip well below the baseline on either side of the narrowest resonances in our spectrum.

Resolution enhancement. The application of an apodization function that emphasizes the later portions of the digitized FID (at the expense of the signal-to-noise ratio) that, upon Fourier transformation, will generate a spectrum with peaks whose widths are narrower than their natural line widths.

3.10 **FORWARD LINEAR PREDICTION**

When we collect a 2-D NMR data set, we fill in one column of the 2-D data matrix at a time. Each column (or pair of columns) contains a digitized FID collected with a unique t_1 evolution time. Sometimes when we carry out a 2-D experiment, we are not able to fill in as many columns (digitize as many FIDs) as we would like (perhaps because of instrument time constraints). When we cut short our filling up of the columns of the data matrix with digitized FIDs, we find that some or all of the resonances in our resulting spectrum (following Fourier transformation) will show truncation error parallel to the f_1 frequency axis. Instead of broadening the spectrum in the f_1 frequency domain with a severe apodization function, we can use linear prediction to extend the data set by adding fictitious columns to our data matrix where we would have placed digitized FIDs instead.

Linear prediction uses a computer algorithm to first examine the periodicity of the signal in the time domain and then make an educated guess as to how the time domain signal would behave if we (1) digitized the FID for a longer period of time; (2) digitized a few more FIDs (for a 2-D data set); or (3) started digitizing our FID sooner than we did (there is a delay called the ringdown delay in the pulse sequence after the last RF pulse but before the FID is digitized). The first two cases are called forward linear prediction; the third case is called backward linear prediction.

Linear prediction. A mathematical operation that generates new or replaces existing time domain data points with predicted ones. Linear prediction can add to the end of an array of digitized FID data points, can extend the number of t_1 time domain data points in a 2-D interferogram, or can replace the initial points in a digitized FID that may have been corrupted by pulse ringdown.

Forward linear prediction. The addition of data points past the last point in time that was digitized. Forward linear prediction can be used to add points on to the end of a digitized FID, or it can be used to add to a 2-D data matrix additional digitized FIDs corresponding to those that would be obtained if longer t_1 evolution times were used.

Backward linear prediction. The use of a mathematical algorithm to predict how the time domain signal would appear during early portions of the digitized FID. Backward linear prediction is normally used to compensate for the corruption of the first few points of a digitized FID by pulse ringdown.

The following example is a simple illustration of how forward linear prediction works. Suppose we start with a single sinusoidal frequency 100 Hz from the spectrum center (the frequency of the rotating frame) that is only slightly damped by a T_2^* of 2 s. If we stop digitizing the FID after 4 s, we find that the net magnetization has decayed only by e^{-2} or by 86.5%, so there is still 13.5% of the original signal intensity remaining at the right side of the digitized FID when we stop collecting data points.

If we Fourier transform the digitized FID without apodization, we will see a very pronounced manifestation of truncation error and, upon showing this data to our superior, he or she will tell us that the data is of poor quality. Applying an apodization function to the digitized FID prior to carrying out the Fourier transformation will smooth out the corduroy of the truncation error but will result in a visually obvious and perhaps aesthetically displeasing increase in the line width of the peaks in the spectrum. However, if we use linear prediction to make an educated guess as to how that single sinusoid would have behaved if we had continued to digitize the FID, we find that the resulting spectrum has a peak that is narrow (0.5 Hz, i.e., $1/T_2^*$) and lacks the corduroy associated with truncation error.

Exponentially damped sinusoid. A sine wave whose amplitude decays exponentially with time. The analog signal induced in the receiver coil of the NMR instrument will consist of one or more exponentially damped sinusoids.

In a way, linear prediction feels like cheating, in that we are getting something for nothing. In fact, though, it is perfectly reasonable to predict the outcome of something as well defined as an exponentially damped sinusoid, so long as we do not try to predict behavior of the signal too far into the future. The general rule of thumb is that forward linear prediction should not be used to more than double the number of data points.

3.11 PULSE RINGDOWN AND BACKWARD LINEAR PREDICTION

Pulse ringdown. The lingering effects of an RF pulse applied just before digitization of the NMR signal.

Linear prediction can also be used to compensate for the corruption of the first few points in a digitized FID caused by pulse ringdown (the lingering effect of an RF pulse applied before digitization commences). In that case, the first 10 or so points of a digitized FID may be discarded and the rest of the points of the digitized FID may be used to predict how the first 10 points would have looked had there not been any observable pulse ringdown. Pulse ringdown is relatively easy to recognize if we display the digitized FID and examine its first few points. Instead of a relatively smooth series of points characteristic of a normal digitized FID, the ringdown-tainted digitized

FID will often show a number of points with an anomalously large amplitude, and may also contain large jumps or discontinuities between adjacent points.

After the application of an RF pulse, the amplitude of the RF pulse will not immediately drop to 0 as it theoretically should. The higher the power of the pulse, the greater the likelihood that the ringdown-induced spectral artifacts will be significant. The negative effects of pulse ringdown are often minimized by inserting a delay between the end of the RF pulse and the start of the digitization of the FID. For solid-state NMR applications intended for the observation of extremely wide NMR signals (e.g., solid-state ^{2}H work on rigid solids), we often choose a pulse sequence that generates a spin echo well after the last RF pulse has been applied to the sample. We may then begin to digitize the FID at the moment the spin echo is at its maximum.

If we wish to observe a spectrum featuring one or more broad resonances, we will find that there is little margin for error in the choice of the ringdown delay (note that the ringdown delay is called alfa in Varian instruments, or it may be called rof2—rof2 occurs after most pulses in a pulse sequence, whereas alfa is only included once before the start of the collection of the FID). A broad NMR peak will have a very short $T_2{}^*$ relaxation time; thus the signal in the **xy** plane (the signal being detected) will die away very quickly. Waiting any extra time after the end of the last RF pulse and before starting to digitize the FID may result in a drastic decrease in the signal-to-noise ratio we obtain. An excessive ringdown delay in our pulse sequence may also complicate the phasing of the spectrum—especially the first order phase correction term.

If we find that the ringdown delay required to produce an artifact-free spectrum yields an unacceptably low signal-to-noise ratio (this is a subjective decision), then we can use backwards linear prediction to correct the first few corrupted points in the FID. Linear prediction is generally accepted in the NMR community as a reasonable method for avoiding choosing between a spectrum with a low signal-to-noise ratio and a spectrum marred by ringdown artifacts.

3.12 **PHASE CORRECTION**

A spectrum can only be phased when the data is collected in a phase-sensitive receiver detection mode. For a spectrum to be phase sensitive, we require either two orthogonal receiver channels digitizing

Spin echo. Magnetization allowed to dephase in the **xy** plane will, following an appropriately phased RF pulse, refocus to give a sharp spin echo. Sampling of the spin echo maxima for a series of refocusing events allows the determination of T_2, the spin-spin relaxation rate constant. Interestingly, the signal we collect in this experiment (see CPMG) is not technically a FID, because we perturb the magnetization periodically in between the collection of data points for a each single passage through the pulse sequence.

simultaneously or one receiver channel collecting data twice as fast and sending every other point to a second channel.

If a particular net magnetization **M** is on-resonance, it will not precess in the rotating frame—even if it has a component in the **xy** plane—and therefore its amplitude will not vary as a damped sinusoid but instead will only decay exponentially as we digitize our FID. If it is off-resonance with respect to the frequency of the rotating frame of reference, however, the second receiver channel will distinguish between a net magnetization precessing at a positive versus a negative rate relative to the rotating frame.

If we are carrying out a simple one-pulse experiment and use a 90° pulse, then immediately after our pulse, when detection is about to commence, all the net magnetization vectors arising from the various chemically distinct species in the sample will point in the same direction. After a short amount of time, these signals will fan out in the **xy** plane. As the detection of the FID continues, the various **M**'s will continue to precess, each at its own rate (positive or negative), and will therefore generate a complex beat pattern caused by constructive and destructive interference. Following the Fourier transformation, we will likely observe a spectrum featuring resonances with various mixtures of absorptive and dispersive character. Phase correction (phasing) of the spectrum may be required to ensure that our observed resonances will be absorptive (and usually positive).

Let us suppose that we have two chemically distinct ^{1}H sites in our molecule (A and B) and that the resonances from these sites are anisochronous (they have different chemical shifts). We can divide the net magnetization vector into the two components M_A and M_B. If we set our transmitter frequency so that M_A is on-resonance, then M_A will generate a signal in the digitized FID that will be a simple exponential decay. Because M_B cannot simultaneously be on-resonance (it precesses at a different frequency), M_B will superimpose a sinusoidal signal on top of the simple exponential decay of M_A in the digitized FID. Technically, both signals are exponentially damped sine waves with a 90° phase shift (damped cosine waves), but the wavelength of M_A's cosine function is infinite because M_A is on-resonance. Let us further stipulate that following the application of the 90° pulse, M_A points exactly along one of our two receiver axes, meaning that we detect all of the signal from M_A in just one of our receiver channels.

Phase correction. The balancing of the relative emphasis of two orthogonal data arrays (or matrices for a 2-D spectrum) to generate a frequency spectrum with peaks that have a fully absorptive (or, in some cases, fully dispersive) phase character.

Anisochronous. Two spins are anisochronous when they undergo transitions between their allowed spin states at different frequencies. That is, two spins are anisochronous if their resonances appear at different points in the frequency spectrum.

To avoid the deleterious effects of pulse ringdown, we always wait a small amount of time between applying our 90° pulse and the commencement of FID digitization. Because of this delay, M_B will not point in the same direction as M_A at the moment we begin to digitize the FID. Therefore, the digital signal we obtain from M_B detected in the channel in which M_A is found will not start at its maximum amplitude. We use first-order phase correction to compensate for this problem. This correction is appropriate whenever we have to compensate for a pulse ringdown delay. In practice, some first-order phase correction is always required. This type of phase correction is called first-order phase correction because there is a linear relationship between the amount of phase correction needed and the precession frequency of each chemically distinct resonance in the rotating frame. A resonance that is nearly on-resonance will require only a tiny bit of first-order phase correction, whereas a signal very far from being on-resonance requires a greater amount of phase correction.

Zero-order phase correction is even easier to understand. It lines up the phase of the receiver with the phase of the transmitter so that a resonance that is exactly on resonance will appear purely absorptive—i.e., with no dispersive character. Zero-order phase correction affects every resonance in the entire frequency spectrum by the same amount.

Figure 3.6 shows two NMR spectra that illustrate the difference between proper and improper phasing of resonances. At the top is a well-phased spectrum (Figure 3.6a); at the bottom, a poorly phased spectrum (Figure 3.6b).

When we collect a 2-D NMR spectrum, both the second frequency dimension data (f_1 or F_1) and the first frequency dimension data (f_2 or F_2) may be phase sensitive. (Note that f_1 and f_2 appear to be reversed but this naming convention derives from the order of their time domain precedents, t_1 and t_2, in the NMR pulse sequence.) Zero- and first-order phasing of the second dimension of a 2-D NMR data set is required in many cases. Some experiments, most notably the gradient-selected heteronuclear multiple bond correlation (gHMBC) experiment, use the absolute value of the signal and hence do not require phasing.

When we phase a spectrum, it is more important to equalize the height of the baseline on both sides of the peak being phase corrected than to obtain a symmetric peak shape. Peak symmetry is less important than baseline alignment because peak symmetry is often

First-order phase correction. Syn. first-order phasing. The variation in the proportion of amplitude data taken from two orthogonal arrays (or matrices) wherein the proportion varies linearly as a function of the distance from the pivot point.

Zero-order phase correction. Syn. zero-order phasing. The variation in the proportion of amplitude data taken from two orthogonal arrays (or matrices) that is applied evenly to each point in the spectrum.

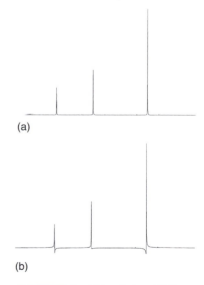

(a)

(b)

■ **FIGURE 3.6** (a) A well-phased NMR spectrum; (b) a poorly phased NMR spectrum.

z⁴ hump. An asymmetric peak shape, usually consisting of a low-intensity peak offset but overlapping with the main peak (a shoulder), that results from a poorly set z^4 RT shim current. The presence of z^4 humps are particularly troublesome when one wishes to observe a weak peak whose chemical shift is near that of an intense peak. ^{1}H shifts near 4.65 ppm when the solvent is 10% D_2O/90% H_2O and suffers particularly from the presence of a z^4 hump.

Baseline correction. The flattening or zeroing of the regions of a frequency spectrum that are devoid of resonances.

Phase cycling. The alternation of the phase of applied RF pulses and/or receiver detection on successive passes through a pulse sequence without the variation of pulse lengths or delays. Data collected when only RF pulse phases are varied is often added to data collected with different RF pulse phases to cancel unwanted components of the detected signal, or to cancel artifacts either inherent in single executions of a specific pulse sequence or inherent to unavoidable instrumental limitations.

Rotamer. Syn. rotational isomer. An isomer generated by rotation (usually 120°) about a chemical bond.

Rotational isomerism. Interconversion of rotational isomers or rotamers.

imperfect when the shims are not adjusted well. Nonoptimal settings of one or both of the higher order even power shims z^4 and z^6 generate asymmetric peak shapes, as do combined missettings of the odd power shims z^3 and z^5. The term "z^4 hump" may sound comical, but over time we will encounter an asymmetric peak shape when our z^4 shim is adjusted poorly often enough that this term should become familiar, especially if we are using 90% H_2O/10% D_2O as our solvent. The z^4 hump may be especially prevalent when we use an instrument that lacks the ability to carry out automatic gradient shimming.

3.13 BASELINE CORRECTION

Even in the absence of ringdown effects, we often find that the baseline of the spectrum is not flat. Baseline distortion results from a number of instrument hardware attributes and experimental methods. Imperfect phase cycling and the application of filters at the edge of the spectral window can both generate an imperfect baseline. In practice, we simply acknowledge that a perfectly phased spectrum may still exhibit a rolling or tilted baseline. Because of the prevalence of baseline distortion, it is a commonly accepted practice to apply baseline correction to virtually all NMR spectra. Each NMR processing software package will feature a protocol for carrying out baseline correction.

Judicious application of baseline correction is important. First and foremost, before we apply baseline correction to our spectrum, we must be sure that we are not wiping out a broad resonance. This problem can be avoided by Fourier transforming our data set, phasing it on the narrow resonances, and then increasing the vertical scale of the spectrum. Atoms (usually ^{1}H's) undergoing chemical exchange generate broader resonances than do their nonexchanging counterparts. Broadening of resonances is observed when portions of our molecule move slowly on the NMR time scale, perhaps as the result of the interconversion of rotamers (rotational isomerism).

Another problem arising from the correction of baseline distortion stems from the presence of minor components, including the satellite peaks observed in ^{1}H NMR spectra caused by ^{13}C spins. Whereas some baseline correction algorithms work automatically, others require the user to define what is baseline and what is not. When defining baseline, we take care not to include any peaks, however weak, in the baseline region. Failure to exclude small peaks from regions of supposed baseline may result in a skewing of the baseline-corrected spectrum, thereby producing strangely shaped resonances and introducing systematic errors in calculated peak integrals.

When carrying out baseline correction as part of a 2-D spectrum, we first ensure that the baseline correction is working properly on a 1-D spectrum of our sample collected with the same spectral window. Then we use those same parameters for the 2-D spectrum.

3.14 **INTEGRATION**

One of the most pleasing aspects of NMR spectroscopy is that, in many cases, the areas under the curves that are the NMR signals or resonances correspond to the relative abundances of each chemically distinct species in the sample. The lack of a varying extinction coefficient (in contrast to IR spectroscopy, for example) in the determination of NMR signal strength is one of the pillars of the practice of NMR. The spectral peak corresponding to the resonance arising from a given chemical species can be integrated to determine the abundance of that chemical species. Integrals help identify and assign various moieties in the solute molecule or molecules. Figure 3.7 shows the 1-D ^{1}H NMR spectrum of 1-ethyl-2-pyrrolidone (in $CDCl_3$) with integrals.

To maximize the accuracy of the integrals we measured from our sample, we pay attention to several items. We make sure that the sample is fully relaxed between scans. Full relaxation back to equilibrium requires that we wait a minimum of five times the longest T_1 between the application of the read pulse (the RF pulse applied just before digitization of the FID commences) and the initial excitation pulse of the following scan (for a garden variety 1-D ^{1}H or ^{13}C spectrum, the read pulse is the first pulse). Recall that the amount of time

Accuracy. How close an experimentally determined value is to the true or actual value.

Read pulse. The RF pulse applied just before the digitization of the FID.

■ **FIGURE 3.7** The 1-D ^{1}H NMR spectrum of 1-ethyl-2-pyrrolidone in $CDCl_3$; the values of integrals of the resonances appear underneath each resonance.

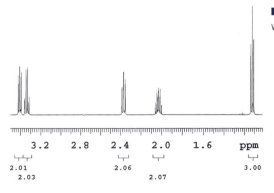

Carr-Purcell-Meiboom-Gill (CPMG) experiment. An experiment wherein the net magnetization is tipped into the **xy** plane, and subjected to a series (or train) of RF pulses and delays to refocus the net magnetization. Maintaining the net magnetization in the **xy** plane allows the measurement of the T_2 relaxation time.

Acquisition time, at (Varian), or AQ (Bruker). Syn. detection period. The amount of time that the free induction decay (FID) is digitized to generate an array of numbers; denoted "at" on a Varian instrument, or "AQ" on a Bruker instrument.

we wait prior to application of the first pulse in a pulse sequence is called the relaxation delay. It is often denoted RD; on a Varian NMR instrument it is often called d1.

The NMR signal begins to relax as soon as the last RF pulse in the pulse sequence is applied in the pulse sequence, so the relaxation delay need not be five times the longest T_1 of interest. That is, in most cases relaxation is already taking place while the FID is being digitized (exceptions are rare, but the Carr-Purcell-Meiboom-Gill or CPMG experiment that is used to measure T_2 is one example). Thus, we subtract the amount of time spent acquiring (digitizing) the FID from the five T_1's to determine the appropriate relaxation delay. The time spent digitizing the FID is referred to as the acquisition time and is called AQ on a Bruker NMR instrument, or "at" on a Varian NMR instrument.

When the T_1's for the various resonances of interest are not known (this is almost always the case) and we wish to obtain accurate integral values, we can either measure the T_1's prior to collecting our spectrum to be integrated, or we can estimate the longest expected T_1 and multiply that number by five to arrive at the total amount of time we will program into the pulse sequence to allow for relaxation.

In practice, the measurement of T_1's plus the collection of the 1-D spectrum to be integrated often requires more instrument time than just going ahead and collecting the 1-D experiment with even the most conservative (i.e., long) estimate of the longest T_1 value. If we elect to skip the T_1 measurement step, we must remember that an insufficient relaxation delay in our pulse sequence will attenuate some of the measured integrals. Thus we should watch for systematic errors when examining integrals. If we collect a 1-D ^1H spectrum and we notice that the integral values we obtain for ^1H's bound to sp^2-hybridized carbons are lower than we expect, we may elect to repeat collection of the 1-D ^1H spectrum with a longer relaxation delay.

Impurities in the sample can also affect measured T_1's. The oxygen molecule O_2 is paramagnetic, and this paramagnetism supplies the various spins in our sample with an efficient relaxation pathway. Thus, dissolved oxygen shortens T_1's. If we want to accurately measure T_1's, we bubble nitrogen or argon gas through our sample solution or use a freeze-pump-thaw cycle in order to remove dissolved oxygen. If, however, we are interested in obtaining accurate integral data as quickly as possible, then having dissolved oxygen in the sample is desirable.

The presence of other paramagnetic species (those with unpaired electrons)—in particular metal ions—also shortens T_1's. In general, however, any chemical species that can move in and out of the solute's solvation shell is likely to perturb measured T_1's. Changes in temperature may also affect the T_1's of the various resonances in the spectrum of our sample.

When we observe ^1H, we often find that a methyl group, if present, will be the moiety with the shortest T_1 (it will relax the fastest). All other integrals should be compared with the area of three protons for a single methyl group in the spectrum. In general, the further downfield (to the left on the ppm scale) the ^1H resonance, the more slowly it will relax. Aldehyde and aromatic protons relax particularly slowly, and we should exercise great caution when we base conclusions on integrals of downfield protons.

Data from the observation of other nuclei can also be integrated, but special precautions are usually required to ensure the collection of quantitative integral data. One-dimensional ^{13}C spectra are almost never integrated, for several reasons. First, ^{13}C's typically relax very slowly and waiting five times the longest ^{13}C T_1 often makes accumulation of a spectrum with a reasonable S/N ratio prohibitively time-consuming for samples with ^{13}C present at its 1.1% natural abundance. A second problem stems from the NOE enhancement experienced by ^{13}C in the present of ^1H decoupling. Switching (gating) the ^1H decoupler off during the relaxation delay but on during the digitization of the ^{13}C FID (to collapse multiplet structure from J-coupling to ^1H's) is a prerequisite for obtaining quantitative ^{13}C peak integrals.

3.15 MEASUREMENT OF CHEMICAL SHIFTS AND J-COUPLINGS

Data processing software is generally very good at determining the location of NMR peaks. Nonetheless, there are several points worth mentioning with regard to peak picking. Peak picking is measuring the chemical shifts of the resonances we observe in our spectrum.

Measurement of the positions (chemical shifts) of the peak maxima allows us to determine approximate chemical shifts and to calculate J-couplings. We obtain the ^1H 1-D spectrum and measure the chemical shifts of the resonances in all cases. If our sample is dilute, we

Paramagnetic species. A molecule containing unpaired electrons.

Downfield. The left side of the chemical shift scale. This corresponds to higher frequency, and the higher resonant frequency in turn indicates a lack of electron density.

A resonance is downfield if it is located on the left side of the spectrum or if it is observed to appear to the left of its expected value. By convention, lower frequencies (also lower chemical shifts values in ppm) appear on the right side of the spectrum and higher frequencies appear on the left side of the spectrum. Because the first generation of NMR instruments (CW instruments) operated with a constant RF frequency and a variable applied magnetic field strength, lower fields (downfield) were required to make higher frequency resonances come into resonance. Therefore, when an NMR spectrum is plotted as a function of field strength, the lower values appear on the left and higher values appear on the right.

Peak picking. The determination of the location of peak maxima in a frequency spectrum, often done by software with minimal user guidance.

Quartet. Four evenly spaced peaks in the frequency spectrum caused by the splitting of a single resonance by J-coupling to three identical spin-½ nuclei to give a multiplet with four peaks with relative intensity ratios of 1:3:3:1.

Splitting. The coupling-induced division of the resonance from an NMR-active atom in a single atomic site in a molecule into two or more peaks, legs, or lines.

Splitting pattern. The division by J-coupling of a resonance into a multiplet with a recognizable ratio of peak intensities and spacings.

Complex data point. A data point consisting of both a real and an imaginary component. The real and imaginary components allow an NMR data set to be phase sensitive, insofar as there can occur a partitioning of the data depicted between the two components. An array of complex data points therefore consists of two arrays of ordinates as a function of the abscissa.

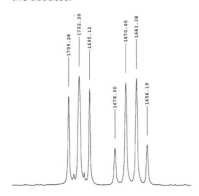

■ **Figure 3.8**   Peak-picking information for two ^{1}H NMR resonances. The triplet on the left shows a splitting of 7.1 Hz, whereas the quartet on the right shows a splitting of 7.3 Hz.

may not be able to obtain a 1-D ^{13}C spectrum with a sufficient S/N to allow the measurement of the ^{13}C chemical shifts. Figure 3.8 shows an expansion of a 1-D ^{1}H spectrum containing two resonances split into multiplets because they are J-coupled to other resonances outside the displayed region of the spectrum. That is, the spin states of NMR-active nuclides one or more bonds distant may divide (split) the resonance from a single chemically distinct species into two or more peaks; these peaks may be well-resolved or they may overlap. The difference in the frequencies of two peaks arising from one resonance is called a splitting. A splitting pattern we measure may correspond to a single J-coupling value, or it may be the sum or difference of two or more J-couplings. Above the two multiplets in the spectrum shown in Figure 3.8 are the digital readouts or "peak-picking" information. When we measure J-couplings, we express our results in Hz; when we measure chemical shifts, we report our values in ppm.

It is important to know the digital resolution of the spectrum whose peak positions we measure. The digital resolution is the sweep width in Hz divided by the number of points of data in one channel of the spectrum. Recall that in order to make our spectrum phase-sensitive, we have two channels (we call them the real and the imaginary channels) to allow us to phase correct our spectrum. Therefore half of the points in the spectrum will lie in the real channel and span the full sweep width. A sweep width of 5 kHz containing 64 k complex data points has 32,768 points distributed evenly along the 5 kHz of frequency in each channel, therefore giving us a digital resolution of 5000 Hz/32,768 points, or 0.15 Hz/point. If we want to measure J-couplings with a precision greater than 0.1 Hz, the size of the Fourier transformed data set must be greater than 64 k (complex). In practice, the fast speed and data handling capabilities of modern computers allow us to always transform a 1-D data set by using a 256 k or larger array.

We continue to increase the size of the data set using zero filling until we no longer observe a connect-the-dots appearance when we examine expansions of the spectrum that reveal the spectrum's fine structure; NMR peaks should appear to be smooth (not jagged) when expanded.

The digital resolution is the lower limit of the uncertainty associated with the measurement of J-couplings and chemical shifts. Increasing the size of the Fourier-transformed spectrum does not reduce the uncertainty of a frequency measurement without bound. At some point, further increasing the number of points per unit frequency fails to further improve the accuracy of frequency measurements, even though the digital resolution continues to shrink. The accuracy of a frequency measurement is assessed by measuring the same spectral feature(s) (chemical shifts or J-couplings) multiple times from different spectra of our sample. We can then average our measured values and calculate their standard deviation. Three is the minimum number of measurements we require to allow computation of a standard deviation.

The values of J-couplings obtained from 1-D spectra are frequently reported without uncertainties or with the digital resolution given as the uncertainty. Using the digital resolution as the uncertainty is not strictly correct; the uncertainty is the accuracy, which is how close a measurement is to its actual value. The precision is the smallest amount a measurement may change from one reading to the next.

Precision. The magnitude of the smallest discernable variations in an experimental measurement.

For 2-D spectra, the problem of having too large a digital resolution along the frequency axis being used to measure chemical shift and/or J-coupling can be difficult to overcome because of the necessity of limiting the size of the data matrix that is the data set. In certain cases we adjust the dimensions of the 2-D data matrix to minimize digital resolution along one axis at the expense of digital resolution along the other. Keeping the product of the two matrix dimensions below a certain value is of paramount importance so that the computer processing time is not excessive. For example, a $4\,k \times 4\,k$ 2-D data set can be converted to a $16\,k \times 1\,k$ data set without drastically increasing processing time because the total size of the data matrix will be the same. With this matrix dimension readjustment, the frequency range of one of the 2-D axes will now be spanned by 8192 points instead of 2048 (for a $4\,k \times 4\,k$ matrix). When we measure chemical shifts and J-couplings from 2-D NMR spectra—which is often the case when we encounter overlapping resonances—the digital resolution is often larger (expressed in Hz/pt) than the accuracy we might calculate by measuring a spectral feature multiple times on a well-isolated resonance under similar laboratory conditions in a 1-D spectrum. When the digital resolution is large, citing the digital resolution as the uncertainty is acceptable.

A resonance arising from a single chemical species may be split into multiple lines (legs) to form a multiplet because of J-coupling of the spins generating the observed resonance to other nearby spins. Nonfirst-order effects may skew the relative intensities of the individual peaks comprising a multiplet from those predicted on the basis of spin-state statistics. For example, the signal from the three [1]H's of a methyl group adjacent to a methylene group may not appear exactly as a 1:2:1 triplet if the chemical shift of the methylene group is close to that of the methyl group. That is, the leg of the triplet nearer to the chemical shift of the adjacent methylene group will be more intense than the leg farther away. In this and similar cases, we must take great care when measuring the chemical shift of the group in question. The chemical shift is defined as the point on the chemical shift axis at which the integral line of the multiplet has risen to 50% of its total value. This skewing of multiplet intensity presents special challenges when we observe that it overlaps with other resonances, because the integral contains contributions from other resonances. Therefore, the point where the integral has risen by 50% may not be readily discernable.

In some cases, we employ 2-D methods to measure chemical shifts and J-couplings. Resonance overlap is often observed in a 1-D ^{1}H spectrum, making accurate chemical shift and J-coupling determination impossible. Resonance overlap often requires that we use 2-D methods to isolate the chemical shifts or J-couplings. That is, the cross peaks we observe in certain 2-D spectra allow us to determine chemical shifts and J-couplings that we cannot measure directly from the 1-D spectrum. The combination of overlapping resonances and nonfirst-order intensity skewing of the various legs of a particular multiplet may make the accurate measurement of a chemical shift difficult. Nonetheless, it is often possible to measure chemical shifts to within 0.01–0.02 ppm for most ^{1}H resonances.

When we attempt to measure ^{13}C chemical shifts, we may encounter a poor S/N in the directly-detected 1-D ^{13}C NMR spectrum. Therefore, we may have to resort to the use of indirect detection methods—especially the ^{1}H-detected HMBC experiment—to obtain the chemical shifts of ^{13}C atoms lacking directly attached ^{1}H's.

3.16 DATA REPRESENTATION

NMR data can be shown in a number of different formats. In some cases, a table containing shifts, multiplicities, integrals, etc. may suffice to describe the information contained in a simple 1-D NMR spectrum. In other cases, however, we may wish to show a colored contour plot to help capture the complex nature of a particular 2-D NMR spectrum.

Assuming we have employed some apodization function and baseline correction in massaging the data set to be as aesthetically pleasing as possible, we may still be confronted with a number of decisions to be made regarding the depiction of our data. For 1-D data sets, we choose both a vertical and a horizontal scale. Expansions are often nice to include in plots of 1-D NMR spectra, because they allow both the inclusion of the entire spectrum (perhaps to show product cleanliness) and an enlarged region showing detail (perhaps a distinctive ^{1}H multiplet). As the operating frequencies of the NMR instruments used for collection of everyday 1-D ^{1}H spectra continue to increase, the widths of multiplets arising from J-couplings relative to spectral window sizes is decreasing, thus making the inclusion of expanded regions increasingly necessary.

If we wish to include peak-picking information above the 1-D NMR spectrum, we may choose to reduce the vertical intensity of the plot to allow room for that extra information (either in Hz or ppm, depending on need).

For 2-D NMR data sets, we normally generate monochrome contour plots. Just as a topographical map shows contour lines that denote specific elevations above sea level as a function of longitude and latitude, a 2-D contour plot shows spectrum intensity as a function of the f_1 and f_2 frequency axes.

When we generate a plot of a 2-D spectrum, we must consider the plot threshold (an intensity cutoff below which no data is shown), the contour spacing (how much the intensity must increase before the next contour line is drawn), and the maximum number of contour lines. In some cases (e.g., the DQFCOSY), we also plot negative contours. Figures 3.9, 3.10, and 3.11 show three plots of the same region of a ^{1}H-^{1}H 2-D gCOSY spectrum. Figure 3.9 shows a plot with the threshold set too high; Figure 3.10 shows a plot with the threshold set too low; Figure 3.11 shows a plot with the threshold set just above the noise floor.

Setting the plot threshold too high can create the misleading impression that a particular cross peak is weak or absent when in fact it is not (see Figure 3.9). This error can occur when one or both of the resonances participating in the cross peak is/are spread out, perhaps as a result of a large number of J-couplings. The true measure of a cross peak's intensity (integral) is not its height but rather its volume or the sum of the absolute values of the volumes of its positive and negative constituents.

If we select a plot threshold that is too low, we will find that the entire region being plotted will contain many more or less random lines, and these lines will distract the eye from the cross peaks of

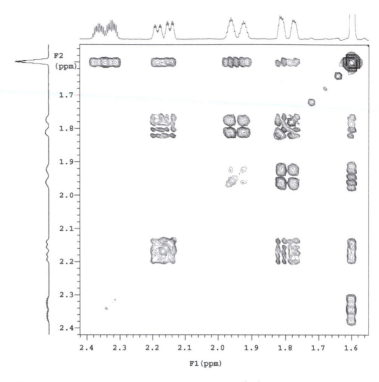

■ FIGURE 3.9   The first of three plots of the same region of a ^{1}H-^{1}H 2-D gradient-selected correlation spectroscopy (gCOSY) spectrum showing the effect of plot threshold variation. In this plot, the plot threshold is set too high (too little information appears in the 2-D plot).

t$_1$ ridge. Syn. t$_1$ noise. A stripe of noise often observed in a 2-D spectrum for a given chemical shift value of the f$_2$ chemical shift axis that runs parallel to the f$_1$ chemical shift axis. A t$_1$ ridge occurs with an f$_2$ chemical shift corresponding to one or more of the most intense peaks in the spectrum of the 1-D spectrum of the f$_2$ nuclide.

interest (see Figure 3.10). There are two distinct schools of thought when it comes to selecting plot threshold. One school holds that we must set the threshold low enough to capture just a little of what is called the noise floor—that portion of the 2-D plot containing only random noise and no cross peaks or prominent artifacts (such as t$_1$ ridges, see Section 9.7). The second school of thought with regard to 2-D plot thresholds dictates that the plot should be made to look as clean as possible; thus the threshold should be set high enough to eliminate noise peaks, leaving only cross peaks of interest. Whereas the former school of thought may be considered to be more intellectually honest, there is clearly some value in drawing to the eye only the cross peaks of interest, thereby eliminating the obvious tangents that may arise if additional information not central to the present argument appears in a plotted region. The choice of the appropriate plot threshold is therefore context dependent and largely subjective.

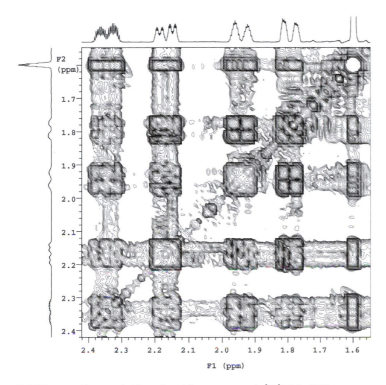

■ **FIGURE 3.10**   The second of three plots of the same region of a ^{1}H-^{1}H 2-D gCOSY spectrum showing the effect of plot threshold variation. In this plot, the plot threshold is set too low (too much information appears in the 2-D plot).

The spacing of successive contour lines is also a subjective choice. Contour lines in 2-D NMR plots are normally done on a logarithmic scale rather than on the linear scale we are used to seeing when we view topographical maps. In a logarithmic intensity plot, each successive line is some multiple (greater than one) of the previous line's intensity. Generating each successive contour line at 110% of the previous line's intensity will usually result in cross peaks that have closely spaces lines and resemble nearly completely filled in shapes (Figure 3.12). A 150% successive contour line intensity plot, however, would appear a much less dark plot (Figure 3.13). Figure 3.12 shows a 2-D NMR cross peak from a gradient-selected heteronuclear single quantum coherence (gHSQC) spectrum with (up to) 40 contour lines spaced 110% apart; Figure 3.13 shows the same cross peak with (up to) 20 contour lines spaced 150% apart; and Figure 3.14 shows the cross peak with only 10 contour lines spaced 110% apart.

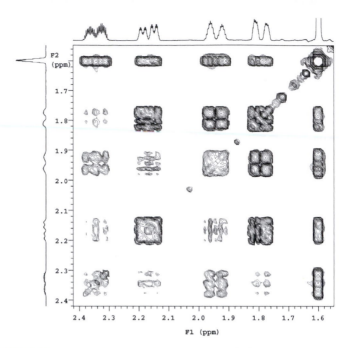

■ **FIGURE 3.11** The third of three plots of the same region of a ^{1}H-^{1}H 2-D gCOSY spectrum showing the effect of plot threshold variation. In this plot, the plot threshold is set properly (the right amount of information appears in the 2-D plot).

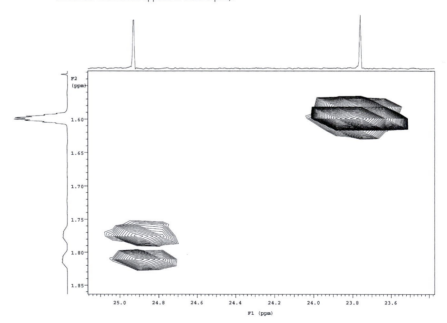

■ **FIGURE 3.12** The first of three plots of a ^{1}H-^{13}C 2-D gHSQC NMR spectrum, showing the effect of spacing and number of contour lines. This plot shows (up to) 40 lines spaced 110% apart.

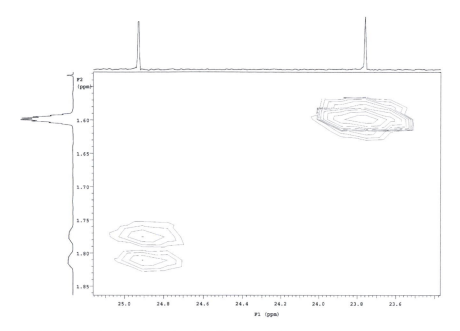

■ FIGURE 3.13   The second of three plots of a ¹H-¹³C 2-D gHSQC NMR spectrum, showing the effect of spacing and number of contour lines. This plot shows (up to) 20 lines spaced 150% apart.

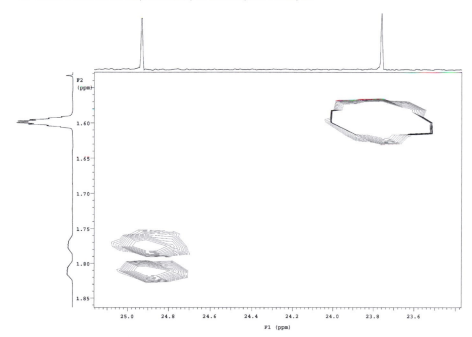

■ FIGURE 3.14   The third of three plots of a ¹H-¹³C 2-D gHSQC NMR spectrum, showing the effect of spacing and number of contour lines. This plot shows 10 lines spaced 110% apart.

Limiting the number of contour lines to just 5 or 10 likely will result in a plot with open, circular (or elliptical) cross peaks (Figure 3.14). This technique is often used by those in the bioNMR subdiscipline because bioNMR plots are often so full of cross peaks that having a large number of filled-in cross peaks is displeasing to the eye.

Limiting the number of contour lines to 15, 20, or 30 can also be used to help illustrate which cross peaks are most intense. This enhancement occurs because a weaker cross peak will have its middle completely filled in by the contour lines, but a stronger cross peak will appear as an open circle, much like a snow-capped mountain (Figure 3.14). That is, a strong cross peak will have an intensity at its center that rises above the maximum intensity denoted by the highest contour line. Therefore, the centers of the stronger cross peaks will be devoid of contour lines.

Chapter 4

^{1}H and ^{13}C Chemical Shifts

4.1 THE NATURE OF THE CHEMICAL SHIFT

As explained briefly in Chapter 1, variation in how much an NMR-active nucleus is shielded by its electron cloud from the applied magnetic field determines the frequency of the photons that induce transitions between allowed spin states. Because the resistance of the electron cloud to the applied field varies in direct proportion to the applied field strength, the frequency units cancel and we are left with a unitless quantity called the chemical shift and denoted δ.

In fact, the chemical shift is quoted in parts per million (ppm), but this modifier is really just a multiplier (10^6) and not a unit proper. The use of the ppm multiplier allows us to bandy about much more aesthetically pleasing shifts in the ranges of one, ten, hundreds, and even more ppm—although shift differences of a hundredth of a ppm are sometimes discussed.

The greater the electron density around an NMR-active nucleus, the more shielded the nucleus and, hence, the lower its NMR frequency relative to the same nuclide in other, less-shielded environments. The Pauli electronegativities are useful as the starting point for understanding chemical shifts. Recall that the electronegativity of an atom reflects how strongly that atom will withdraw electron density from other atoms to which it is bonded or connected via common bonds. Table 4.1 shows the electronegativities of the atoms most commonly found in the garden-variety molecules encountered in the application of NMR methods to organic compounds.

Electronegativity. A number reflecting the affinity of an atom for electron density.

Qualitative arguments about more- versus less-shielded NMR-active nuclei do have their place, but the choice of a standard is required to put the discussion of chemical shifts on firmer footing. For ^{1}H, ^{13}C, and ^{29}Si, the standard is tetramethylsilane (($CH_3)_4$Si), or TMS.

Table 4.1 Electronegativities [1] of elements found in organic compounds.

Element	Atomic number	Symbol	Electronegativity
Hydrogen	1	H	2.20
Boron	5	B	2.04
Carbon	6	C	2.55
Nitrogen	7	N	3.04
Oxygen	8	O	3.44
Fluorine	9	F	3.98
Sodium	11	Na	0.93
Magnesium	12	Mg	1.31
Aluminum	13	Al	1.61
Silicon	14	Si	1.90
Phosphorus	15	P	2.19
Sulfur	16	S	2.58
Chlorine	17	Cl	3.16
Selenium	34	Se	2.55
Bromine	35	Br	2.96
Iiodine	53	I	2.66

By definition, the chemical shift of the resonance of the 12 ^{1}H's in TMS is 0 ppm. Likewise, the resonance of the four ^{13}C's in TMS has a chemical shift of 0 ppm, as does the resonance of the single ^{29}Si atom. To clear up any potential confusion, it should be noted here that each different atom in TMS has its own unique NMR frequency for a given field strength—that is, the ^{1}H's in TMS do not resonate at the same frequency as the ^{13}C's in TMS, etc. Nonetheless, the resonance of each nucleus in TMS defines 0 ppm for the resonances of the same nuclide.

Because most organic compounds we will encounter will contain hydrogen (electronegativity = 2.20) and carbon (2.55), and may also contain nitrogen (3.04) and/or oxygen (3.44), we can appreciate that the relatively low electronegativity of silicon (1.90) means that the silicon atom in TMS does a poor job of withdrawing electron density from the four methyl groups around it when compared with more electronegative atoms. In fact, the opposite occurs; the silicon atom is an electron density donor to the four methyl groups; thus, those groups are insulated with additional electron density and hence experience additional shielding from the effect of the applied field (recall that electron density screens or shields the nucleus). The additional electron density found in the four methyl groups of TMS serves to place the shift of the resonances of ^{1}H's and ^{13}C's in TMS to the right of the range of shifts found in nearly all organic compounds.

Functional groups that withdraw electron density are called electron-withdrawing groups (EWGs). Functional groups that donate electron density are called electron-donating groups (EDGs).

By convention, 0 ppm occurs at the right side of the chemical shift scale in the NMR spectrum, and higher shifts are listed to the left of 0 ppm. This representation is counterintuitive, in that lower frequency (and lower chemical shift) is on the right and higher frequency (and higher chemical shift) is on the left.

The methyl groups in TMS receive extra electron density from the silicon atom, so they are said to be shielded. Methyl groups next to an electronegative atom like oxygen (3.44) lose electron density to the electronegative oxygen and are said to be deshielded. That is, moieties near electron-withdrawing groups have less electron density to shield them from the effect of the applied field and thus resonate at a higher frequency (for a given field strength).

In much of the literature, chemical shifts are referred to as being upfield or downfield. This terminology is left over from the days of continuous wave (CW) instruments wherein the frequency was held constant and the applied magnetic field was varied over time to bring nuclei with one chemical shift into resonance at a time. At a constant frequency, the magnetic field would have to be raised (hence upfield) to bring more shielded groups into resonance. Similarly, a lowering of the magnetic field (downfield) will bring the more exposed (deshielded) atoms into resonance.

For the ^{1}H and ^{13}C chemical shift scales, Table 4.2 should help sort out the various terms used interchangeably in much of the discussion found in the NMR literature.

Electron-withdrawing group, EWG. A functional group that withdraws electron density through chemical bonds from other groups or atoms nearby. Electron density withdrawal may occufr either directly from groups that are alpha to the EWG, or from groups that are more distant through the phenomenon of (chemical) resonance. Effective electron withdrawers include atoms with large electronegativities, and functional groups that can, through (chemical) resonance, assume a negative formal charge (and still comply with the octet rule).

Electron-donating group, EDG. A functional group that donates electron density through chemical bonds to other groups or atoms nearby. Electron density donation may occur either directly to groups that are alpha to the EDG, or to groups that are more distant through the phenomenon of (chemical) resonance. Effective electron donors include atoms or functional groups with lone pairs on their attachment points and functional groups containing atoms with low electronegativities.

Shielded group. A functional group with extra electron density.

Deshielded group. A chemical functional group deprived of its normal complement of electron density.

Table 4.2 Correlation of terms used to describe relative chemical shift locations.

δ_H (ppm)	δ_C (ppm)	Description of chemical shift	Description of nuclear shielding	Relative frequency for a given B_0
10	250	downfield	deshielded	higher
0	10	upfield	shielded	lower

Upfield. The right side of the chemical shift scale, corresponding to lower frequency, and the lower resonant frequency in turn indicates additional electron density. A resonance is upfield if it is located on the right side of the spectrum or if it is observed to appear to the right of its expected value. By convention, lower frequencies (also lower chemical shifts values in ppm) appear on the right side of the spectrum and higher frequencies appear on the left side of the spectrum. Because the first generation of NMR instruments (CW instruments) operated with a constant RF frequency and a variable applied magnetic field strength, higher fields (upfield) were required to make lower frequency resonances come into resonance. Therefore, when an NMR spectrum is plotted as a function of field strength, the lower values appear on the left and higher values appear on the right.

In discussions of NMR spectra, we often anticipate where along the ppm scale we will observe the chemical shift of the resonance arising from a particular functional group. However, surprising differences and marked deviations can be ascribed to the shielding or deshielding of functional groups, atoms, and molecular moieties, and the resonances can be said to be shifted upfield or downfield. For example, we can state that an electronegative atom alpha to (one bond away from) a methyl group deshields the methyl group and shifts the resonance of the methyl group downfield from its normally observed chemical shift.

The chemical shift (δ) for either ^{1}H, ^{13}C, or ^{29}Si is calculated as

$$\delta = [(\nu_{obs} - \nu_{TMS})/\nu_{TMS}] \times 10^6 \qquad (4.1)$$

where ν_{obs} is the frequency observed from the resonance of the group in question and ν_{TMS} is the frequency observed for the resonance of the TMS standard. Notice that the units of frequency (Hz) cancel in Equation 4.1. The multiplier of one million is used to bring the numbers up to a reasonable value so that, for example, we do not have to talk about a shift of 0.00000104 but can instead discuss a shift of 1.04 ppm.

4.2 ALIPHATIC HYDROCARBONS

Consider aliphatic hydrocarbons. A methyl group (CH_3) at the end of a long aliphatic chain will show a ^{1}H resonance with a shift of about 0.75 ppm. Figure 4.1 shows the ^{1}H shifts for the resonances of the aliphatic side chain of 5-hydroxydodecanoic acid δ-lactone. The methyl ^{1}H's (position 12) that are nine bonds removed from the oxygen atom of the carbonyl group are sufficiently well isolated to

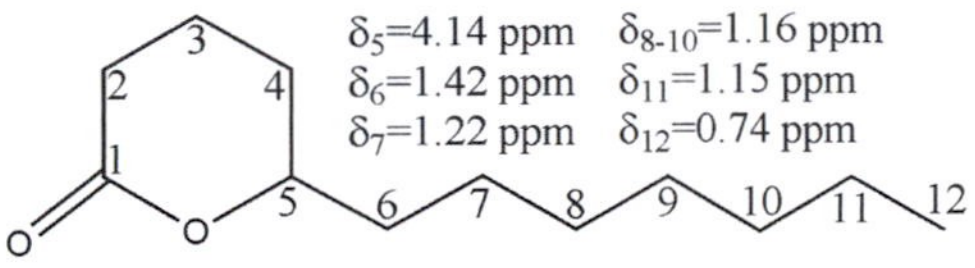

δ_5=4.14 ppm δ_{8-10}=1.16 ppm
δ_6=1.42 ppm δ_{11}=1.15 ppm
δ_7=1.22 ppm δ_{12}=0.74 ppm

■ **FIGURE 4.1** [1]H chemical shifts of the aliphatic side chain of 5-hydroxydodecanoic acid δ-lactone.

show a resonance with a shift of 0.74 ppm. This shift is as upfield as we are likely to see from any methyl [1]H's in the absence of a ring current or other effects, including those stemming from chemical shift anisotropy (see Section 4.4). Therefore, we should memorize that 0.7 ppm marks the right edge of the aliphatic ^{1}H chemical shift range. The presence of any more electronegative atom nearby will pull the ^{1}H shift of a methyl group's resonance downfield (to the left).

A methylene (CH_2) group in the middle of a long chain of hydrocarbons will generate a ^{1}H resonance with a chemical shift of around 1.15 ppm. The chemical shifts of the methylene ^{1}H resonances on carbons 8–11 of 5-hydroxydodecanoic acid δ-lactone are found at 1.15–1.16 ppm.

Consider a CH_2—R fragment. If R = H, less electron density will be pulled away from the CH_2 than if R = C, because the electronegativity of carbon is greater than that of hydrogen. Therefore we see the chemical shift of the methylene group further downfield than that of the methyl group.

A methine (CH) group surrounded by methyl or methylene groups will produce a resonance with a ^{1}H chemical shift of 1.5–1.6 ppm. Again we can apply the argument that the more electronegative substituents withdraw more electron density. For the CH of the R_1—CH—R_2 fragment, the shifts of its ^{1}H and ^{13}C resonances will be more downfield if R_1 and R_2 are both carbon, intermediate if one is hydrogen and one is carbon, and more upfield if R_1 and R_2 are both hydrogen.

The ^{13}C shifts of methyl group resonances in aliphatic hydrocarbons normally fall in the range of 13 to 24 ppm, whereas the ^{13}C shifts of methylene group resonances in the same class of compounds will lie in the range of 22 to 35 ppm, with 29 ppm being the shift of the ^{13}C resonance of the methylene group in the middle of the pendant seven-carbon chain of 5-hydroxydodecanoic acid δ-lactone. [13]C shifts from methine group resonance normally appear in the range of 30 to 42 ppm.

Electronegative groups and atoms only affect nearby atoms in molecules. If we are observing the resonance of a particular ¹H or ¹³C in a molecule and an electronegative atom or group is many bonds distant, we are unlikely to observe much of an effect on the chemical shift of the ¹H or ¹³C resonance in question. The presence of oxygen or other electronegative heteroatom or group within two or three bonds will move observed shifts noticeably downfield.

4.3 SATURATED, CYCLIC HYDROCARBONS

If a molecule has one of more saturated rings in it, the shifts of the ¹H and ¹³C resonances in the molecule will often be observed to be upfield from what we would otherwise expect. The greater the ring strain, the greater the effect. Three-membered rings show this effect to the greatest extent, whereas six-membered rings typically only show this effect to a small degree.

Chemical shifts from six-membered rings are often easy to understand. Most six-membered rings are found to be in the chair conformation; axial ¹H resonances are commonly found 0.4 ppm upfield relative to their equatorial counterparts. This difference is ascribed to variations in electron density at the two positions.

Five-membered rings, on the other hand, are particularly troubling. The main problem is that they tend to adopt a conformation in which four of the five atoms in the ring lie in a single plane, with the fifth atom out of the plane. To visualize this conformation, imagine four atoms on the corners of a postage envelope with the fifth atom occupying the tip of the envelope's partially open flap. The problem lies in determining which atom (or atoms) is (are) on the point of the envelope's flap, and also how long it (they) resides (reside) there.

4.4 OLEFINIC HYDROCARBONS

The introduction of double bonds complicates our discussion of hydrocarbon shifts because of the asymmetry of the carbon–carbon double bond. As we might expect, a terminal double-bonded ¹³C ($H_2C=$) will be deshielded relative to a methyl group because instead of being bonded to three protons, it is single-bonded to two protons and double-bonded to a carbon (recall the electronegativity argument presented in Section 4.1); the chemical shift of the ¹³C resonance of the $H_2C=$ fragment is about 114 ppm. ¹H's attached to doubly bonded carbons typically give resonances with shifts in the range of 4.9 to 5.7 ppm. ^{13}C's participating in carbon-carbon double bonds produce resonances in the range of 114 to 138 ppm, but other factors can influence shift.

We should not spend too much time memorizing exact ranges for chemical shifts of the resonances arising from the various functionalities we are likely to encounter. Effort aimed at understanding and developing an appreciation for shift trends due to adjacent substitution pays higher dividends in the long run. If we assign the ^{1}H and ^{13}C resonances to the atoms in the molecules in Chapters 11 and 12, we will be well on our way.

Besides seeing a downfield shift of ^{1}H and ^{13}C resonances in the vicinity of the region of the molecule deprived of electron density by the double bond, we will also see an effect from the bulky π-electron cloud above and below the C$=$C bond. The π cloud is much less confined than the σ bonds, and the lack of confinement allows the electron density in the π cloud to circulate to various degrees as a function of the orientation of the double bond relative to the z-axis of the applied field. Because the circulation of the π-electron density varies with molecular orientation, the shielding provided to nearby nuclei by the π electrons will also vary. Stated another way, because the π electrons generate a circulating charge whose extent of circulation varies with double bond orientation, the tiny magnetic field resulting from the circulating charge will also vary with orientation.

The variation in shielding resulting from double bond orientation is one example of chemical shift anisotropy (CSA). A CSA is said to be observed when the chemical shift varies as function of the orientation of a molecule relative to the direction of $\mathbf{B_0}$. When molecular tumbling rates in solutions become slow enough, the CSA and also the dipolar interaction provide a very efficient relaxation mechanisms and drastically broaden resonances through shortening of the $T_2{}^*$ ($T_2{}^*$ cannot be greater than T_1). In the limit of no movement— i.e., in rigid solids—we often find that resonances become very broad, and special hardware is required to acquire a spectrum.

Alternating single and double bonds (conjugation) relax the restriction on the quick drop-off of the effects of electronegative groups in molecules. That is, an EWG many bonds distant may exert a profound deshielding effect on a given ^{1}H or ^{13}C if resonance structures can be drawn to link the two.

4.5 ACETYLENIC HYDROCARBONS

Carbon–carbon triple bonds are relatively rare and not often found in the molecules of nature. ¹H's at the end of a carbon–carbon triple bond will produce resonances with shifts in the range of 1.8 to 1.9 ppm, and ¹³C's in carbon–carbon triple bonds will resonate from 70 to 85 ppm.

Just as the π cloud of the carbon–carbon double bond is affected by molecular orientation, so too is the π cloud of the carbon–carbon triple bond. What is interesting here is that the ¹H and ¹³C resonance shifts we observe with triple bonds are lower than with double bonds. That is, the argument that the more electron-density deprived the system, the greater the observed chemical shift fails to predict what we observe. The explanation for this apparent anomaly is that the extensive amount of electron density surrounding the triple bond serves to provide additional shielding to the ¹H and ¹³C nuclei; thus, these nuclei feel less of the applied field than we might predict on the basis of the level of unsaturation in the molecular fragment.

4.6 AROMATIC HYDROCARBONS

Aromaticity imparts a special magic on chemical shifts as a result of the ring currents that occur when the vector normal to the plane of the aromatic ring is parallel to the z-axis of the applied field. When rapid molecular tumbling takes place, some averaging takes place, but a very significant effect is still observed. Instead of being confined to just a single π bond, π electrons in aromatic systems circulate around the ring. This charge circulation is called a ring current, and it generates an important magnetic field. Because all magnetic field lines travel in circular arcs (expressed in Maxwell's equations), the resonances arising from the ¹H's on the outside of the ring will be shifted in one direction (downfield) and the resonances from the ¹H's in (or near) the inside of the ring will be shifted in the opposite direction (upfield). Location of ¹H's in the interior (shielding) region of an aromatic π system is common in proteins but rare in small molecules. Nonetheless, [18]-annulene possesses these types of ¹H's, as well as ¹H's with their resonances shifted to the more commonly encountered downfield end of the chemical shift range. The downfield shift of the resonances of aromatic ¹H's is caused by their location on the outside of the aromatic π system. Figure 4.2 shows both the downfield chemical shift for the resonances of the ¹H's on the outside of the aromatic ring and also the upfield chemical shift for the resonances of the ¹H's in the interior of the aromatic ring.

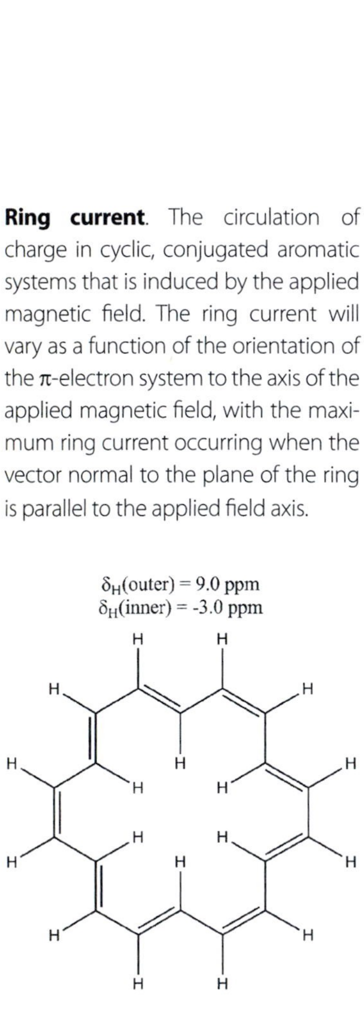

Ring current. The circulation of charge in cyclic, conjugated aromatic systems that is induced by the applied magnetic field. The ring current will vary as a function of the orientation of the π-electron system to the axis of the applied magnetic field, with the maximum ring current occurring when the vector normal to the plane of the ring is parallel to the applied field axis.

δ_H(outer) = 9.0 ppm
δ_H(inner) = -3.0 ppm

■ **FIGURE 4.2**   [18]-Annulene shows both upfield and downfield shifts for the ¹H's located outside and inside (respectively) the aromatic π-electron system.

An analogy may help us visualize how the aromatic ring current works. Imagine a rock in the middle of a swiftly flowing stream. Directly downstream from that rock, there will be an eddy wherein the effect felt by the current is reduced; at the edges of the rock, however, the water must flow more rapidly to get around the rock. If the current is slow in one spot, it must increase somewhere else to compensate. The current is like the applied field and the rock is like the delocalized electrons: At the edge of the rock (ring) the current (field) is greater, whereas behind the rock (in the middle of the ring) the current is minimal.

The circulation of aromatic electrons generates a field that opposes the applied magnetic field at the center of the ring; but at the outside edge of the ring (and even at the ring), the applied field is actually enhanced. Because ^{1}H's on aromatic rings are most commonly found on the outside of the ring, the applied field is enhanced by the circulating aromatic π-electron density; hence, we observe lower chemical shifts for the resonances of aromatic ^{1}H's than for the resonances of ^{1}H's on isolated sp^2-hybridized (olefinic) carbons. That is, the ring current induced by the applied field adds to the applied field and makes aromatic ^{1}H's and ^{13}C's resonate at lower ppm values than their olefinic counterparts. Typical aromatic ^{1}H resonance chemical shifts are 7.2–7.7 ppm, and typical aromatic ^{13}C resonance chemical shifts are 120–145 ppm. In polyaromatic hydrocarbons, the ^{1}H resonances may be observed as far downfield as 8.7 ppm.

The presence of ring substituents that can donate and withdraw electron density will move aromatic ^{1}H and ^{13}C resonances upfield and downfield, respectively. The effect is generally observed only at positions on the ring that are *ortho* and *para* to electron-donating groups or electron-withdrawing groups. Alkyl groups, through a bizarre incantation called no-bond resonance, are electron-donating groups.

No-bond resonance. An extension of traditional resonance structure formulations wherein one violates the octet rule with a two-electron deficit being placed on an alkyl group to explain the electron-donating nature of alkyl groups.

Strongly electronegative atoms with lone pairs (e.g., O, N, F) will shift the resonance of the ^{13}C to which they are attached downfield (they will withdraw electron density through the σ bond) but will shift the resonances of *ortho* and *para* ^{13}C's and ^{1}H's upfield because of their electron donation through the resonance structures involving π electrons.

4.7   HETEROATOM EFFECTS

The introduction of nitrogen and oxygen into this discussion of ^{1}H and ^{13}C shifts introduces a great deal of complexity, but with a little

Table 4.3 ^{1}H and ^{13}C chemical shifts in and near heteroatom-containing functional groups.

Group name	Structure	^{1}H shift (ppm)	^{13}C shift (ppm)
Hydroxyl	R—OH	0.5–3.0	–
Methoxy	R—OCH$_3$	3.2–3.3	55–61
Primary ether	R$_1$—O—CH$_2$—R$_2$	3.3–3.4	72–75
Secondary ether	R$_1$—O—CH—R$_2$R$_3$	3.5–3.6	66–73
Tertiary ether	R$_1$—O—C—R$_2$R$_3$R$_4$	–	71–75
Aldehyde	HC=O	9–10	191–206
Ketone	R$_1$—(C=O)—R$_2$	–	192–215
Carboxylic acid	R$_1$—CO$_2$H	11–12	177–185
Ester	R$_1$—CO$_2$—R$_2$	–	165–179
Primary amino	R—NH$_2$	0.5–2.0	–
Secondary amino	R$_1$R$_2$—NH	1.8–4.0	–
Primary amide	R$_1$—(C=O)—NH$_2$	5–7	168–177
Secondary amide	R$_1$—(C=O)—NH—R$_2$	6–8.5	161–177
Tertiary amide	R$_1$—(C=O)—N—R$_2$R$_3$	–	169–173

patience we can readily grasp the key concepts. Nitrogen and oxygen, with their large electronegativities, deshield nearby NMR-active nuclei. The resonances from ^{1}H's directly attached to heteroatoms sometimes reveal chemical exchange. We know when exchange is taking place because it (1) broadens the ^{1}H resonance, and (2) suppresses the appearance of fine structure even though we might expect to observe a multiplet indicating J-coupling to nearby spins. Table 4.3 shows typical chemical shifts for a wide array of functional groups and molecular fragments containing heteroatoms.

Examination of the data in Table 4.3 reveals that the chemical shifts of the resonances of ^{1}H's and ^{13}C's cannot always be explained with arguments accounting for which and how many electronegative atoms are a certain number of bonds distant. Bond lengths, the proximity of nearby π-electron density, and other factors also influence chemical shifts. Recall that placing electronegative substituents near a group (or farther away if conjugation lies between the two)

Table 4.4 ^{13}C chemical shift trends in tetrasubstituted methanes.

Compound	Ligand electronegativity	^{13}C chemical shift (ppm)
Methane, CH_4	2.20 (H)	-2.3
Tetrachloromethane, CCl_4	3.16 (Cl)	96.5
Tetrabromomethane, CBr_4	2.96 (Br)	-28.5
Tetraiodomethane, CI_4	2.66 (I)	-292.5

will usually shift that group's NMR resonance downfield. But this is not always the case. For example, the ^{13}C chemical shift of the resonance arising from a carbonyl ^{13}C in a ketone is further downfield than that from an ester even though the ester carbonyl ^{13}C has an additional oxygen atom. This shift discrepancy occurs because the lone pair on the oxygen singly bonded to the carbonyl ^{13}C in the ester donates electron density through resonance, thereby providing additional shielding, thus moving the ester carbonyl ^{13}C resonance upfield (to the right).

Hydrogen bonding serves to increase electron density around hydroxyl and amino ^1H's, thus providing shielding and placing hydroxyl and amino ^1H resonances unexpectedly far upfield. That is, despite the fact that nitrogen and oxygen have greater electronegativities than carbon has, the shifts of the resonances of ^1H's bound to nitrogen and oxygen relative to those of ^1H's bound to carbon are not especially far downfield.

In fact, any soft atom (a polarizable atom, usually the one in the molecule with the highest atomic number) may, despite its electronegativity relative to hydrogen and carbon, help to shield any NMR-active nuclide in a molecule from the applied field. A good example of this effect is the ^{13}C chemical shift trend shown by methane and the halogenated methanes (see Table 4.4). Between chlorine and bromine, the sheer number of electrons plus the softness of the halogens begins to dominate the chemical shift through shielding of the ^{13}C at the center of the molecule. Once iodine is introduced, the shielding from the iodine atom's many polarizable electrons far outweighs the greater electronegativity of iodine relative to that of hydrogen, thus resulting in a ^{13}C chemical shift well outside the range normally observed for carbon compounds.

Soft atom. An atom with a low electronegativity, and often with a higher atomic number (Z). These atoms have electron clouds that are more easily distorted by external forces and, as such, they are often the source of electron density that shields unexpectedly (alters the chemical shifts to values that are anomalously upfield) nearby atoms.

There are many fine and more detailed discussions of ^{1}H and ^{13}C chemical shifts to be found in the literature. Also, a number of software packages predict ^{1}H and ^{13}C chemical shifts in user-supplied molecular structures. Despite the obvious temptation to rely heavily on predictive tools, we should develop our own intuition in this important area. To understand ^{1}H and ^{13}C chemical shifts, we assign molecules. Assignment is the correlation of observed chemical shifts to the ^{1}H's and ^{13}C's (or other NMR-active nuclides) present in a molecule. We first assign simple molecules and then incrementally advance to increasingly complex and diverse molecules. Chapters 11–14 in this book contain problems that can form the basis of more extensive discussions of chemical shifts.

■ REFERENCE

[1] A. L. Allred, *J. Inorg. Nucl. Chem.*, **17**, 215 (1961).

Chapter 5

Symmetry and Topicity

If a large ensemble of an NMR-active nuclide (e.g., ^{1}H) experience the same chemical environment, solvent, temperature, pressure, concentration, and field(s), they form a single net magnetization that, when excited with RF, generates a resonance (peak) in the spectrum with a particular chemical shift. For example, all the ^{1}H's in chloroform generate a resonance with a shift of 7.27 ppm relative to the resonance of the ^{1}H's in TMS.

In molecules exhibiting symmetry, NMR-active groups (e.g., methyls) related by an inversion center, a mirror plane, a rotational symmetry axis, or rotation plus inversion will have the same chemical shift in the absence of chiral molecules. Chemical species showing the same chemical shift are said to be isochronous. All the ^{1}H's in benzene are isochronous because they are all identical to one another. That is, they are all related by symmetry through a sixfold (60°) rotation normal to the plane of the molecule. Therefore, all the ^{1}H's in benzene resonate at 7.16 ppm. Groups with anisochronous (different) shifts are never related by a true symmetry operation. We must take care never to assume that two isochronous groups *must* be related by a symmetry operation. Sometimes groups are isochronous by coincidence.

Topicity (from the Greek root topos meaning place or local) is important because it allows us to prediction the number of unique chemical shifts we expect to see from the resonances of a particular molecule. Groups or atoms related by symmetry come in three varieties: homotopic, enantiotopic, and diastereotopic.

5.1 HOMOTOPICITY

Two ^{1}H's are said to be homotopic if, by marking only one atom or group—e.g., by isotopic substitution (substitute ^{2}H for ^{1}H), we produce

Isochronous. Two atoms or functional groups are isochronous when they generate NMR resonances with the same chemical shift.

Symmetry operation. A geometrical manipulation involving rotation, inversion, reflection, or some combination thereof.

Homotopic. Two or more atoms or groups in a molecule are homotopic if labeling one generates the same molecule as labeling the other. Homotopic atoms or groups always generate resonances with the same chemical shift.

the same product regardless of which ^{1}H is marked. Other NMR-active nuclides and even functional groups (e.g., methyls) can be homotopic with respect to each other. Homotopic atoms and groups are always isochronous. Figure 5.1 shows four molecules containing homotopic atoms and groups. *Tert*-butylcyclohexane has three homotopic methyl groups (Figure 5.1a); labeling any of the three methyl groups still produces the same molecule (through rotation about the single bond attaching the *t*-butyl group to the cyclohexane ring). Mesitylene (Figure 5.1b) has three homotopic methyl groups, as well as three homotopic ring protons. In addition, the three nonprotonated carbon atoms also form a homotopic set, as do the three protonated aromatic carbons. 3-Pentanone (Figure 5.1c) has two homotopic ethyl groups on either side of the carbonyl carbon-oxygen pair. Malonic acid (Figure 5.1d) has two homotopic carboxylic acid groups, and also the two protons of the center methylene group are homotopic.

The surest way to test for homotopicity is to build two identical molecules with a molecular model kit. Then we can label (mark) the two groups in question and quickly determine whether a simple symmetry operation—for example, rotation about a single bond or flipping the molecule over—gives the same molecule. With a little practice, we can usually determine topicity with just one model of the molecule in question.

We should actually build models of the molecules whose topicity we are exploring, because it is challenging to visualize molecules in three dimensions. Discussions are also much easier to carry out when we can point to specific parts of molecular models. Model kits range in size, complexity and price. One model kit that the author regards with particular esteem is the Molecular Visions™ model kits available from Darling Models™ (if searching online for "Darling Models," be sure the include the term "Chemistry").

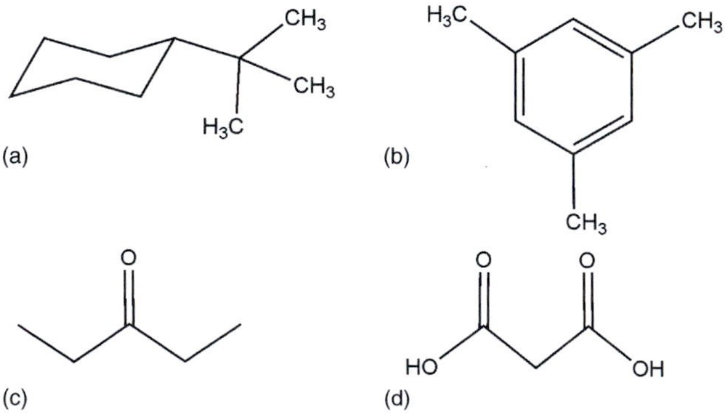

■ **FIGURE 5.1**   Four organic molecules containing homotopic groups or atoms. (a) *t*-butylcyclohexane; (b) mesitylene; (c) 3-pentanone; (d) malonic acid.

5.2 **ENANTIOTOPICITY**

Two atoms or groups are said to be enantiotopic if labeling one of these groups produces a molecule that is the mirror image of the molecule produced when the other atom or group is labeled. Enantiotopic atoms and groups will always give the same chemical shift in an achiral solvent or environment. If the solvent is chiral, however, enantiotopic atoms or groups may become anisochronous. To resolve enantiotopic groups, we can introduce a chiral compound to our sample. Resolving distinct chemical shifts from enantiotopic groups may require that the chiral additive associate intimately with our solute molecule (in the first solvation shell). Figure 5.2 shows molecules containing enantiotopic atoms and groups. Isopropylcyclohexane (Figure 5.2a) has two methyl groups that are enantiotopic; labeling one of the methyls renders a molecule that is not superimposable on a second molecule in which the other methyl is labeled. 1,3-Dimethylcyclohexane (Figure 5.2b) has two enantiotopic methyl groups and two enantiotopic methine groups. Two of the four methylene groups are also enantiotopic with respect to each other (the two that lie directly below the methine groups as the molecule appears in the figure). Butanone (Figure 5.2c) contains two enantiotopic protons on its methylene group. 2-Methyl malonic acid (Figure 5.2d) has enantiotopic carboxylic acid groups.

Enantiotopic atoms and groups are said to be prochiral, meaning that labeling one of the atoms or groups generates a chiral molecule (a molecule that is not superimposable on its mirror image). Enantiotopic atoms and groups are designated pro-R and pro-S through a simple evaluation algorithm. Labeling one atom or group so that it has a higher priority in our evaluation relative to its

Enantiotopic. Two atoms or functional groups in a molecule are said to be enantiotopic if marking one of the atoms or groups renders a molecule that is the mirror image of the molecule obtained when the other atom or group is marked. The resonances of NMR-active enantiotopic groups or atoms will be isochronous in the absence of a chiral compound.

Prochiral. An atom or functional group that is part of an enantiotopic pair which, upon (isotopic) labeling, generates a chiral compound.

Chiral molecule. A group of atoms bonded together that cannot be superimposed upon its mirror image.

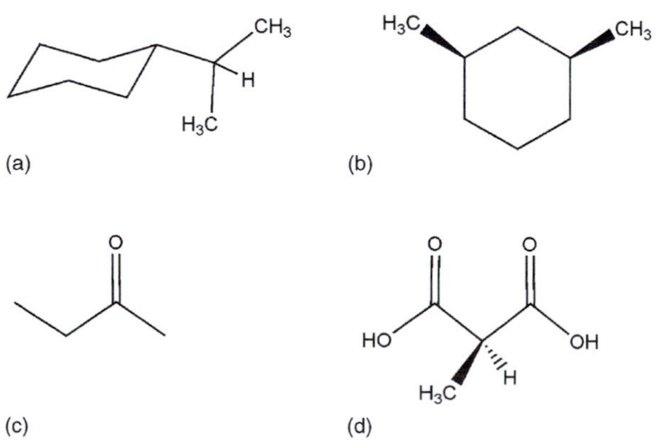

(a)

(b)

(c)

(d)

■ **FIGURE 5.2**   Four organic molecules containing enantiotopic groups or atoms. (a) isopropyl-cyclohexane; (b) 1,3-dimethyl-cyclohexane; (c) butanone; (d) 2-methyl malonic acid.

Pro-R. An atom or functional group that is part of an enantiotopic or diastereotopic pair that, upon (isotopic) labeling such that it acquires a higher precedence, generates an R chiral center.

Pro-S. An atom or functional group that is part of an enantiotopic or diastereotopic pair that, upon (isotopic) labeling such that it acquires a higher precedence, generates an S chiral center.

Diastereotopic group. Two chemical moieties that appear to be related by symmetry but are not. Isotopic labeling of one of two diastereopic groups produces a molecule that cannot be superimposed on the molecule produced when the other group is labeled (even following energetically plausible rotations about single bonds).

Newman projection. A graphical image useful for exploring rotational conformation wherein one sights down the bond about which rotation occurs.

Alpha position. One bond removed from the chemical substituent (atom or functional group) in question.

Beta position. Two bonds distant from the chemical substituents (atom or functional group) in question.

enantiomeric counterpart allows us to assign the molecular framework connecting the two enantiomeric atoms or groups as an R versus S center. For example, we consider the two methylene protons of glycine. Changing one proton to a deuteron will result in the R compound, whereas changing the other proton will give the S compound. An atom or series of atoms in a molecule that bears enantiotopic groups or atoms is said to be a prochiral center.

The different chemical shifts that arise from enantiotopic atoms and groups in chiral solvents can be explained by assuming that the chiral solvent or additive must be part of the solvation shell of the solute molecule. Thus, the presence of chiral molecules nearby will result in different intermolecular interactions depending, on the right- or left-handedness of the prochiral atoms or groups.

5.3 DIASTEREOTOPICITY

Two atoms or groups are said to the diastereotopic if the two atoms or groups appear to be similar but are not actually related by a symmetry operation. Selectively labeling one diastereotopic atom or group will not produce the mirror image of the molecule in which the other atom or group is labeled. For example, the methylene protons in phenylalanine are diastereotopic. The resonances from diastereotopic atoms and groups will normally be anisochronous, unless they happen to fortuitously have the same chemical shift.

Newman projections provide an excellent method of illustrating how two atoms or groups are diastereotopic. Newman projections show that, as we rotate atoms or groups in our molecule about a chemical bond, the exact chemical environment of the two diastereotopic atoms or groups is not reproduced. Figure 5.3 shows Newman projections of the three lowest energy rotational isomers (rotamers) of phenylalanine as rotation takes place when looking down the C_α—C_β bond. Recall that the alpha position is one bond distant from the highest precedence functional group—in this case, the carboxyl group—and the beta position is two bonds distant.

The two protons on the beta carbon (C_β) are labeled $H_{\beta1}$ and $H_{\beta2}$, and the proton on the alpha carbon (C_α) is labeled H_α. Although $H_{\beta1}$ and $H_{\beta2}$ may at first appear to be identical—especially if we draw the molecule in two dimensions—the three Newman projections showing the rotamers resulting from rotation about the C_β—C_α bond reveal that the beta protons do not experience the same chemical environment.

In the discussion of rotamers, it is first important to define the dihedral angle. Starting with basic geometry, we recall that it takes three noncolinear points to define a plane. The dihedral angle is the angle between the two planes defined by four atoms, the middle two of which share a common bond. If we consider the four atoms H_α—C_α—C_β—$H_{\beta 1}$ in the context of the lowest energy rotational states of the C_β—C_α bond, we can see that the dihedral bond angles (the angles between the planes defined by H_α—C_α—C_β and by C_α—C_β—$H_{\beta 1}$) can be 180°, +60°, and −60°. Figure 5.3 shows this graphically.

If a group or atom has a ±60° dihedral angle with respect to a group or atom three bonds distant (e.g., from $H_{\beta 1}$ to C_β to C_α to H_α) then $H_{\beta 1}$ and H_α are said to be *gauche* to each other. If the dihedral is 180°, then the two are said to be *trans*.

In Figure 5.3, we can see that in rotamer 1 (Figure 5.3a) the phenyl group of phenylalanine is *gauche* to the carboxylic acid group and *trans* to the amino group. This conformation places $H_{\beta 1}$ and $H_{\beta 2}$ *gauche* to the amino group, and seemingly in the same environment. But, in fact, $H_{\beta 1}$ and $H_{\beta 2}$ are not in the same environment. $H_{\beta 1}$ is closer to the carboxylic acid group, whereas $H_{\beta 2}$ is closer to H_α. If we proceed to rotamer 2 by rotating the back of the molecule 120° counterclockwise around the C_β—C_α bond, we see that $H_{\beta 1}$ is now *gauche* to the carboxylic acid group and H_α, while $H_{\beta 2}$ is *gauche* to the carboxylic acid and amino groups. Rotamer 3 puts $H_{\beta 2}$ *gauche* to the carboxylic acid group and H_α, but never can $H_{\beta 1}$ and $H_{\beta 2}$ exchange positions while keeping all other groups in the same relative orientation. Although $H_{\beta 1}$ and $H_{\beta 2}$ can both sample all three rotameric positions depicted in Figure 5.3, $H_{\beta 1}$ and $H_{\beta 2}$ can never be in exactly the same environment. Once $H_{\beta 2}$ rotates over to a position formerly occupied by $H_{\beta 1}$, the phenyl group is not in the same place, so the environment is not exactly the same. The fact that the three rotamers have different energies and are therefore not present in a 1:1:1 ratio is immaterial.

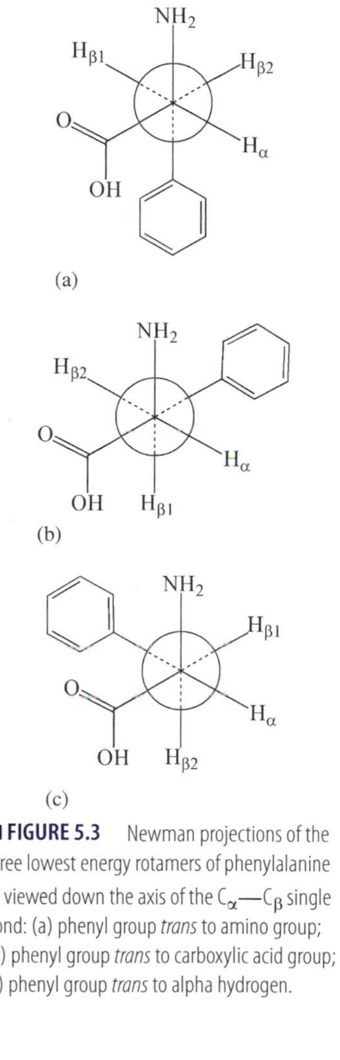

5.4 CHEMICAL EQUIVALENCE

Two atoms are said to be chemically equivalent if they are in the same chemical environment, meaning that they are related by symmetry. Homotopic atoms are thus chemically equivalent.

5.5 MAGNETIC EQUIVALENCE

Two atoms are said to be magnetically equivalent if they have the exact same geometrical relationship to every other NMR-active atom

Dihedral angle (Φ). The angle between two planes defined by four atoms connected by three bonds, the middle two of which share a common bond.

(a)

(b)

(c)

■ **FIGURE 5.3** Newman projections of the three lowest energy rotamers of phenylalanine as viewed down the axis of the C_α—C_β single bond: (a) phenyl group *trans* to amino group; (b) phenyl group *trans* to carboxylic acid group; (c) phenyl group *trans* to alpha hydrogen.

Chemically equivalent. Atoms or functional groups in the same chemical environment.

Magnetically equivalent. Two nuclei are magnetically equivalent when they are of the same nuclide and when both have exactly the same geometrical relationship to every other NMR-active nucleus in a molecule.

in the molecule. Magnetic equivalence is quite rare because its definition is so restrictive. Nonetheless, being able to recognize magnetic equivalence is important. When magnetically equivalence nuclei couple with other spins such that the magnitude of the chemical shift difference is similar in magnitude to the coupling, a very complex multiplet pattern may be observed. This is less often the case using higher field instruments, so instances of observing very complex splittings are rare.

Through-Bond Effects: Spin-Spin (J) Coupling

Besides integrals (Chapter 3) and the chemical shift (Chapter 4), NMR spectroscopy provides another key piece of information in the form of how spins affect one another through the electrons of the molecule. This intramolecular communication is the phenomenon known as J-coupling, scalar coupling, or spin-spin coupling.

J-couplings can be used to determine many things, including bond linkage and local molecular conformation. One or more J-couplings can be calculated from the splittings (in Hz) of the peaks that comprise a multiplet.

6.1 ORIGIN OF J-COUPLING

J-coupling occurs because NMR-active nuclei (spins) are tiny magnets. When a spin changes its spin state, its orientation changes; the reorientation of the spin is communicated through the electron cloud to other spins in the form of a resonant frequency shift. For a single spin-½ nucleus in a molecule, half of the frequency shifts communicated to neighboring spins are positive and half are negative. The shifts caused by J-coupling thus split resonances into two or more peaks but do not affect chemical shift. The magnitude and the sign of the J-coupling-induced splitting give us important information on bond connectivity and molecular conformation.

J-coupling does not take place through empty space, but through a molecule's electron cloud, and thus occurs almost exclusively through the chemical bonds we draw when we satisfy the octet rule. J-couplings are written with a leading superscript to indicate the number of bonds between the two coupled spins. For example, geminal ^{1}H's are separated by two bonds and so we write their J-coupling

J-coupling. Syn. coupling, scalar coupling, spin-spin coupling. The alteration of the spin-state transition frequency of one spin by the spin state of a second spin. J-coupling is a through-bond effect and is, for a given system (molecule), invariant as a function of the applied magnetic field strength. When a J-coupling is described in writing, a leading superscript denotes how many bonds separate the two spins, and a trailing subscript may denote the identities of the two coupled spins.

Geminal. Two atoms or functional groups are geminal if they are both bound to the same atom. Geminal atoms or functional groups discussed in the context of NMR spectroscopy are usually but not always the same species.

Vicinal. Two atoms or functional groups are vicinal if they are separated by three bonds, meaning that they are bound to two atoms sharing a common bond. Vicinal atoms or functional groups discussed in the context of NMR spectroscopy are usually but not always the same species.

^{13}C satellite peaks. Syn. ^{13}C satellites. Small peaks observed in the NMR frequency spectrum found on either side of the resonance of an NMR-active nuclide (most often ^{1}H) bound to carbon. Because ^{13}C is 1.1% abundant relative to ^{12}C, each ^{13}C satellite peak will be 0.55% of the central peak's intensity.

as ^{2}J, whereas vicinal ^{1}H's (^{1}H's on two atoms sharing a common bond) are separated by three bonds and therefore denoted by ^{3}J. A trailing pair of subscripts may be added to describe the participants in the coupling. A one-bond coupling, written ^{1}J$_{CH}$, is observed to exist between a ^{1}H and a ^{13}C to which it is bonded. The value of ^{1}J$_{CH}$'s is usually about 140 Hz if the ^{13}C is sp^3 or sp^2 hybridized, while it may be about 220 Hz if the ^{13}C is sp hybridized.

The ^{1}J$_{CH}$ can be measured directly from a given resonance in the ^{1}H 1-D spectrum if other resonances do not obscure what are called the ^{13}C satellite peaks. Because ^{13}C is 1.1% abundant, 1.1% of the total integrated intensity from the resonance of a carbon-bound ^{1}H will be split such that two peaks (each 0.55% of the total peak area) will occur approximately equidistant from and on either side of the ^{1}H-^{12}C center peak. One peak will be from ^{1}H's bound to ^{13}C's in the α spin state, and the other from ^{1}H's bound to ^{13}C in the β spin state. The distance (in Hz) between the two ^{13}C satellite peaks is ^{1}J$_{CH}$.

If a spin-½ nucleus is in the α spin state, it will slightly perturb the strength of the applied field felt by its nearby neighbors in the molecule. If the spin-½ nucleus is in the β spin state, there will also be a perturbation of the field felt by its neighbors, but the sign if the β perturbation will be opposite that of the α perturbation. Put another way, if spin A is J-coupled to spin X, then spin A will alter spin X's precession frequency such that spin X's near spin A's in the α spin state will be shifted in one direction, and spin X's near spin A's in the β spin state will be shifted in the other direction. Spin A is said to split the resonance arising from spin X because the resonance for spin X is now observed to be split into two equal peaks.

The amount of the splitting is measured in cycles per second (Hz), is independent of the applied field strength, and is called the J-coupling. The constancy of J-couplings exists because NMR-active spins are tiny magnets with fixed strengths. But variations in the electron cloud between two NMR-active spins affect the magnitude (and sign) of the J-coupling. ^{2}J's are often negative in sign, whereas ^{1}J's and ^{3}J's are often positive. If a J-coupling is positive and both coupling partners have positive gyromagnetic ratios, then having one spin in the α spin state will lower the frequency of its coupled neighbor.

In contrast to the chemical shift, which varies in direct proportion to the strength of the applied magnetic field, the J-coupling is independent of field. This invariance stems from the fact that the size of the nuclear dipole or quadrupole moment does not vary with the applied field. Therefore, any effect exerted by one spin on another through the electron cloud depends only on properties such as local electron density and molecular bonding geometry—but NOT on the strength of the applied magnetic field.

The simplest observable splitting consists of a signal from an ensemble of atoms in a unique chemical environment split (divided) into two peaks (or lines or legs). This splitting pattern is called a doublet. Any resonance arising from a single species that is split into two or more lines is termed a multiplet. A resonance that is unsplit is a singlet.

Leg. Syn. line. One individual peak within a resonance split into a multiplet through the action of J-coupling.

Singlet. A resonance that appears in the frequency spectrum as a single peak.

The discussion of the Boltzmann equation in Chapter 1 shows that under normal conditions (i.e., if the sample is not hyperpolarized), spin A will be nearly equally distributed between the α and β spin states, so we expect to see the resonance from spin X split into two peaks with equal integrals (areas). If spin A were 75% α and 25% β, we would see the resonance of spin X split into two lines with relative intensities of 75:25. Although we do often see a skewing of the intensities of doublets, the reason for this asymmetry is due not to unequal populations of the spin states of nearby NMR-active nuclei, but to selection rules with a complex mathematical basis.

6.2 SKEWING OF THE INTENSITY OF MULTIPLETS

Fortunately, the skewing of the intensities of the individual legs of multiplets can be partially explained by using a qualitative argument. The argument is really just an offshoot of the following two facts about J-coupling of spins in NMR.

> J-coupling Fact One: If spin A couples to spin X, then spin X couples to spin A by the same amount.
> J-coupling Fact Two: If spin A and spin X have the same chemical shift, then they cannot show coupling with each other.

J-coupling fact one is just a restatement of "for every action there is an equal and opposite reaction." Put another way, if spin A can feel the spin state of spin X, then spin X can feel the spin state of spin A just as well. This effect applies even if spin A is a ^{1}H and spin X is a ^{13}C. The J-coupling we measure when looking at the resonance of spin A (in Hz) will be the same as the J-coupling we measure when we look at the resonance of spin X.

J-coupling fact two is not as obvious. The inability of two isochronous spins to display coupling with each other greatly simplifies the appearance of many NMR spectra, and for this we should be grateful.

Spin-tickling experiment. The low power CW RF irradiation of one resonance to observe the effect this irradiation has on the resonances generated by other spins, which may be partial or complete multiplet collapse or a change in integrated intensity.

Continuous wave RF. Syn. CW RF. A sinusoidally varying electrical current or photons with a frequency in the radio-frequency region of the electromagnetic spectrum, usually with a low amplitude, applied for a relatively long period of time (typically one or more seconds) to the sample.

For example, the three equivalent methyl protons in methyl iodide do not show coupling with one another. Instead, their resonance appears as a singlet because these three ^{1}H's are homotopic and therefore their resonances are isochronous all the time.

Perhaps the best explanation for why two isochronous spins cannot show coupling with each other (J-coupling fact two) comes from considering a spin-tickling experiment wherein we apply a very soft (low power) RF pulse (or even continuous RF, also called continuous wave or CW RF) in the vicinity of two nearly identical resonances from spin-½ nuclei of the same species, e.g., ^{1}H. In this experiment, our detection method will not be tipping the net magnetization vector for all spins into the **xy** plane and digitizing the FID that results, but rather detecting a change in the complex impedance of the receiver coil as various chemically distinct spins come in and out of resonance as we sweep the applied field and apply RF with a constant frequency. For spin A to split spin X via J-coupling, spin A's α and β spin states cannot be rapidly exchanging with each other while we observe the signal from spin X. If we have to excite spin X to observe it, we cannot avoid exciting spin A if spin A's chemical shift (δ_A) is very close or identical to that of spin X's chemical shift (δ_X). Because observation of spin X's resonance requires irradiation of spin X at δ_X, we end up scrambling the spins of spin A with the applied RF (the α spins rapidly become β spins and vice versa), so spin X can no longer be split by spin A's that are themselves rapidly flipping back and forth between the α and β spin state. That is, the various spin A's do not stay in one or the other spin state long enough to split spin X; instead, we only see the average (a singlet instead of a doublet).

The transition from the state in which different spins exhibit J-coupling with each other to the state in which the spins show no coupling because they are isochronous is a smooth one. As two resonances that split each other into doublets through J-coupling approach each other on the chemical shift scale, we observe that the outer legs of both doublets drop in integrated intensity while the inner legs increase in integrated intensity because the total area for each resonance must remain constant. As the difference in chemical shifts ($\Delta\delta$) approaches zero (as is the case when the applied field strength is reduced), observation of J-coupling between the two must no longer be possible. But the demise of the observable J-coupling does not involve variation of the J-coupling (recall that J-couplings are constant). Instead, the demise of the observable J-coupling as $\Delta\delta$ goes to zero takes place through the diminution

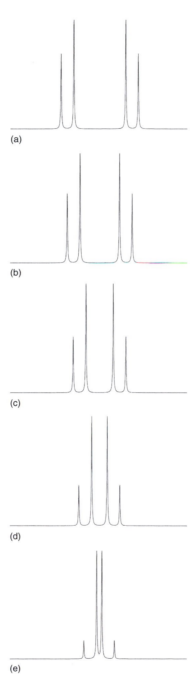

(a)

(b)

(c)

(d)

(e)

■ **FIGURE 6.1** The Dach effect showing how the outer legs of the doublets from two spins lose intensity as the separation between the two chemical shifts ($\Delta\delta$) becomes comparable to the J-couplings between the two spins: (a) $\Delta\delta/J = 5$; (b) $\Delta\delta/J = 4$; (c) $\Delta\delta/J = 3$; (d) $\Delta\delta/J = 2$; (e) $\Delta\delta/J = 1$.

Nonfirst-order behavior. Syn. Dach effect, intensity skewing. The deviation of the intensity of the individual peaks in a multiplet from those intensities predicted using Pascal's triangle.

Dach effect. Syn. roof effect. The skewing of the intensities of the individual peaks (legs) of a multiplet caused by the close proximity (in the spectrum) of another resonance to which the resonance in question is coupled. The Dach effect is due to nonfirst-order coupling behavior.

of the intensity of the outer leg of the doublets. This skewing of the intensities of the individual legs of multiplets is called nonfirst-order behavior. Only when the ratio of the chemical shift difference ($\Delta\delta$) to the J-coupling is large (10 or more) do we observe multiplets free of this skewing.

The intensity skewing of the legs of multiplets is called the Dach effect (Dach means "roof" in German). The name stems from the pitched roof we can imagine if we draw lines connecting the tops of the legs of two nearby J-coupled doublets. Figure 6.1 shows simulations of how the Dach effect progresses as a function of the ratio $\Delta\delta/J$. The Dach effect also is readily apparent for higher order multiplets, e.g., triplets and quartets.

At a certain point two nearby doublets may take on the appearance of another multiplet called a quartet. Careful measurement of the alternate spacing of the four legs of the quartet may reveal that the difference between legs 2 and 3 is not the same as that between 1 and 2 (and 3 and 4). If the spacing is exactly identical, we can still determine unequivocally whether we are observing two intensity skewed doublets or a quartet by running our sample in an NMR instrument with a different magnetic field strength. Lacking access to a second NMR instrument of different field, we can also change solvents and/or solute concentration to possibly alter the chemical shift difference of the two doublets slightly.

J-couplings do not vary with applied field strength, so the multiplets observed with a lower field instrument will proportionally take up more space on the ppm scale than they do with a higher field instrument. For this reason, high-field instruments are preferred when we have a complicated molecule with many resonances showing extensive J-coupling, because a higher field instrument disperses our observed multiplets over a bigger frequency range. For a ^1H spectrum, a multiplet that is 10 Hz wide will be proportionally narrower when there are 500 Hz per ppm (on a 500 MHz instrument) than when there are only 200 Hz per ppm (on a 200 MHz instrument). That is, on a 500, the 10 Hz multiplet will occupy $\frac{1}{50}$th of a ppm, whereas on a 200, the same multiplet will occupy $\frac{1}{20}$th of a ppm.

6.3 PREDICTION OF FIRST-ORDER MULTIPLETS

In the absence of Dach effects, we can predict the appearance of multiplets by counting the number of nearby NMR-active spins. Consider the molecule ethanol (Figure 6.2) with its methyl group

■ **FIGURE 6.2** The structure of ethanol.

and methylene group (we neglect the OH group with its exchanging hydroxyl ^{1}H for the time being in our discussion of J-couplings). The methyl protons are homotopic, and the methylene protons are enantiotopic; so, in an achiral solvent, the methylene protons will be chemically equivalent and therefore isochronous. That is, we will not observe the homonuclear ^{2}J's between the methylene ^{1}H's. Rapid rotation about the C—C bond in ethanol ensures that the ^{3}J-coupling of each methyl ^{1}H to each methylene ^{1}H is the same. Furthermore, the proximity of the oxygen atom next to the methylene group ensures that the spectrum will be relatively free of non-first-order intensity skewing if we are using a 200 MHz instrument or higher, because the chemical shift difference between the methyl ^{1}H's and the methylene ^{1}H's will be much greater than the ^{3}J-couplings.

Now we need only consider the statistics for each spin in both groups. The methylene ^{1}H's will be 50% in the α spin state and 50% in the β spin state. But, to fully predict the relative intensities and spacing of the legs of the methyl's ^{1}H resonance, we must consider every spin state possibility for the ^{1}H's of the methylene group. Four possibilities exist for the two methylene ^{1}H's, ^{1}H$_α$ and ^{1}H$_β$: αα, αβ, βα, and ββ. Because the presence of a nearby spin in the α spin state shifts a resonance in one direction whereas the presence of the nearby spin in the β spin state shifts a resonance in the other direction, we can see that the βα and αβ methylene spin state combinations will not shift the resonance of the methyl ^{1}H's in either direction from their original chemical shift or from where we would observe the methyl's ^{1}H resonance if we were to selectively irradiate (decouple) the methylene ^{1}H's. The αα methylene spin state combination will exert an additive effect, however; and the resonance from the methyl ^{1}H's near a methylene group with both ^{1}H's in the α spin state will be shifted twice in one direction, whereas the resonance from the methyl ^{1}H's near the ββ methylene ^{1}H's will be shifted twice in the opposite direction. Summing up all the possibilities, we expect to see three peaks in the multiplet of the methyl ^{1}H's, with the middle peak being twice as intense (due to the methylene ^{1}H's being αβ and βα) as each of the outer peaks (one from the methylene ^{1}H's being αα, and the other from them being ββ). This multiplet is a 1:2:1 triplet, or simply a triplet.

Note that the splitting of the methyl resonance into three peaks reflects that there are two ^{1}H's close to it and has nothing to do with the fact that the methyl group contains three ^{1}H's itself. The area of the 1:2:1 methyl triplet, however, will correspond to that of three ^{1}H's.

Quintet. Five evenly spaced peaks in the frequency spectrum caused by the splitting of a single resonance by J-coupling to four identical spin-½ nuclei to give a multiplet with five peaks with relative intensity ratios of 1:4:6:4:1.

Sextet. Six evenly spaced peaks in the frequency spectrum caused by the splitting of a single resonance by J-coupling to five identical spin-½ nuclei to give a multiplet with six peaks with relative intensity ratios of 1:5:10:10:5:1.

Septet. Seven evenly spaced peaks in the frequency spectrum caused by the splitting of a single resonance by J-coupling to six identical spin-½ nuclei to give a multiplet with seven peaks with relative intensity ratios of 1:6:15:20:15:6:1.

Multiplicity. Into how many peaks of what relative intensities the resonance from a single NMR-active atomic site in a molecule is divided.

Table 6.1 Prediction of relative intensities of multiplet peaks as a function of number of nearby spin-½ coupling partners assuming identical ^{3}J-couplings.

Number of nearby spins	Relative peak intensity	Name
0	1	singlet
1	1:1	doublet
2	1:2:1	triplet
3	1:3:3:1	quartet
4	1:4:6:4:1	quintet
5	1:5:10:10:5:1	sextet
6	1:6:15:20:15:6:1	septet

We can apply similar statistics when considering the effect of the spin states of the methyl ^{1}H's on the methylene's ^{1}H resonance. The spin states of the three methyl ^{1}H's can be ααα, ααβ, αβα, βαα, αββ, βαβ, ββα, or βββ. Summing the contributions of the α and β spin states, we predict a 1:3:3:1 quartet. Again, the four peak splitting pattern we observe is caused by J-coupling to three nearby ^{1}H's; the number of lines into which this methylene ^{1}H resonance is split is unrelated to the integrated intensity of the methylene group's resonance itself.

The general rule for predicting multiplets in cases where all J-couplings are the same and where the spins producing the observed splitting are spin-½ is that the number of nearby spins is one fewer than the number of observed peaks in the multiplet. Normally, the prediction of resonance multiplicities is limited to ^{3}J's. But consider 3-pentanol. We expect the methine ^{1}H on carbon 3 to be split into a quintet by the four adjacent methylene ^{1}H's on carbons 2 and 4 (assuming that all ^{3}J-couplings between the methine ^{1}H and the two pairs of enantiotopic methylene ^{1}H's are the same). We predict an intensity pattern of 1:4:6:4:1 for this resonance. Pascal's triangle (a.k.a. the magic triangle) can be used to predict the multiplicity of any multiplet when all the ^{3}J-couplings are the same (see Table 6.1). In some cases, however, ^{2}J's from diastereotopic geminal ^{1}H's may also be comparable to vicinal ^{3}J's between ^{1}H's.

Similar predictions can be made if a nearby spin is spin-1 instead of spin-½, but we must keep in mind that a spin-1 nucleus has three allowed spin states instead of two, so coupling to two or three spin-1 nuclei quickly makes for very complicated multiplets. Deuterated chloroform and deuterated benzene both show a 1:1:1 triplet in their ^{13}C NMR spectrum, centered at 77.23 ppm and 128.39 ppm, respectively. The single ^{2}H splits the resonance of the directly bonded ^{13}C into three peaks because ^{2}H is spin-1 and thus has three allowed spin states. No amount of ^{1}H decoupling will collapse the 1:1:1 triplet seen at 77.23 ppm when observing the ^{13}C NMR spectra with chloroform-*d* (or at 128.39 ppm for benzene-d_6) as the solvent.

Observation of a ^{13}C 1-D spectrum without the use of ^{1}H decoupling shows how many ^{1}H's are directly attached to each chemically distinct ^{13}C. When predicting ^{13}C multiplets, we consider the number of directly attached ^{1}H's and not the number of ^{1}H's on adjacent carbons. For example, the ^{1}H-coupled ^{13}C resonance from a methyl group will be a 1:3:3:1 quartet.

In practice, however, we rarely collect a ^{1}H-coupled ^{13}C spectrum because of the poor sensitivity associated with this experiment. More elegant experiments that require less instrument time are available to provide similar information. These experiments include the attached proton test (APT) [1] and the distortionless enhancement through polarization transfer (DEPT) [2] experiments. Careful examination of data generated by the ^{1}H-^{13}C heteronuclear multiple quantum correlation (HMQC) and ^{1}H-^{13}C heteronuclear single quantum correlation (HSQC) NMR experiments can also allow us to determine the number of attached ^{1}H's. In one variation of the HSQC experiment, the sign of a cross peak indicates whether an even or an odd number of ^{1}H's is directly bonded to a given ^{13}C (just like a DEPT-135 1-D ^{13}C experiment).

Decoupling of ^{1}H's is common when observing ^{13}C, whereas decoupling of other NMR-active nuclei is rare. The presence of ^{19}F's, ^{31}P's, and other NMR-active nuclei in a molecule will usually split some of the observed ^{1}H and ^{13}C resonances. Recognition of heteronuclear splittings from other NMR-active nuclei in a molecule is often important in assigning resonances, because the observed heteronuclear splittings will help place ^{1}H's and ^{13}C's near the NMR-active heteroatom in the molecule.

In many cases, we find that J-couplings from one spin to its neighbors are not equal in magnitude. For example, if a particular ^{1}H has a 5 Hz coupling to a second spin and a 9 Hz coupling to a third spin, we observe a 1:1:1:1 multiplet that is called a doublet of doublets

Attached proton test, APT. An experiment that determines whether or not a given ^{13}C resonance corresponds to a protonated carbon site in a molecule.

Distortionless enhancement through polarization transfer, DEPT. A ^{13}C-detected one-dimensional experiment that can determine the number of ^{1}H's bound to a ^{13}C.

Doublet of doublets. A single resonance split into a multiplet containing four (roughly) equal intensity peaks. Two coupling constants can be extracted from this multiplet by numbering the peaks from left to right and obtaining the shift differences (in Hz) between some of the peak pairs. The 1–2 and 3–4 differences give the smaller coupling, and the 1–3 and 2–4 differences give the larger coupling.

(not a quartet). Complex multiplets may appear daunting upon initial inspection, but in many cases we can systematically extract coupling constants and readily explain a multiplet's appearance on the basis of the number of nearby spins. Much of the discussion of J-couplings observed in ^{1}H NMR spectra found in the literature centers around homonuclear coupling (coupling to other ^{1}H's), but other NMR-active species are often found to contribute to observed multiplicities in ^{1}H NMR spectra, most notably ^{13}C, ^{19}F, ^{31}P, ^{2}H, and ^{15}N. A ^{13}C with both ^{1}H's and ^{19}F's bound to it will show a complicated multiplet, because the $^1J_{CH}$ and $^1J_{CF}$ differ. Although ^{14}N is spin 1, we rarely observe couplings between ^{14}N's in our molecule and other spins; this absence of couplings is because the ^{14}N nuclei undergo rapid quadrupolar relaxation in an asymmetric environment. Only in a rare instances is the nitrogen atom's chemical environment sufficiently isotropic to allow the ^{14}N nuclei to persist in a given spin state and thus to generate observable couplings to other spins.

Karplus relationship. A mathematical function based on orbital overlap that relates the magnitude of J-coupling as a function of the geminal bond angle or the vicinal dihedral angle.

6.4 THE KARPLUS RELATIONSHIP FOR SPINS SEPARATED BY THREE BONDS

The dihedral angle between coupled NMR-active spins separated by three bonds modulates the magnitude of the observed ^3J. Figure 6.3 shows this relationship graphically. There are two maxima observed as the dihedral angle varies from 0° to 180°. One maximum occurs

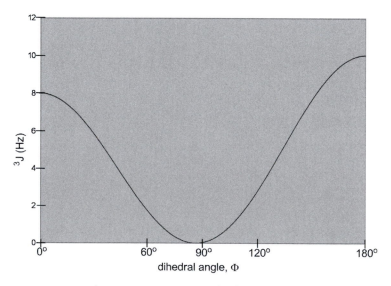

■ **FIGURE 6.3** The ^3J Karplus relationship for vicinal ^{1}H's. ^3J varies as a function of the dihedral angle Φ.

at 0° and the second at 180°. A minimum occurs near 90°, actually occurring around 85°. The 0° ^3J maximum is typically 8–10 Hz, and the 180° ^3J maximum is typically 10–17 Hz.

For ^1H's on sp^3-hybridized carbon atoms, the energy minima associated with rotation about a single bond will normally ensure that the dihedral angle between ^3J-coupled spins will be ±60° and 180°. For *gauche* (±60°) couplings, the observed ^3J will typically be 5–6 Hz; and for the *trans* (180°) coupling, the ^3J will be about 12 Hz. In certain cases the dihedral angle between ^1H's that are three bonds apart may approach 90°, in which case the coupling may approach 0 Hz.

The magnitude of the ^3J-coupling results from the overlap of the occupied molecular orbitals connecting the coupling participants to the atoms between them. When the two orbitals are orthogonal, the overlap of the orbitals is minimized; thus the coupling is likewise minimized.

The presence of additional electron density between the spins separated by three bonds serves to enhance the amount of coupling that takes place. ^1H's that are *trans* across a C=C bond exhibit a ^3J usually on the order of 17 Hz.

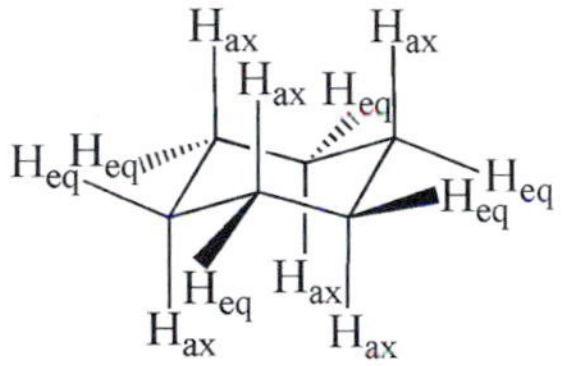

^3J's in cyclohexane rings provide an amazing wealth of information. Whenever we encounter a six-membered ring, we initially assume that the ring adopts the chair conformation and not the boat or skew-boat conformations (Figure 6.4). Because of 1,3-diaxial interactions, bulky substituents tend to occupy equatorial and not axial positions for cyclohexane rings in the chair conformation. Axial ^1H's on adjacent carbons in cyclohexane rings are *trans* to each other and, because of this, their resonances show a large ^3J relative to the other possible combinations of ^1H's on adjacent carbons (axial/equatorial and equatorial/equatorial) that are always *gauche*. Newman projections of a cyclohexane ring in the chair conformation with a view down the C_1—C_2 bond axis (Figure 6.5) show that only the axial ^1H on C_1 (H_{1ax}) is *trans* to the axial ^1H on C_2 (H_{2ax}). All other ^3J-couplings between ^1H's on C_1 and C_2—i.e., H_{1ax} to H_{2eq}, H_{1eq} to H_{2eq}, and H_{1eq} to H_{2ax}—are *gauche*. Large ^3J's found in cyclohexane rings are distinctive indicators of 1,2-diaxial (*trans*) couplings.

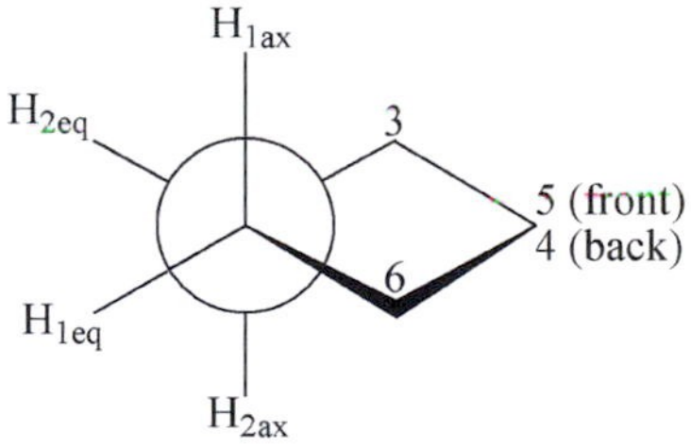

FIGURE 6.4   A cyclohexane ring in the chair conformation has distinct axial and equatorial hydrogens.

FIGURE 6.5   A Newman projection of a cyclohexane ring viewed along the C_1—C_2 bond axis. The subscripts denote whether the hydrogens lie on C_1 or C_2, and also whether the hydrogens are axial (ax) or equatorial (eq). The other ring carbons are numbered 3 through 6.

6.5   THE KARPLUS RELATIONSHIP FOR SPINS SEPARATED BY TWO BONDS

The magnitude of the ^2J produced by geminal ^1H's also varies as a function of bond angle. Just as increasing orbital overlap increases the ^3J-coupling for vicinal ^1H's, so too does the increase of orbital overlap increase ^{2}J-coupling for geminal ^{1}H's. If we liken the two bonding orbitals to the individual blades of a pair of scissors, closing the pair of scissors increases the overlap of the two blades as the angle between the blades is reduced. Figure 6.6 shows the ^{2}J Karplus relationship as a function of bond angle.

Bond angle. If one atom is bonded to two other atoms, the bond angle is the geometrical angle between the vectors connecting the common atom to the two other atoms.

Valence shell electron pair repulsion, VSEPR. A mental tool used to predict deviations from standard hybridized bond angles based on how much of the surface area of an atom a given electron pair in its outer shell occupies.

The two principal bond angles we will most often encounter between geminal ^{1}H's are 120° from sp^2 hybridized atoms and 109.5° from sp^3 hybridized atoms. The ^{2}J for ^{1}H's sharing a 120° bond angle—in a terminal double bond, for example—is 2–4 Hz. The ^{2}J for ^{1}H's sharing a 109.5° bond angle is about 12 Hz.

Distortions in predicted molecular geometry due to the presence of rings in a molecule affect observed ^{2}J's. Valence shell electron pair repulsion (VSEPR) theory can be used to account for deviations in ^{2}J's resulting from unusual bond angles. Probably the best example of this effect occurs in a cyclopropane ring, in which geminal protons splay apart because of the additional room afforded by the compression of the C—C—C bond (the bond angle goes from 109.5° to 60°;

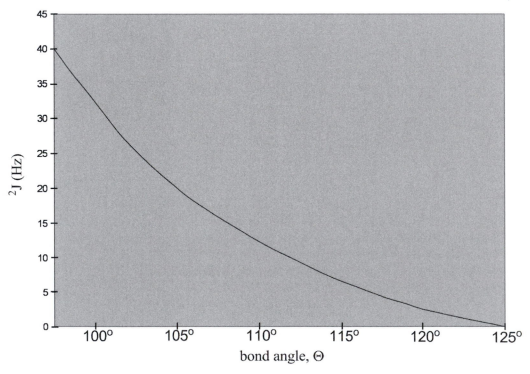

■ FIGURE 6.6 The ^{2}J Karplus relationship for geminal ^{1}H's. The ^{2}J varies as a function of the bond angle Θ.

see Figure 6.7). That is, bringing together two C—C bonds makes more room for the hydrogens, increasing the bond angle between the geminal hydrogens (and reducing orbital overlap), thereby increasing the H—C—H bond angle and decreasing $^2J_{HH}$. The $^2J_{HH}$ we observe in the resonances of geminal ^1H's in a cyclopropane ring are about 5 Hz.

6.6 LONG RANGE J-COUPLING

J-couplings are sometimes observed when spins are separated by four or more bonds, especially when extra electron density lies between the spins. The most commonly observed ^4J is called a W-coupling because the four bonds in between are all coplanar and form the shape of a W (Figure 6.8). ^4J's are often seen between ^1H's on aromatic rings, but other molecular geometries arising from cyclic and steric constraints generate these couplings. ^4J's seldom are observed to be greater than 2–3 Hz, but with the ability to shim a sample so that its narrowest peaks have line widths of less than 0.5 Hz, the ^4J can be readily detected and quantified. In some instances, ^4J's as large as 7 Hz are observed.

6.7 DECOUPLING METHODS

Decoupling is the practice of irradiating one set of NMR-active spins for the purpose of altering and often improving the quality of the NMR signal detected from a second set of spins. Decoupling a particular set of NMR-active spins results in saturation of these spins. Saturation is the equalization of the populations of the various non-degenerate spin states. Decoupling can be either homonuclear or heteronuclear, meaning that the nuclide we observe may be the same or it may differ from the nuclide being irradiated. Heteronuclear decoupling is by far the more commonly practiced form of decoupling.

As stated previously, ^{13}C is rarely observed without using ^1H decoupling (Section 6.2). Applying low-power RF to cover the entire chemical shift range of ^1H's will cause all ^1H's in the sample to interconvert between the α and β spin states such that the ^{13}C resonances will all appear as singlets instead of multiplets. The collapse of a ^{13}C multiplet down into a singlet has three principal advantages over the collection of ^1H-coupled ^{13}C data. One, ^1H decoupling consolidates all

■ FIGURE 6.7 Compression of the C—C—C bond angle in a three-membered (cyclopropane) ring expands the H—C—H bond angle.

W-coupling. A four-bond J-coupling occurring when two NMR-active spins are separated by four bonds held in a static, planar conformation that forms the letter "W".

Homonuclear decoupling. The simultaneous decoupling and observation of spins of the same nuclide, normally accomplished with the application of low-power, single frequency RF that is gated off only long enough to digitize each point of the FID. Because the homonuclear decoupling RF is low power, the pulse ringdown delays normally associated with the application of high power (hard) pulses can be foregone. The location of the decoupled resonance may lie within the limits of the observed spectral window, so great care must be taken to avoid overloading the receiver.

Heteronuclear decoupling. The decoupling of spins of one nuclide to favorably affect the signal observed from a second nuclide, e.g., to the multiplets observed in the resonances of a second nuclide.

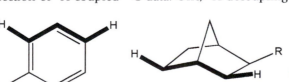

■ FIGURE 6.8 Two molecular systems in which ^4J's (W-couplings) are expected; the R-groups disrupt the molecular symmetry.

the ^{13}C signal intensity into a single peak, so the signal-to-noise ratio we obtain is better. Two, ^1H decoupling greatly simplifies the appearance of ^{13}C NMR spectra that may contain many closely spaced resonances. Three, ^1H decoupling generates an additional improvement in the magnitude of the ^{13}C NMR signal through a phenomenon known as the nuclear Overhauser effect (NOE). The NOE will be discussed more fully in Chapter 7.

Decoupling works by inducing transitions between the spin states of one nuclide (e.g., ^1H) to prevent those nuclei from persisting in a particular spin state long enough to affect the frequency of the nuclide that is being observed. If the power of the applied RF reaching the spins in the sample is too low to induce the transitions at a sufficiently rapid rate, the resonances from the nuclei being observed will show only partial collapse of the multiplets. This problem is often the result of poor tuning of the decoupler channel of the probe. A high salt content in the sample solution may also reduce decoupling efficacy.

Continuous wave decoupling, CW decoupling. The application of a single frequency of RF to a sample for the purpose of selectively irradiating (saturating) a particular resonance, thus perturbing other resonances from spins that interact with the spins corresponding to the irradiated resonance.

Gated decoupling. Gated decoupling occurs when the decoupling RF is turned on and off (gated on or off) at particular points in the pulse sequence. The most common use of this method arises when one wishes to acquire a quantitative 1-D spectrum with decoupling. The decoupler is gated off during the relaxation delay and on during the acquisition time (the time during which the FID is digitized). This protocol prevents varying amounts of NOE enhancement (which can vary from site to site within a molecule) from skewing the relative intensities of the components of the net magnetization generated by each site in a molecule, but at the same time preserves the more lucid presentation of resonances as singlets (multiplets collapsed into a single peak), not to mention the better signal-to-noise ratio associated with the placement of all the intensity of a given resonance into a single peak.

The simplest type of decoupling, continuous wave (CW) decoupling, is also the least efficient decoupling method. The inefficiency of CW decoupling stems from the undesired selectivity of irradiating the sample with RF of a single frequency. If we instead apply shorter bursts of RF whose frequency is modulated periodically, we find that the power required to decouple a given range of frequencies is greatly reduced. Minimizing the power put into the sample helps reduce RF heating; RF heating can degrade our sample, detune the probe, create sample convection currents, and cause a host of other undesirable effects. Modern NMR instruments use decoupling methods such as Waltz-16, GARP, and other broadband decoupling pulse sequences.

Decoupling can be turned on during the relaxation delay of the pulse sequence and off during the digitization of the FID to yield an NOE-enhanced yet coupled spectrum (this practice is rare). In the gated decoupling experiment, decoupling is conversely turned (gated) off during the relaxation delay but on during the acquisition of the FID; this experiment yields quantitative ^{13}C spectra whose resonances are singlets instead of multiplets. The limitation associated with waiting five times the longest T_1 relaxation time still applies if we seek to obtain quantitative integrals from the resonances in our spectrum. The T_1's of ^{13}C's can be 60 seconds or more.

Homonuclear single-frequency decoupling can also be used to simplify the multiplicities of other resonances. Homonuclear decoupling, however, can potentially overload the receiver, because irradiation may take place during digitization of the FID and because the frequency of the RF we apply is the same as that we detect (before mixing). This potentially serious problem can be avoided if the appropriate variables are properly set (on a Varian instrument, type homo = 'y') by interleaving the timing of irradiation and the collection of the individual points that comprise the digitized FID. Interleaving of the decoupling RF pulses and digitization events is possible only if the duration of the decoupling pulses is shorter than the dwell time (the time interval between successive points in the FID). For a 10 kHz sampling rate giving a 5 kHz or 10 ppm sweep width when observing ^{1}H on a 500 MHz instrument, the dwell time is $1/(10 \text{ kHz})$ or 100 μs. This dwell time is sufficient to allow the application of a series of small angle pulses at a lower transmitter power because the typical pulse width for a hard ^{1}H 90° pulse is 8–10 μs. Even 18 dB below the transmitter power for a hard pulse, the factor of 8 times the 8–10 μs pulse gives an 64–80 μs pulse (the pulse width for a given tip angle doubles every time the transmitter power is reduced by 6 dB).

Homonuclear decoupling is presently not as popular as it used to be because of the relative ease with which we can now carry out 2-D experiments to obtain comparable information.

6.8 ONE-DIMENSIONAL EXPERIMENTS UTILIZING J-COUPLINGS

Two important 1-D experiments that make use of J-couplings are the attached proton test (APT) and the distortionless enhancement through polarization transfer (DEPT) experiment. Both the APT and DEPT experiments provide information about whether or not the observed nucleus—virtually always ^{13}C—is protonated, and if so, with how many ^{1}H's.

The APT experiment can be explained by using spin gymnastics to track the motion of the ^{13}C net magnetization vectors in the rotating frame at various points in the pulse sequence. Figure 6.9 shows the APT pulse sequence. The APT pulse sequence works by exploiting the nearly constant $^1J_{CH}$ (this assumption is a poor one only when the molecule has a terminal ^{1}H on a triple-bonded ^{13}C).

First, imagine a methine ^{13}C signal that is on-resonance. Now split the methine ^{13}C's into two populations; Let $^{13}CH_\alpha$ be those methine ^{13}C's with attached ^{1}H's in the α spin state, and let population $^{13}CH_\beta$ be those methine ^{13}C's with attached ^{1}H's in the β spin state. If we apply

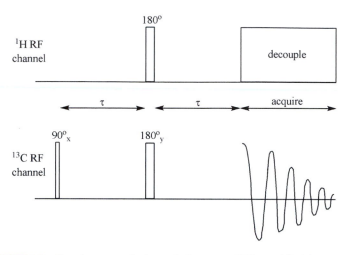

■ FIGURE 6.9 The pulse sequence for the attached proton test (APT). $\tau = (2\,^1J_{CH})2^1$.

a 90° ^{13}C RF pulse along the +x-axis of the rotating frame, the net magnetization vectors for ^{13}CH$_\alpha$ and ^{13}CH$_\beta$ will both point in the same direction in the xy plane, i.e., along the +y-axis of the rotating frame.

If we then wait $\tau = (2\,^1J_{CH})^{-1}$, the ^{13}CH$_\alpha$ and ^{13}CH$_\beta$ vectors will fan out and accumulate a phase difference of 180° (because they must refocus at $^1J_{CH}^{-1}$, by definition). Why we pause here is not important in this case; but once we are no longer on-resonance, the importance of pausing at this point becomes apparent. If we wait an additional τ of the same length, the ^{13}CH$_\alpha$ and ^{13}CH$_\beta$ vectors will refocus along the $-$y-axis. Now turning on the ^{1}H decoupler allows us to detect the methine ^{13}C signal as a singlet that is 180° out of phase with respect to the normal receiver phase. In this way we can distinguish whether or not an on-resonance ^{13}C signal has an attached proton.

Because only one ^{13}C resonance can be on-resonance, a slight modification to the pulse sequence is in order if we are to determine whether or not each ^{13}C is protonated. This modification allows the experiment to work for resonances that are off-resonance, and is accomplished with what is called a chemical shift refocusing pulse. If, after the first τ, we simultaneously apply both a 180° ^{1}H and 180° ^{13}C pulse, we will (1) change the spin state of the ^{1}H's bound to the ^{13}C's, and (2) flip the ^{13}C vectors so that after the τ period they will refocus along either the +y- or $-$y-axis of the rotating frame.

When we use the APT pulse sequence, methylene groups behave like the nonprotonated (C_{np}) groups, and methyl groups behave like the methine groups.

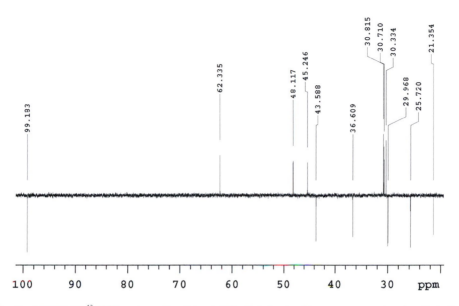

■ FIGURE 6.10   The 1-D DEPT-135 ^{13}C NMR spectrum of longifolene in CDCl$_3$. Methyl and methine resonances are phased positively (they point up), whereas methylene resonances are phased negatively (they point down). Nonprotonated ^{13}C resonances are not observed.

An extensive amount of NMR literature is devoted to explaining the intricacies of NMR pulse sequence design and implementation. The purpose of the preceding example is to highlight the powerful nature of NMR pulse programming and to provide a small taste of the various manipulations possible with NMR pulse sequences.

The DEPT experiment is another 1-D NMR experiment that can be used to distinguish the number of attached ^{1}H's on the various ^{13}C resonances we observe. The DEPT experiment is used more commonly than the APT experiment; however, the DEPT experiment employs a strategy similar to that used in the APT experiment. A DEPT experiment is often used when molecules are relatively simple or when the resonances found in the ^{13}C spectrum have already been mostly assigned. Figure 6.10 shows the DEPT spectrum of longifolene in CDCl$_3$ (longifolene also appears as Problem 12.1).

6.9 TWO-DIMENSIONAL EXPERIMENTS UTILIZING J-COUPLINGS

The J-coupling allows us to obtain extremely useful information about solute molecules. The list of 2-D NMR experiments that employ J-couplings for correlating chemical shifts continues to expand; only a few key experiments will be discussed here. The major 2-D NMR experiments employing J-couplings can be split into homonuclear and heteronuclear groups.

The ever-present but small variations in the strength of the applied magnetic field in the detected region of the sample stipulate that we carry out all 2-D experiments without spinning our sample (for 1-D spectrum collection, we normally spin the sample). Sample spinning causes our solute molecules to travel in circles (except those on the spinning axis); as the molecules move about because of the sample rotation, their NMR-active spins experience a range of field strengths caused by the residual field inhomogeneities we were unable to eliminate with shimming. Because the pulse sequences for all 2-D spectra involve the fanning out of magnetization and then its subsequent refocusing at a later point in time, variation of the field strength experienced by individual molecules is avoided. Obviously, diffusion and convection cannot be prevented, but sample spinning is omitted to yield greater amounts of signal.

6.9.1 Homonuclear Two-Dimensional Experiments Utilizing J-Couplings

The homonuclear 2-D NMR experiments that use J-coupling include the correlation spectroscopy (COSY, and variants including gradient-selected COSY or gCOSY, double-quantum filtered COSY or DQF-COSY) experiment, the total correlation spectroscopy (TOCSY) experiment, and the incredible natural abundance double quantum transfer experiment (INADEQUATE) [3].

6.9.1.1 COSY

The COSY pulse sequence does not lend itself to explanation with visual images and spin gymnastics. To arrive at a better understanding of coherence selection in NMR pulse sequences, we can study the product operator formalism. Fortunately, an intimate grasp of how the COSY pulse sequence works is not required to use the method.

Coherence selection. The isolation of a particular component of the total magnetization, often accomplished with the application of pulsed field gradients (PFGs).

A homonuclear 2-D COSY NMR spectrum shows chemical shifts from the same nuclide on both the f_1 and f_2 axes. Signal normally appears on the diagonal where the analogous 1-D spectrum contains resonances ($\delta_1 = \delta_2$). A signal off the diagonal is called a cross peak ($\delta_1 \neq \delta_2$). The COSY cross peaks appear whenever the spins with resonances at δ_1 and δ_2 are coupled to each other. The intensity of COSY cross peaks varies in direct proportion to the magnitude of the J-coupling between the two resonances.

The COSY experiment can be either phase sensitive or nonphase sensitive (a.k.a. absolute-value COSY). The total integrated volume of a phase-sensitive COSY cross peak will always be zero because the cross peak will be composed of an equal distribution of positive and negative components. Phase-sensitive COSY spectra are useful for the measurement of J-couplings between spins.

6.9.1.1.1 **Phase Sensitive COSY**

The simplest phase-sensitive COSY cross peak shows coupling between two doublets not coupled to any other spins. In this case, the cross peak will consist of two positive peaks and two negative peaks. Peaks of the same sign will be diagonally opposed. The spacing of peaks of opposing sign will show the magnitude of the coupling constant.

If a cross peak arises from the coupling of two spins whose multiplicities are more than just doublets, the extraction of the couplings is slightly more involved. Only the J-coupling responsible for the generation of the cross peak will modulate the alternation in sign of the components of the cross peak.

Whenever we attempt to measure J-couplings accurately using a 2-D COSY data set, we must ensure that we have sufficient digital resolution (see Chapter 3) to give the desired precision, usually along the f_2 axis. This matrix dimensional asymmetry may entail reprocessing the data set with array dimensions set at $16\,k \times 1\,k$ instead of $4\,k \times 4\,k$.

In some instances, we may find that examination of a cross peak close to the diagonal (δ_1 is almost equal to δ_2) is difficult because of distortion of the cross peak due to intensity on the diagonal of the spectrum. The double-quantum filtered COSY (DQF-COSY) experiment fortunately provides a method for suppression of most of the signal intensity found on the diagonal.

Any phase-sensitive COSY experiment has a minimum phase cycle (usually 4 scans per t_1 time increment), and calibration of the 90° pulse is strongly recommended. Setting the receiver gain in any COSY experiment should be carried out when collecting the 1-D spectrum (and not when we digitize the first FID of the 2-D COSY, because the signal intensity in the COSY experiment builds up and reaches a maximum only after a number of FIDs have been digitized.

t_1 time increment. A discreet variation (normally an increase by a fixed amount) in the t_1 time delay in the NMR pulse sequence.

6.9.1.1.2 **Absolute-Value COSY, Including gCOSY**

Absolute-value COSY (avCOSY) experiments, in contrast to phase-sensitive COSY experiments, generate diagonal and cross peaks with the same sign. The absolute-value COSY experiment cannot be used to measure J-couplings directly, but cross peak intensities are in proportion to J-couplings. Cross peaks that are spread out because one or both participating resonances arise from spins that are J-coupled to many other spins may give the false impression that the cross peak is weak. The volume of a cross peak is the true measure of its intensity.

In many cases, we find that our sample is concentrated enough to obviate the need to collect more than one scan per t_1 time increment, yet the minimum phase cycle dictates the collection of at least four scans per t_1 time increment. Fortunately, the use of pulsed field gradients allows the collection of COSY spectra with just one scan per t_1 time increment through a process called coherence selection. The gradient-selected COSY (gCOSY) experiment often allows us to collect a 2-D ^{1}H-^{1}H gCOSY experiment with 256 FIDs (256 t_1 time increments) in 6 minutes. Another favorable feature of the gCOSY experiment is that it is tolerant of poorly calibrated pulses.

6.9.1.2 *TOCSY*

The total correlation spectroscopy (TOCSY) experiment has become an indispensable tool for unraveling the spectra of complicated molecules, especially those of biological origin. The TOCSY experiment utilizes a mixing time during which the net magnetizations of the various spins in the sample are all held near a common axis in the **xy** plane of the rotating frame by using what is called a spin lock. During the spin lock, a series of relatively soft 180° pulses (typically 6–10 dB below the hard 90° transmitter power) are used to prevent the divergence of the various net magnetization vectors of the chemically distinct spins. Because the various net magnetizations spend most of their time along a common axis, they exchange phase information through spin-spin (T_2) relaxation that serves to generate the TOCSY cross peaks.

Spin lock. The placement of magnetization into the **xy** plane, followed by the application of CW RF with a 90° phase shift. The spin lock can also be accomplished with a series of 180° pulses to generate a repeated spin echoes. If the frequency of the spin lock RF is in the center of the spectral window and the 90° pulse width is around 30 μs, the spin lock gives optimal TOCSY mixing. If the frequency of the spin lock RF is well outside the spectral window and the 90° pulse width is around 90 μs, then the spin lock gives optimal ROESY mixing. Using a longer 90° pulse width helps minimize the development of TOCSY cross peaks in the ROESY spectrum.

The TOCSY spin lock can sometimes cause RF heating of the sample, especially when the mixing time gets long and the sample contains ions. Recall that RF heating can cause sample degradation—especially protein denaturation—as well as other problems. If the sample temperature is not held constant during the acquisition of a 2-D NMR data set, some of the chemical shifts—especially those associated with ^{1}H's participating in hydrogen bonding—may vary enough over the course of the experiment to create artifacts that will render the data useless. The problem of changing chemical shifts can be eliminated by scanning (applying the pulse sequence) many times before we scan and save the data. Scans applied to allow the sample to reach thermal equilibrium are called steady-state scans or dummy scans. Steady-state scans can be applied either just at the beginning of an experiment (common), or before the collection of every group of scans corresponding to a particular t_1 time increment and phase cycle (rare). Additionally, a few scans can be collected from each FID so that the sample does not reach different equilibrium states as the t_1 evolution time increases. For example, if we are collecting 64 FIDs and each FID is comprised of 16 scans, we collect scans 1 through 4 for FID number 1, then scans 1 through 4 for FID number 2, and so on until we have collected scans 1 through 4 for FID number 64. Then we collect scans 5 through 8 for FID number 1, and so on. Because the power put into the sample per unit time is lower when we have a longer t_1 evolution time, we do not collect all 16 scans from each FID and then move on to collecting the next FID, but instead collect 4 scans from each FID in order, then we collect 4 more scans for each FID, and so on. This process is called interleaving, and it serves to average the amount of power put into the sample per unit time.

The TOCSY experiment allows the generation of cross peaks between virtually all spins in the same spin system. A spin system is a group of spins that all couple to one or more other members of the group. A molecule may contain one spin system, or it may contain a hundred spin systems. There can be no coupling between spins from two different spin systems: if there is a nonzero coupling between two spins then those two spins are part of the same spin system.

The efficiency with which cross peak intensity is generated in the TOCSY experiment is a function of the magnitude of the J-coupling, the spin lock mixing time, and the relative distance of the coupled resonances from the frequency of the RF used for the spin lock. Resonances close to each other in chemical shift will generate larger cross peaks than resonances that are further apart, unless the resonances that are widely separated are both comparable distances from the frequency of the spin lock RF. This relationship generates an X-shaped profile for maximum cross peak generation with the center of the X at the frequency of the spin lock RF (normally that

Dummy scans. Syn. steady-state scans. Executions of the pulse sequence carried out prior to the saving of any of the data for the purpose of ensuring that the net magnetization has reached a constant magnitude following the relaxation delay at the start of each execution of the pulse sequence. Steady-state scans also help ensure that the sample reaches a constant temperature, which is especially important in carrying out any 2-D experiment involving X-nucleus decoupling (HMQC and HSQC) or a spin-lock (TOCSY and ROESY).

Spin system. A group of spins within a molecule that all couple to one or more other members of the same group.

of the transmitter). This X-shaped profile stems from the synchrony of the components of the net magnetization vectors for the various resonances as they dephase and rephase during the spin lock's refocusing pulses. Swapping of the phase information encoded by the t_1 evolution time is most efficient when J-coupled spins precess with one or both of their **xy** components in synchrony.

The TOCSY experiment is perhaps most useful in the study of proteins and nucleic acids. In proteins the amide linkage connecting amino acid residues is an effective barrier to J-couplings; thus, each residue constitutes a separate spin system.

The TOCSY experiment is normally carried out in a phase-sensitive manner. For large molecules (MW $>1000\,g\;mol^{-1}$), the sign of the cross peaks will be opposite that of the signal found on the diagonal of the spectrum. Cross peaks with the same sign as that of the diagonal peaks (i.e., δ along $f_1 = \delta$ along f_2) are not due to J-coupling but to chemical exchange, usually with the solvent.

The power of the TOCSY experiment can be illustrated with the following theoretical example. Suppose we have a spectrum with four spins, A, B, B', and C, and that spins B and B' have the same chemical shift (they are isochronous, and that is why we didn't give them separate letters). Let us further suppose that the COSY spectrum shows us that spin A is coupled to a spin with the chemical shift of B and B'. Because we are assigning arbitrary labels, we say that A is coupled to B and not to B'. If the COSY spectrum also shows that spin C is coupled to B or B', we can use the TOCSY experiment to determine whether spin A and spin C are both coupled to the same spin (B) or whether A is coupled to B and B' is coupled to C.

During mixing in the TOCSY experiment, spin A will pass its phase information to spin B, meaning that B will be detected with phase information from both A and B and therefore will show a cross peak to spin A. If spin B is also coupled to spin C, then as long as the mixing time is of sufficient duration (and relaxation does not destroy the signal), spin B will pass whatever phase information it has on to spin C. Because spin B will have some of spin A's phase information, spin C will end up as the recipient of some of this information and we will thus observe a cross peak between spins A and C. If, on the other hand, spin B' is coupled to spin C, then we will not see a cross peak between spin A and spin C. In the first case, spin B acts as an intermediary between spins A and C because they are all part of the same spin system. In the second case, however, spins A and B are in one spin system and spins B′ and C are in another spin system. Note that the coupling between spins A and C is zero in both cases.

The longer the mixing time, the greater the likelihood that we will observe a cross peak from a spin that is many bonds away. That is, even though the direct coupling may be zero between spins separated by many bonds, the TOCSY experiment passes phase information between spins separated by many bonds as long as they are both members of the same spin system. A short TOCSY mixing time is 30–50 ms, an average mixing time is 80 ms, and a long mixing time is 120 ms or greater. TOCSY mixing times cannot be lengthened indefinitely because of $T_2{}^*$ relaxation, which decreases the overall amount of signal detected. Choice of the appropriate mixing time depends on the $T_2{}^*$'s for the spins in the molecule being examined and on the sizes of the J-couplings between members of the spin system of interest. A small J-coupling in the middle of a spin system can act as a bottleneck, severely limiting the passage of cross peak-generating phase information.

6.9.1.3 *INADEQUATE*

As a last resort, we use the incredible natural abundance double-quantum transfer experiment (INADEQUATE) to determine the carbon connectivities in a molecule. The INADEQUATE experiment is a ^{13}C-detecting experiment that relies on the proximity of two ^{13}C's to generate signal. The experiment works by exploiting the $^1J_{CC}$'s present. The principal problem with the experiment is that the acquisition of the data set requires a large amount of instrument time in the absence of a sample isotopically enriched with ^{13}C at the site or sites of interest. This large instrument time requirement stems from the low natural abundance of ^{13}C (1.1%), thus making the probability of having two ^{13}C's adjacent to each other at any particular site in a molecule only 1 in 8300 (0.011^{-2}).

To maximize the detected signal, our sample should be made as concentrated as possible, just so long as our high sample concentration does not cause viscosity broadening or generate microscopic aggregates, both of which will decrease the signal-to-noise ratio we attain. We carry out the INADEQUATE experiment using a probe with a normal coil configuration (with the coil closest to the sample tuned to the ^{13}C frequency, as opposed to an inverse probe).

In some cases, we may find that it is impossible to make a sample that is concentrated enough to generate a spectrum with a reasonable signal-to-noise ratio in the amount of instrument time available to us. If our sample is limited and we absolutely have to have this data, we should consider synthesizing, isolating, or otherwise acquiring more sample. If solubility and/or viscosity broadening hinder detection, we might consider increasing temperature slightly. Excessive sample heating, however, may degrade our sample; heating will also reduce the magnitude of the net magnetization vector as shown by Boltzmann distribution statistics and will decrease the spectrum's signal-to-noise ratio because it will increase the resistance of the receiver coil.

Another possible solution is to find a higher field instrument for this experiment. It is good to have friends at other sites with NMR instruments that we can turn to from time to time. Many NMR laboratories receive government funding so that they can serve as a regional NMR resource. Not personally knowing the people at a particular laboratory should not prevent us from contacting them.

The Fourier-transformed INADEQUATE data set will have the magnitude of the J-coupling as the abscissa (vertical axis) and the ^{13}C chemical shifts as the ordinate (horizontal axis). A cross peak indicates that a ^{13}C at a particular shift has a J-coupling that can be read off the ordinate scale. To find the ^{13}C resonance to which the first ^{13}C is bonded, look for another cross peak at the same height in the plot and read the second chemical shift from the abscissa scale.

6.9.2 Heteronuclear Two-Dimensional Experiments Utilizing J-Couplings

In the heteronuclear experiment category, the experiments of interest are the heteronuclear multiple quantum correlation (HMQC) experiment, the heteronuclear single quantum correlation (HSQC) experiment, and the heteronuclear multiple bond correlation (HMBC, including the gradient-selected version gHMBC) experiment. Both the HMQC and HSQC produce similar results, but each has its own unique advantages and disadvantages.

6.9.2.1 *HMQC and HSQC*

The heteronuclear multiple quantum correlation (HMQC) and heteronuclear single quantum correlation (HSQC) experiments both correlate ^{1}H resonances with the resonances of some other nuclide, usually ^{13}C or ^{15}N. Both the HMQC and HSQC experiments can be run in the phase-sensitive or nonphase-sensitive mode. As in other 2-D experiments, field gradient pulses can be used at the start of the pulse sequence to dephase residual magnetization left over from the previous scan. Both the HMQC and HSQC experiments are ^{1}H-detected experiments, in contrast to the older heteronuclear correlation (HETCOR) experiment that detects the X nucleus' magnetization.

An HMQC or HSQC experiment carried out with an inverse probe will provide the same information as a HETCOR experiment carried out with a normal probe, but will do so in far less time. Even carrying out the HMQC or HSQC experiment on a normal probe is more efficient than the HETCOR experiment. This advantage stems from the higher signal-to-noise ratio we obtain when we detect signal from the ^{1}H (with its higher gyromagnetic ratio) instead of from ^{13}C or ^{15}N.

The principal technical hurdle that must be overcome when using ^{1}H detection to correlate ^{1}H and ^{13}C (or ^{15}N) resonances is that the bulk of the ^{1}H signal being detected corresponds to ^{1}H's bound to ^{12}C (or ^{14}N). Only through cancellation of 98.9% of the ^{1}H signal is the ^{1}H-^{13}C HMQC or HSQC experiment able to generate useful data. Because a large fraction of the total detected signal must be removed to leave the small fraction of the signal of interest, RF stability is critically important. The use of bandpass RF filters in the ^{1}H and ^{13}C channels improves the quality of the data by preventing the ^{2}H lock channel and other RF sources from disturbing the ^{1}H and ^{13}C spin populations.

After we shim our sample, we increase the amount of lock power (decreasing the lock gain to keep the lock intensity on scale) to just below the point where the lock level decreases and/or becomes unstable. Increasing the lock power improves the quality of HMQC and HSQC data sets by increasing the signal-to-noise ratio in the lock channel. The greater signal-to-noise ratio in the lock channel makes the instrument more sensitive to applied field fluctuations, and this increased sensitivity to field variations allows the instrument to better maintain the constancy of the applied field. Thus, the ^{1}H and ^{13}C (or ^{15}N) RF becomes more stable relative to a jittery spectral window; this greater stability improves the cancellation efficiency of the signal from ^{1}H's bound to ^{12}C (or ^{14}N).

Besides concerns related to optimizing cancellation of the unwanted signal from ^{1}H's on ^{12}C's, another important feature of the HMQC and HSQC experiments is the requirement for decoupling of the X nucleus (^{13}C or ^{15}N) during detection of the ^{1}H signal. That is, we only turn on the X-nucleus decoupler while we are amplifying, mixing, and digitizing (all three occur simultaneously) the signal from the FID of the ^{1}H's. If we fail to turn on the X-nucleus decoupler during the acquisition time in the pulse sequence, each ^{1}H resonance we observe will be shifted to higher or lower frequency by roughly ½ of the ^{1}J$_{CH}$ (or ^{1}J$_{NH}$). That is, when we select only the signal from the ^{1}H's bound to ^{13}C's (or ^{15}N's), each ^{1}H resonance will be split by the ^{1}J$_{CH}$ (or ^{1}J$_{NH}$). In the case of ^{13}C and ^{15}N, the problem posed by having to excite a large frequency range (recall the discussion of pulse roll-off in Chapter 2) for the purpose of decoupling requires careful calibration of the X decoupler by using what is called the γH$_2$ calibration procedure on a standard sample that is labeled with ^{13}C or ^{15}N. The decoupler duty cycle is normally no larger than 15–20% (the decoupler duty cycle is the percentage of time the X decoupler is on relative to the time of executing the pulse sequence one time). In many instances, we will find that the 2k (2048) data point rule of thumb (to keep the size of the 2-D data set manageable) is less restrictive than the more stringent requirement of keeping the X decoupler duty cycle below 15–20%. Collection of 1k (1024) or even 0.75k (768) data points is often more appropriate. Failure to limit the X decoupler duty cycle can heat the X nucleus coil in the probe to the point where the solder joining the X coil to the wires leading to the coil may melt (the author has done this). Once the solder in the probe melts, the probe will most likely require expensive and lengthy repairs.

As stated earlier, the HMQC and HSQC NMR experiments work by detecting ^{1}H magnetization and canceling the signal from ^{1}H-^{12}C (or ^{1}H-^{14}N) pairs, leaving only the signal from ^{1}H-^{13}C (or ^{1}H-^{15}N) pairs. Cancellation of the unwanted signal is accomplished through alternation of the phase (phase cycling) of one of the pulses in the pulse sequence. Figure 6.11 shows an abbreviated ^{1}H-^{13}C HMQC pulse sequence, and Figure 6.12 shows the fate of the ^{1}H magnetization for ^{1}H's bound to ^{12}C's and ^{13}C's for the abbreviated HMQC pulse sequence.

Consider an isolated methine group whose ^{1}H resonance is slightly downfield. That is, the ^{1}H net magnetization vector of this methine group will, when tipped into the **xy** plane with an RF pulse, precess at a higher frequency than the rotation rate of the rotating frame. Initially, the net magnetization vector of these ^{1}H's can be split into three components: ^{1}H-^{12}C, ^{1}H-^{13}C$_\alpha$, and ^{1}H-^{13}C$_\beta$ where the subscripts denote the spin states of the ^{13}C's (point 1 in Figure 6.12). Assuming that ^{13}C is 1.10% abundant, the relative amounts of the three populations will

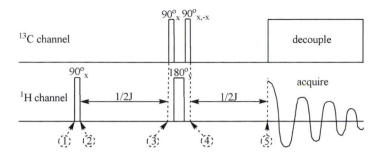

■ FIGURE 6.11   An abbreviated version of the heteronuclear multiple quantum correlation (HMQC) pulse sequence. The state of the net magnetization is discussed for five points corresponding to the five numbers in the dashed circles.

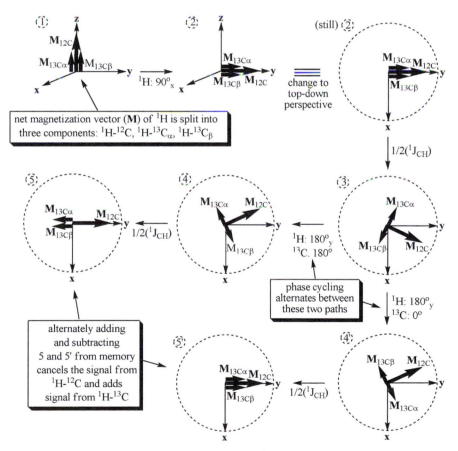

■ FIGURE 6.12   The net magnetization vector of the ^{1}H magnetization is followed through the abbreviated HMQC pulse sequence shown in Figure 6.11.

then be 98.90%, 0.55%, and 0.55%, respectively. If a 90° ^{1}H pulse is applied along the +**x**-axis of the ^{1}H rotating frame (to be strictly correct, we should write this as the +**x'**-axis, but in virtually all the NMR literature the prime is implied and not explicitly written), the vectors associated with all three ^{1}H spin populations will be tipped into the **xy** plane and will point along the +**y**-axis (point 2 in Figure 6.12).

If we then wait for a period of time $\tau = (2^1J_{CH})^{-1}$, we find that the ^{1}H-^{12}C vector will have moved only slightly clockwise during this delay as a result of chemical shift evolution, because the ^{1}H-^{12}C vector precesses just a little faster than a spin that is on resonance. In the same period of time, however, the ^{1}H-^{13}C$_\alpha$ vector will have moved 90° counterclockwise relative to the ^{1}H-^{12}C vector, and the ^{1}H-^{13}C$_\beta$ vector will have moved 90° clockwise relative to the ^{1}H-^{12}C vector. That is, the phase difference between the ^{1}H-^{13}C$_\alpha$ vector and the ^{1}H-^{13}C$_\beta$ vector will be 180°. At this point the three ^{1}H vectors make up the parts of a capital letter "T" (point 3) in Figure 6.12.

If we next apply a 180° pulse to the ^{13}C spins, this pulse will flip (convert) every ^{13}C in the α spin state to the β spin state and vice versa. For the ^{1}H net magnetization vectors, however, the 180° ^{13}C pulse will only swap the α and β labels on the two ^{1}H-^{13}C vectors. To get our magnetization to refocus after the next τ, we apply a ^{1}H pulse to rotate the three ^{1}H vectors 180° about the **y** axis. The ^{1}H 180° pulse rotates the "T" about the **y** axis and the ^{13}C 180° pulse makes the α and β labels exchange (point 4 in Figure 6.12).

Waiting another $\tau = (2^1J_{CH})^{-1}$ causes the ^{1}H-^{12}C vector to precess clockwise because its precession frequency is slightly faster than on resonance; thus, the ^{1}H-^{12}C vector returns to its starting point in the **xy** plane: the +**y**-axis. Meanwhile, the ^{1}H-^{13}Ca vector continues its counterclockwise precession and winds up pointing along the −**y**-axis. Similarly, the ^{1}H-^{13}Cb vector precesses clockwise and also winds up pointing along the −**y**-axis (point 5 in Figure 6.12).

Turning on the ^{13}C decoupler at this point causes the ^{1}H doublet (the combination of ^{1}H-^{13}C$_\alpha$ and ^{1}H-^{13}C$_\beta$) to collapse back into a singlet that is 180° out of phase with respect to the ^{1}H-^{12}C vector, but this singlet will precess at the same frequency as the ^{1}H-^{12}C signal (there may be a slight chemical shift difference between ^{1}H's bound to a ^{12}C's and that of ^{1}H's bound to ^{13}C's). Therefore, the ^{1}H signal we detect from the methine will have an amplitude that is 97.8% of the total possible signal (98.9% − 1.1% = 97.8%). Not such a profound effect (yet).

Next, we repeat the above pulse sequence. But, instead of applying the 180° ^{13}C pulse after the first τ delay, we apply a +90° ^{13}C pulse followed immediately by a $-90°$ ^{13}C pulse to give a composite pulse with a net tip angle of 0°, whose duration and power dissipation in the sample is the same as the ^{13}C 180° pulse. For the 0° ^{13}C pulse, the α and β labels of the two ^{1}H-^{13}C vectors do not get swapped (point 4′ in Figure 6.12). The 180° ^{1}H pulse is still applied, and we still wait a second τ delay (point 5′ in Figure 6.12). In this second pass through the pulse sequence, the ^{1}H-^{13}C$_\alpha$ and ^{1}H-^{13}C$_\beta$ vectors also return to their original starting point in the **xy** plane: along the +**y**′-axis. Turning on the ^{13}C decoupler and digitizing the signal from the ^{1}H FID captures ^{1}H signal from the methine whose amplitude is 100% (98.9%+1.1%=100%).

If we subtract the signal collected during the second scan from that of the first, we are left with a signal due only to the ^{1}H's on ^{13}C's. All the ^{1}H-^{12}C signal cancels, whereas the ^{1}H-^{13}C signal grows. This method of selecting a subset of the total signal is a simple but important example of phase cycling. Instrumentally, the subtraction is accomplished by alternating the phase of the receiver, although the phase of the ^{1}H 90° pulse can also be used to control along which axes (+**y** versus $-$**y** in this case) the various net magnetization vectors will ultimately point.

The preceding discussion shows how alternation of the ^{13}C pulse between 180° and 0° serves to alternate the phase of the signal from ^{1}H's bound to ^{13}C's and thus how alternatively adding and subtracting the digitized signals from the FIDs to and from computer memory cancels the ^{1}H-^{12}C signal while accumulating the ^{1}H-^{13}C signal. What we are still missing is how the second dimension—the ^{13}C chemical shift axis—is introduced.

Figure 6.13 shows the unabbreviated ^{1}H-^{13}C HMQC pulse sequence. By introducing a pair of delays after the first and before the second of the two ^{13}C 90° pulses (that combine to make the ^{13}C 180° or 0° pulses), the ^{13}C chemical shift axis is introduced. The sum of this pair of delays is the t_1 ^{13}C chemical shift evolution time, or more simply the t_1 evolution time. Note that in the middle of the t_1 evolution time, the ^{1}H 180° pulse is applied to refocus the ^{1}H chemical shifts; thus, the t_1 evolution time is split into two equal parts of $t_1/2$.

After the first ^{13}C 90° pulse is applied, the ^{13}C net magnetization vector will lie in the **xy** plane—say along the +**y**-axis of the ^{13}C rotating frame—and will begin to process. As the ^{13}C magnetization

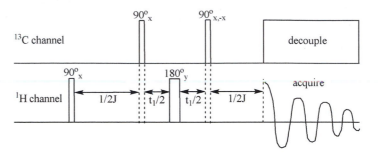

■ **FIGURE 6.13** The 2-D HMQC pulse sequence. Note that the t_1 time increment is split into two equal parts that sandwich the ^{1}H chemical shift refocusing (180°) pulse.

precesses, the projection of the ^{13}C vector along the +y-axis will vary sinusoidally for each chemically distinct ^{13}C. Following the second ^{13}C 90° pulse, the component of the ^{13}C vector tipped to either the −z- or +z-axis will depend on the combination of how long the t_1 evolution time period was and how much any particular ^{13}C chemical shift deviated from being on resonance. In other words, during the t_1 evolution time, the component of the ^{13}C magnetization traces out a sine wave and the frequency of the sine wave depends on the ^{13}C chemical shift. For a given ^{13}C chemical shift, varying the t_1 evolution time will cause the component of the ^{13}C magnetization tipped down to the −z axis or back up to +z axis to vary according to a cosine function. The net magnetization vector from each chemically distinct ^{13}C precesses a unique amount depending on how far it is from being on resonance and the length of the t_1 evolution time. Digitization of the signal from a number of FIDs—with each subsequent FID resulting from applying the pulse sequence with increasing t_1 evolution times—generates a data set that allows us to mathematically extract how the ^{13}C vectors vary sinusoidally as a function of the t_1 evolution time.

However, the preceding discussion still fails to explain how the ^{13}C chemical shift axis information gets transmitted to the ^{1}H signal we detect. The ^{13}C chemical shift information gets passed to attached ^{1}H's on the basis of the efficiency with which the ^{13}C α and β spin state populations are inverted by the ^{13}C 180° pulse. Because the inversion of the ^{13}C spin populations is modulated by the t_1 evolution time, the strength of the signal for the ^{1}H's bound to ^{13}C's will therefore also be modulated by the t_1 evolution time. The frequency of this modulation will depend on how far from being on resonance each chemically distinct ^{13}C spin population is. In this way, the ^{13}C chemical shift information is transmitted through the $^1J_{CH}$ only to those ^1H's bound to ^{13}C's. Fourier transformation of the 2-D data set with respect to the t_1 evolution time shows how a particular ^1H resonance correlates with a particular ^{13}C resonance.

The HMQC experiment, although less sensitive to the miscalibration of pulses than the HSQC experiment, produces broader cross peaks than the HSQC experiment. We use the HMQC experiment if signal is abundant, if the ^1H and ^{13}C 1-D spectra are not too crowded, and if probe tuning and pulse calibration are not likely and/or practical to carry out. For more exacting work with more demanding samples, we prefer the HSQC experiment. However, ^1H-^{15}N HMQC is generally preferred to ^1H-^{15}N HSQC.

The HSQC experiment cannot be explained readily by using spin gymnastics arguments. To understand the HSQC and other more complex pulse sequences, we should consult other texts to learn the product operator formalism and density matrix theory.

Many variations of the HMQC and HSQC pulse sequences are available. In many cases, we can simply set a few flags in the NMR software to include or exclude this or that feature.

The HMQC and HSQC experiments can be made sensitive to unusual couplings if we change the $(2J)^{-1}$ delay in the pulse sequence. In this way, we can make a particular HMQC experiment sensitive to a ^1H on the end of a triply bonded carbon (a terminal alkyne). Typical delays in the HMQC/HSQC experimental setup are based on $^1J_{CH} = 140\,Hz$ and $^1J_{NH} = 90\,Hz$. If one or more of the ^1J's between the XH spin pair differs significantly from the assumed value, the HMQC/HSQC cross peak may be weak or even absent. Table 6.2 shows how normalized volume integrals taken from the HSQC cross peaks for several molecules vary as a function of the $(2J)^{-1}$ delay ($^1J_{CH}$'s can be obtained by measuring the spacing, in hertz, between the ^{13}C satellite peaks in the ^1H 1-D spectrum). From these data, it should be clear that we must be careful in selecting the appropriate delay. Furthermore, factors other than the $(2J)^{-1}$ delay (e.g., ^1H relaxation, especially for longer delays) may influence cross peak volume.

Plotting of the HMQC/HSQC data sets should include all cross peaks; in lieu of the use of projections of the actual data matrix (with their poor digital resolution), we include 1-D spectra along the

Table 6.2 Normalized volume integrals of HSQC cross peaks as a function of observed $^1J_{CH}$ and the $(2J)^{-1}$ delay in the HSQC pulse sequence.

Observed $^1J_{CH}$ (Hz)	$^1J_{CH}$ upon which $(2J)^{-1}$ HSQC delay is based				
	120 Hz	140 Hz	160 Hz	170 Hz	205 Hz
128/129	1.00	0.31	0.21	0.03	0.25
145	0.51	0.86	0.62	0.06	0.23
160–163	0.59	0.87	0.89	–	–
167–170	0.43	0.84	0.99	0.90	0.80
178	–	0.40	–	0.89	1.00
204	–	0.03	–	0.31	0.86

f_1 projection. The summation or maxima picking of a 2-D data matrix parallel to the f_2 axis. If the f_1 axis is vertical, then the f_1 projection will normally be shown on the left side of the data matrix. The projection is obtained by summing all points or finding the maximum value of each row of the matrix.

f_1 and f_2 axes. If the ^{13}C (or ^{15}N) 1-D is not available, the 2-D data set can be summed parallel to the f_2 axis to give what is called the f_1 projection. The line widths of the resonances in the f_1 projection will very likely be much greater than those found in a ^{13}C (or ^{15}N) 1-D spectrum. A ^{1}H-^{13}C HMQC or HSQC 2-D data set with an acceptable signal-to-noise ratio can typically be collected in 20–30 minutes using a 20 mM sample on a 500 MHz instrument equipped with an inverse probe.

6.9.2.2 *HMBC*

The heteronuclear multiple bond correlation (HMBC) experiment employs nearly the same methodology as the HMQC experiment, except that the HMBC experiment exploits a much smaller J-coupling than the ^{1}J of ^{1}H-^{13}C or ^{1}H-^{15}N. The HMBC experiment is typically tuned to a J-coupling of 8 Hz even though the range of J-couplings of interest is usually 2–12 Hz. We bias the $(2J)^{-1}$ time delay in the HMBC experiment toward larger couplings because T_2^* relaxation effects significantly attenuate the overall amount of signal we observe if the $(2J)^{-1}$ delay is lengthened. Many ^{1}H's have T_2^*'s on the order of several hundred milliseconds, so tailoring the $(2J)^{-1}$ delay to be most sensitive to a J-coupling of 3 Hz results in a $(2J)^{-1}$ delay of $(2 \times 3\,\text{Hz})^{-1}$, or 167 ms, which will result in the loss of a great amount of signal prior to detection. Clearly, losing two-thirds or more of the ^{1}H signal is undesirable. The good news is that the intensity of HMBC cross peaks does not depend critically on the choice of the time delay in the pulse sequence. Consequently, making the HMBC experiment most sensitive to J-couplings of 8 Hz does not preclude the observation of cross peaks arising from ^{1}H's exhibiting much smaller J-couplings. While the $(2J)^{-1}$ delay may be optimized for a coupling of 8 Hz, any cross peak arising from a coupling of larger than 8 Hz will generate roughly the volume integral as a cross peak arising from an 8 Hz coupling. That is, if we wish to differentiate between a cross peak generated by a coupling of 8 Hz and 12 Hz, we must shorten the $(2J)^{-1}$ to correspond to a coupling of 12 Hz.

Another favorable feature associated with the HMBC experiment is that X-decoupling is not employed. Therefore, better actual f_2 resolution (not digital) can be obtained because more data points can be acquired in the collection of the digital array of points we obtain from the signal of the FID. That is, each FID can be digitized for a longer period of time. Leaving the X-decoupler off has the added advantage of allowing cross peaks arising from ^{1}J's to be readily identified, as the HMQC cross peak will now be split into two cross peaks with the same ^{13}C (or ^{15}N) chemical shift and two different ^{1}H shifts.

When collected in a phase-sensitive mode, HMBC cross peaks are found to have a mixed phase character. That is, we cannot phase HMBC cross peaks so that they are purely absorptive. The use of pulsed field gradients for the purpose of coherence selection in the HMBC experiment (gHMBC) renders a nonphase-sensitive 2-D data set. This latter method is generally preferred because phasing of the spectrum is not required.

The HMBC experiment typically requires roughly four times more scans per t_1 time increment (to obtain a spectrum with a comparable signal-to-noise ratio) than do the HMQC and HSQC experiments. Thus, the HMBC experiment takes three or four times longer to acquire than the HMQC or HSQC experiment (involving the same two nuclides, e.g., ^{1}H and ^{13}C). This extensive scanning is required because the sensitivity of the HMBC experiment is poorer than that of the HMQC or HSQC experiments.

It generally holds that the intensity of an HMBC cross peak drops off as the number of bonds separating the two coupled nuclides (normally ^{1}H and ^{13}C) increases because cross peak intensity roughly mirrors the magnitude of the J-coupling between the two species.

Karplus diagram. A plot showing the Karplus relationship.

Strong HMBC cross peaks are often observed between ^{1}H's and ^{13}C's that are two and three bonds distant, provided that the geminal and vicinal (dihedral) bond angles are not such that the coupling is at or near zero (recall the Karplus diagrams presented in Sections 6.4 and 6.5). The exception occurs when we are dealing with a ^{1}H on a ^{13}C that is sp^2 hybridized; in this case, the HMBC cross peak between a ^{1}H bound to an sp^2-hybridized carbon and an adjacent ^{13}C will be weak or may even be absent, although the cross peak between a ^{1}H and a ^{13}C three bonds away will likely be intense. The absence of the HMBC cross peak between a ^{1}H on an sp^2-hybridized carbon and an adjacent ^{13}C stems from the fact that a geminal bond angle of 120° (caused by the sp^2 hybridization) gives a smaller ^{2}J—usually less than 2 Hz. The presence of a heteroatom near the ^{13}C two bonds from the ^{1}H in question may be sufficient to disrupt the conditions that drive the ^{2}J to near zero; hence, the presence of atoms other than H and C may favorably influence HMBC cross peak intensity between ^{1}H's bound to sp^2-hybridized carbons and ^{13}C's two bonds distant.

The ^{1}H-^{13}C 2-D HMBC spectra of aromatic and conjugated compounds often allow us to distinguish between ^{13}C signals that are *cis* versus *trans* to a ^{1}H three bonds away. Because the entire network of atoms in an aromatic or conjugated system is often planar (at least locally, barring steric constraints), there are only two vicinal bond angles from which to choose (0° and 180°). Thus, there are only two values likely to be encountered between a ^{1}H on an sp^2-hybridized carbon and a ^{13}C three bonds away; the *cis*-^{3}J is smaller (5–6 Hz) than the *trans*-^{3}J (8–10 Hz). Consequently, the *cis*-^{3}J generates a weaker HMBC cross peak than the *trans*-^{3}J does.

HMBC cross peaks are sometimes observed between ^{1}H's and ^{13}C's as many as four or five bonds distant, especially if the entire system is planar and there are π-electrons between the two nuclides.

f$_2$ projection. The summation or maxima picking of a 2-D data matrix parallel to the f$_1$ axis. If the f$_2$ axis is horizontal, then the f$_2$ projection will normally be shown on the top of the data matrix. The projection is obtained by summing all points or finding the maximum value of each column of the matrix.

In plotting HMBC data sets, it is important to avoid the use of a projection of the data matrix on the f$_2$ (^{1}H chemical shift) axis. Instead, we use the ^{1}H 1-D spectrum. Use of the f$_2$ projection (like the shadow or profile of the cross peaks when viewed parallel to the f$_1$ axis) will show spurious peaks due to the splitting of the ^{1}J cross peaks.

In some cases, we may find that a ^{13}C 1-D spectrum that we collected overnight will not reveal the weaker ^{13}C resonances, e.g., the resonances from nonprotonated carbons, including carbonyl carbons. Nonetheless, the HMBC data set may reveal the shifts of these peaks. The HMBC experiment is in many cases the only practical means of determining the chemical shifts of weak ^{13}C signals (the precision of the chemical shift measurement will be lower than for a directly observed ^{13}C or ^{15}N 1-D spectrum). Use the f_1 projection in lieu of the ^{13}C (or ^{15}N) 1-D spectrum if direct observation of the X nucleus is not practical.

A 20 mM sample will often yield a reasonable quality ^{1}H-^{13}C 2-D gHMBC using 400 t_1 increments in about 75 minutes on a 500 MHz NMR equipped with an inverse probe.

■ REFERENCES

[1] S. L. Patt, J. N. Shoolery, *J. Magn. Reson.*, **46**, 535–539 (1982).

[2] M. R. Bendall, D. M. Doddrell, D. T. Pegg, *J. Am. Chem. Soc.*, **103**, 4603–4605 (1981).

[3] A. Bax, R. Freeman, S. P. Kempsell, *J. Am. Chem. Soc.*, **102**, 4849–4851 (1980).

Chapter 7

Through-Space Effects: The Nuclear Overhauser Effect (NOE)

The nuclear Overhauser effect (NOE) is a powerful tool that is used effectively by many practicing bioNMR spectroscopists to elucidate the structures of proteins and other biomolecules. Beyond finding application in the area of bioNMR, the NOE also influences favorably the outcome of many routine ^{13}C 1-D spectra. Even those investigators studying small molecules with NMR often find the NOE to be of great utility in making unambiguous stereochemical assignments.

Nuclear Overhauser effect, NOE, nOe. The perturbation of the populations of one set of spins achieved through saturation of a second set of spins less than five angstroms distant.

7.1 THE DIPOLAR RELAXATION PATHWAY

The NOE arises through the dipolar interaction of spins, which is an interaction that occurs between two spin-½ nuclei through space. The dipolar interaction is tensorial in nature, meaning simply that it depends on the relative orientation of the two dipoles in space. It is a through-space effect and not a through-bond effect like J-coupling.

The dipolar interaction arises from the precession of one nucleus next to another, for as it precesses, its nuclear magnetic moment μ will influence the field felt by nearby nuclei. The precessing magnetic moment generates an alternating field in the **xy** plane that is normally greater in magnitude than either J-couplings or the variations in chemical shifts that occur as a result of changing the orientation of a molecule. In the liquid state, rapid isotropic molecular tumbling normally averages out the dipolar interaction, so the dipolar interaction in liquids normally does not cause observed resonances to shift.

The dipolar interaction does, however, often play a role in the relaxation of spins. Even though the net effect of dipolar interactions on the frequencies observed in a given system with rapid isotropic molecular tumbling is zero, the momentary shifts in the transition energies of one spin that is influenced by a nearby magnetic dipole provides an efficient relaxation pathway.

7.2 THE ENERGETICS OF AN ISOLATED HETERONUCLEAR TWO-SPIN SYSTEM

Consider an isolated methine group (again). The ^{1}H-^{13}C duo is a spin pair that can exist with four possible spin state combinations: $\alpha\alpha$, $\alpha\beta$, $\beta\alpha$, and $\beta\beta$. At equilibrium, the population of each possible combination of spin states is governed by the Boltzmann equation. The lowest energy combination is $\alpha\alpha$ and the highest energy combination is $\beta\beta$ as we would expect from consideration of the Zeeman energy diagram (see Figure 1.1). The $\alpha\beta$ and $\beta\alpha$ spin state combinations of course have energies between the $\alpha\alpha$ and $\beta\beta$ extremes. The $\alpha\beta$ combination (^{1}H in the α spin state, ^{13}C in the β spin state) is lower in energy than the $\beta\alpha$ combination because the energy difference at a given applied field strength is greater for ^{1}H than for ^{13}C. Therefore, the spin state of the ^{1}H has a greater effect on the overall energy of the two mixed ($\alpha\beta$ and $\beta\alpha$) spin state combinations.

Each combination is linked to every other combination by a relaxation pathway with a characteristic dipolar relaxation rate constant denoted W_0, W_{1H}, W_{1C}, or W_2. If the populations of the four spin state combinations are disturbed, simple first-order kinetics (rate constant W_X times the deviation from equilibrium) will attempt to return the system to equilibrium. The trailing subscript of the four rate constants denotes which spin or spins must undergo spin flips in that relaxation process. W_0 denotes the zero quantum spin flip rate constant. The term zero-quantum does not mean that nothing happens; rather, it means that one spin goes from the α to the β spin state while the other goes from the β to the α spin state, and thus the simultaneous exchange results in no net change in the number of spins in the α and β spin states. W_2, on the other hand, denotes a double quantum spin flip and therefore links the $\alpha\alpha$ and the $\beta\beta$ spin state combinations. W_{1H} links the $\alpha\alpha$ to the $\beta\alpha$ and the $\alpha\beta$ to the $\beta\beta$ spin-state combinations, whereas W_{1C} links the $\alpha\alpha$ to the $\alpha\beta$ and the $\beta\alpha$ to the $\beta\beta$ spin state combinations (remember that the ^{1}H spin state is the first α or β, and the ^{13}C spin state is the second). Both W_{1H} and W_{1C} are rate constants for single quantum transitions, because only one spin undergoes a spin flip when these pathways are used.

Dipolar relaxation rate constant, W. For a two spin system, W_0 is the rate constant for the zero quantum spin flip, W_1 is the rate constant for the single quantum spin flip, and W_2 is the rate constant for the double quantum spin flip. For short correlation times, the ratio of W_2:W_1:W_0 is 1:¼: ⅙.

Zero quantum spin flip rate constant, W_0. The kinetic rate constant controlling the simultaneous change in both spin states for a two-spin system where one of the spin-½ spins goes from the α to the β state while the other spin-½ spin goes from the β to the α state.

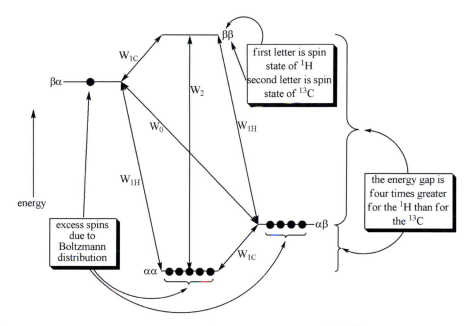

■ FIGURE 7.1 The energetics and excess populations of the four allowed spin state combinations for a ${}^{1}H$-${}^{13}C$ pair at equilibrium.

Because the NMR frequency of the ${}^{1}H$ is roughly four times that of ${}^{13}C$, the number of excess spins (N) in the four combinations will be $N_{\alpha\alpha} = 5$, $N_{\alpha\beta} = 4$, $N_{\beta\alpha} = 1$, and $N_{\beta\beta} = 0$. That is, tracing along W_{1H} will always show a difference in population four times greater than tracing along W_{1C}. Figure 7.1 shows the excess equilibrium populations of the four allowed spin-state combinations for a ${}^{1}H$-${}^{13}C$ pair.

Before considering what happens when we apply ${}^{1}H$ decoupling to this system, we acquaint ourselves with the spectral density function, $J(v)$, and how it relates to the relative magnitude of W_0, W_{1H}, W_{1C}, and W_2.

7.3 **THE SPECTRAL DENSITY FUNCTION**

The spectral density function, $J(v)$, is a mathematical function that describes how energy is spread as a function of frequency. This energy may be what the ancients referred to in their discussions of the ether that permeates all space, but that is more of a philosophical topic. In a liquids sample at a given temperature, the spectral

Single quantum spin flip rate constant, W_1. The kinetic rate constant controlling the change in the spin state of a single spin-½ spin from either the α to the β state or from the α to the β state.

Double quantum spin flip rate constant, W_2. The kinetic rate constant controlling the simultaneous change in both spin states for a two-spin system where both spin-½ spins go from the α to the β state or from the α to the β state.

Spectral density function, $J(v)$, $J(\omega)$. A mathematical function that describes how energy is spread about as a function of frequency.

Correlation time, τ_c. The amount of time required for a molecule to diffuse one molecular diameter or to rotate one radian (roughly ⅙ of a complete rotation).

density function is controlled by a variable called the correlation time τ_c, which is the measure of how long it takes for a molecule to diffuse one molecular diameter or rotate one radian (about one sixth of a complete rotation). The explicit relationship between $J(v)$ and τ_c is relatively simple and is shown in Equation 7.1.

$$J(v) = 2\tau_c/(1+4\pi^2 v^2 \tau_c^2) \tag{7.1}$$

Figure 7.2 shows the spectral density function in the range of 10 to 10,000 MHz on a logarithmic scale for a series of correlation times ranging from 32 ps to 3.162 ns (32, 100, 316, 1000, and 3162 ps). The numerator of Equation 7.1 dominates for small values of the product of v times τ_c and the equation reduces to $J(v) = 2\tau_c$. On the other hand, when v times τ_c gets large, the second term in the denominator becomes dominant and $J(v)$ drops to essentially zero very quickly.

For a transition between any of the various combinations of spin states to be efficient, appreciable spectral density is required at the frequency of the photons whose energy is tuned to complement the difference in energy between the initial and final spin state combinations. Put another way, the spectral density has to make up the energy difference when a transition takes place; if the spectral density is low at the

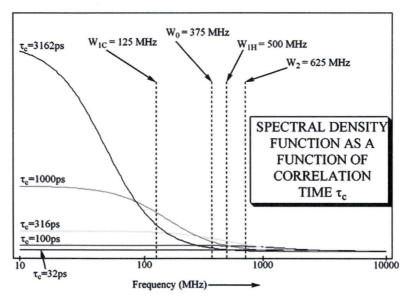

■ **FIGURE 7.2**   The spectral density function, $J(v)$, for five different correlation times as a function of frequency.

frequency needed to drive a particular transition, the transition will not occur rapidly.

Figuring out the frequency of a given transition is simple. On a 500 MHz instrument, the ^{1}H frequency is 500 MHz and the ^{13}C frequency is 125 MHz. For the single quantum W_{1H} transition that involves only flipping the spin of the ^{1}H, the frequency of the photons that will drive this transition is 500 MHz. If the spectral density function is not zero at 500 MHz, then the W_{1H} transition will be efficient and the dipolar relaxation mechanism will be an efficient relaxation pathway for the ^{1}H. For the W_{1C} transition, the frequency is 125 MHz; so in this case, having spectral density at 125 MHz will make the single quantum ^{13}C dipolar relaxation mechanism efficient. For the double quantum W_2 transition that connects the $\alpha\alpha$ and $\beta\beta$ spin state combinations, spectral density at 625 MHz ($\nu_H + \nu_C$) is required. For the zero quantum W_0 transition (also called the flip-flop transition), spectral density at 375 MHz ($\nu_H - \nu_C$) is required to make the transition efficient.

If the spectral density function drops to zero at the frequency of a given transition as the result of an increase in the correlation time τ_c, then the rate constant for the transition decreases. This transition rate constant reduction is important when the size of the molecule increases, when field strength increases, or when the solution viscosity increases (e.g., during a polymerization or as the result of cooling).

In the so-called fast-exchange limit (short τ_c), the ratio of W_0:W_1: W_2 is 2:3:12. Calculation of the relative values of W_0, W_1, and W_2 is complicated and beyond the scope of this text. Nevertheless, it can be argued that W_2 will be largest because it involves the greatest change in spin states. That is, it makes sense that the rate constant for the double quantum spin flip will be the largest of the three because it links the relaxation pathway that allows the greatest change in the sum of the quantum numbers.

Flip-flop transition. Syn. zero quantum transition, W_0 transition, zero quantum spin flip. When two spins undergo simultaneous spin flips such that the sum of their spin quantum numbers is the same before and after the transition takes place. For example, if spins A and B undergo a flip-flop transition, then if spin A goes from the α to the β spin state, then spin B must simultaneously goes from the α to the β spin state.

Fast-exchange limit. The fast-exchange limit is said to be reached when no further increase in the rate at which a dynamic process occurs will alter observed spectral features. Normally, we speak of resonance coalescence as occurring when the fast-exchange limit is reached.

7.4 DECOUPLING ONE OF THE SPINS IN A HETERONUCLEAR TWO-SPIN SYSTEM

When ^{1}H decoupling takes place, the number of excess α spins for the ^{1}H's of the methine group will be eliminated. That is, the B_1 field of the applied RF at 500 MHz equalizes the number of ^{1}H's in the α versus the β spin state. Figure 7.3 shows how the number of excess spins

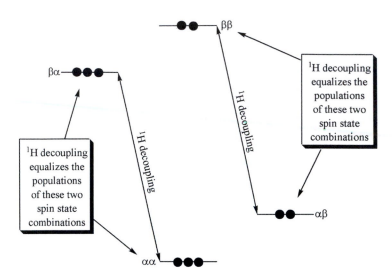

■ **FIGURE 7.3**   The effect of ^{1}H decoupling on the equilibrium population of the ^{1}H-^{13}C two–spin system: the ^{1}H's rapidly flip between the α and β spin states, thus equalizing the number of ^{1}H's in the two states.

changes as a result of saturating (decoupling) the ^{1}H spins. Saturation of the ^{1}H spins makes the populations of the four possible spin state combinations for the ^{1}H-^{13}C pair change from $N_{\alpha\alpha} = 5$, $N_{\alpha\beta} = 4$, $N_{\beta\alpha} = 1$, and $N_{\beta\beta} = 0$, to $N_{\alpha\alpha} = 3$, $N_{\alpha\beta} = 2$, $N_{\beta\alpha} = 3$, and $N_{\beta\beta} = 2$.

7.5 RAPID RELAXATION VIA THE DOUBLE QUANTUM PATHWAY

Saturating the ^{1}H spins initially disturbs only the ^{1}H spin populations by increasing the population of the higher energy β spin state at the expense of the lower energy α spin state; following this disruption, relaxation commences to return the system to equilibrium. Because the double quantum (W_2) spin flip transition has the largest rate constant, it converts the population of the $\beta\beta$ spin state combinations to the $\alpha\alpha$ combination most rapidly. The efficient transfer of the excess $\beta\beta$ combination to the $\alpha\alpha$ combination therefore yields the following spin state combination populations: $N_{\alpha\alpha} = 5$, $N_{\alpha\beta} = 2$, $N_{\beta\alpha} = 3$, and $N_{\beta\beta} = 0$ (Figure 7.4). Because the strength of the ^{13}C signal depends on the relative number of spins in the α versus the β spin state for the ^{13}C nuclei, we can assess the effect on the ^{13}C signal strength by comparing both the relative populations of the $\alpha\alpha$ versus the $\alpha\beta$ and the $\beta\alpha$ versus the $\beta\beta$ spin state combinations. Before

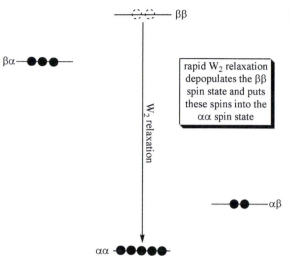

■ **FIGURE 7.4**   Rapid double quantum (W_2) relaxation depopulates the $\beta\beta$ spin state and puts the spin pair into the $\alpha\alpha$ spin state.

saturating the ^{1}H spins $N_{\alpha\alpha} - N_{\alpha\beta} = N_{\beta\alpha} - N_{\beta\beta} = 1$. After saturation of the ^{1}H spins followed by rapid double quantum relaxation $N_{\alpha\alpha} - N_{\alpha\beta} = N_{\beta\alpha} - N_{\beta\beta} = 3$. Thus, the differences between the populations of the $\alpha\alpha$ versus $\alpha\beta$ and $\beta\alpha$ versus $\beta\beta$ spin state combinations before and after saturation of the ^{1}H spins followed by W_2 relaxation shows the ^{13}C signal strength (which depends on the excess of spins in the lower energy spin state) increases by 200%. The signal increase we obtain for the ^{13}C signal when ^{1}H's are irradiated is called the nuclear Overhauser effect (NOE).

In practice, NOE enhancements will vary, with a twofold (200%) signal increase being the theoretical maximum enhancement for ^{13}C observation with ^{1}H decoupling. Other relaxation mechanisms besides that due to the dipolar interaction will diminish the observed enhancement, η.

Similar calculations can be carried out on other spin systems. The general result for the NOE enhancement η is

$$\eta = \gamma_x / (2\gamma_A) \tag{7.2}$$

where η is the maximum theoretical enhancement (in addition to that normally observed), the nucleus being saturated (^{1}H in the preceding example) is the X nucleus with gyromagnetic ratio γ_X, and the nucleus being observed is the A nucleus with gyromagnetic ratio γ_A. In the homonuclear case ($\gamma_X = \gamma_A$), an enhancement of 50% is therefore

Enhancement, η. Syn. NOE enhancement. The numerical factor by which the integrated intensity of a resonance increases as the result of irradiation of a spin that is nearby in space. For the irradiation of nuclear spins, the upper limit for the observation of a nearby spin is on the order of five angstroms.

1-D NOE difference experiment.
The subtraction of a 1-D spectrum
obtained by irradiating a single resonance at low power with CW RF from
a 1-D spectrum obtained by irradiating a resonance-free region in or
near the same spectral window. The
resulting spectrum shows the irradiated resonance phased negatively,
and any resonance that has its equilibrium spin population perturbed
through cross relaxation with the
irradiated resonance shows a positive integral.

also possible. This 50% enhancement is the basis of the homonuclear
1-D NOE difference experiment used often by synthetic chemists.
Note that if one of the gyromagnetic ratios is negative, the enhancement also will be negative.

After having read this far, we may question why discussion of the
spectral density function is relevant. The spectral density function
$J(v)$ becomes critically important when the correlation time τ_c gets
longer. As τ_c gets longer, the point on the frequency axis (v) at which
the second term in the denominator of Equation 7.1 becomes dominant and thus forces $J(v)$ to zero moves further to the left (compare
the 100 ps trace to the 1000 ps trace in Figure 7.2).

Now recall that, for a given transition to be efficient, appreciable
spectral density at the frequency of the transition is required. As the
spectral density drop-off moves further to the left as τ_c increases,
the double quantum (W_2) transition at 625 MHz is the first to shut
down (the rate constant gets small) because there is no longer appreciable spectral density at that frequency. Once the W_2 transition
shuts down, saturating the ^{1}H spins will no longer allow the excess
population of the $\beta\beta$ spin state combination to rapidly relax to the
$\alpha\alpha$ spin state combination; thus, the NOE enhancement we expect
is not observed. This situation normally occurs when the molecular weight of a molecule approaches about 1000 g mol^{-1} in D$_2$O at
25°C. The physical basis for the shutdown of the double quantum
transition is a lengthening of the correlation time τ_c due to slower
molecular tumbling. Spectrometer frequency, solution viscosity, and
molecular weight all play a role in determining at what point the
double quantum (W_2) transition shut downs.

However, the complete cancellation of the NOE is only temporary.
Longer correlation times allow the NOE to be observed (this time
with an opposite sign). The NOE no-man's-land in the intermediate
correlation time range impedes through-space NMR investigations,
but we can circumvent this impediment by changing solvent, temperature, and field, or by changing the NMR experiment itself.

7.6 A ONE-DIMENSIONAL EXPERIMENT UTILIZING THE NOE

The 1-D NOE difference experiment is often preferred to the 2-D
NOESY experiment because the former is more quantitative; that is,
the precision of the 1-D experiment is finer. The 1-D NOE difference
experiment is accomplished by (1) selectively irradiating (saturating)

a single resonance for 2 to 5 seconds and then digitizing the resulting FID; (2) selectively irradiating a resonance-free region in or near the spectral window, digitizing the resulting FID (this is the control FID); and then (3) subtracting the latter from the former (obtaining the difference between the two). This time domain data array, which resembles a digitized FID, is then Fourier transformed. In the absence of any NOE effects, the difference spectrum shows only the irradiated resonance phased fully negatively (the irradiated resonance is absorptive, but it is negative). If saturating one resonance increases the intensities of other resonances, then the affected resonance integrals increase and the difference spectrum reflects these changes.

The 1-D NOE difference experiment is normally carried out on ^{1}H's, and is an excellent way of determining *cis-trans* substitution patterns of groups across carbon–carbon double bonds. The 1-D NOE difference experiment is also useful in determining the proximity of ^{1}H-containing functional groups in cyclic or other molecules locked into specific conformations because of steric constraints.

Irradiation of multiple resonances is possible, but each resonance being irradiated requires the digitization of an additional FID. That is, the control FID can be used over and over by subtracting it from each digitized FID collected when a unique resonance is saturated.

The control FID we digitize to subtract from each of the other FID(s) we digitize is prepared by irradiating a resonance-free point in or near the spectral window. Irradiation of an empty spectral region when we collect our control FID improves the cancellation efficiency for spectator resonances (those whose intensities are unaffected, i.e., those that do not participate). This reproduction of the power delivery timing into the sample is directly analogous to practice we use in the HMQC experiment wherein we alternate using a $+90°/-90°$ pulse with the 180° pulse instead of using a 0° pulse alternating with a 180° pulse for successive scans.

The NOE difference experiment is carried out as follows. First, we collect a regular 1-D spectrum. Next, we record the transmitter offset required to make each resonance to be irradiated on resonance. Third, we find a suitable location in or near the spectral window for dummy (or control) irradiation (perhaps -5 ppm or $+15$ ppm) where no resonances are observed. Fourth, we prepare both FIDs using a long period of single-frequency low power RF irradiation followed by a hard 90° RF pulse, except that in the preparation of the first FID, a resonance of interest is irradiated, whereas in the preparation of the second FID, the dummy location is irradiated instead. If we digitize both FIDs using the same acquisition timing (ringdown delay, receiver frequency, and dwell time), then the difference of the two digitized FIDs is Fourier transformed, phased, baseline corrected, and integrated to yield the NOE difference spectrum. If the irradiated resonance is phased with a negative integral, any enhancements we observe will be positive.

If we irradiate a methine ^{1}H, then we take the difference spectrum and set the integral of the peak being irradiated to -100. If the software does not allow us to set a negative integral, we can phase the irradiated peak so that it is fully absorptive and positive, perform baseline correction, set the integral to $+100$, and finally invert the spectrum with a 180° phase shift. This sequence of operations ensures that any other peaks we integrate will show positive integrals in percent (be sure to normalize if the receiving resonance contains more than one ^{1}H). If we irradiate both ^{1}H's of a methylene group, then the integral of the irradiated peak should be set to -200. If we irradiate a methyl group, we set the integral of the methyl resonance to -300.

If our instrument is appropriately equipped, we can use shaped RF pulses to selectively excite resonances. We might wish to do this if using CW, RF is not selective enough. This practice is required in cases where the resonance to be irradiated is near other resonances in the spectrum (not in space) whose irradiation might yield ambiguous results.

The efficiency with which the NOE is generated depends on the physical distance between the irradiated and the observed spins. The dipolar interaction scales as r^{-6}, thus making the effect diminish severely with increasing internuclear distance r. But this strong distance dependence is what makes the NOE so useful. Only spins that are very close in space will exhibit a strong dipolar interaction that may, upon irradiation of the resonances from one of the spins, increase the integrated intensity of the receiving spin's resonance in the NOE difference spectrum.

Internuclear distance, r. The through-space distance between two nuclei.

A second important consideration to bear in mind when we use the NOE difference experiment concerns competing relaxation mechanisms. If the dipolar relaxation mechanism contributes negligibly to enhancing the overall rate of relaxation for a particular spin because of other efficient and available relaxation mechanisms, then we will find that the NOE enhancements we measure will not be statistically significant. Put another way, if the dipolar interaction is drowned out by other competing relaxation mechanisms, then we will not see an increase in the integral of the resonance from what we expect to be the receiving spin—even if the two spins in question (irradiated and receiving) are nearby. Methyl groups are notoriously poor receivers of NOE enhancements. This aloofness stems from the fact that methyl groups relax efficiently as a result of their rapid rotation about the bond that attaches the group to the rest of the molecule. Whenever we want to look for an NOE between a methyl group and something else, we always irradiate the methyl group and look for the effect on the other spin.

NOE enhancements of 2–5% are normally considered reasonable indicators of close proximity in a molecule, perhaps 0.3–0.4 nm.

7.7 **TWO-DIMENSIONAL EXPERIMENTS UTILIZING THE NOE**

There are two basic 2-D NMR experiments that make use of the NOE: the NOESY and the ROESY [1] experiments. NOESY stands for nuclear Overhauser effect spectroscopy and ROESY stands for rotational Overhauser effect spectroscopy. The ROESY experiment is also referred to in some of the literature as the CAMELSPIN experiment. The principal difference between the NOESY and ROESY experiments lies in the time scale associated with the dipolar relaxation mechanism.

7.7.1 **NOESY**

For the NOESY experiment, the spectral density function's amplitude for the various transitions is the same as that described in the earlier explanation of the origin of the NOE. However, for the ROESY experiment, a spin lock is applied to the various net magnetization vectors of the spins following a 90° pulse to put the net magnetization vectors into the **xy** plane. In the ROESY experiment, the exchange of phase information that leads to cross peak generation occurs as the spins are aligned in the **xy** plane along the magnetic field component of the applied RF. That is, the dipolar interaction occurs as the various spins have components aligned with the B_1 field. Because the B_1 field strength is much weaker than the B_0 field strength, the spectral density actuating the magnetization exchange is always present. In practice, liquid samples never have a correlation time τ_c that

Rotational Overhauser effect spectroscopy, ROESY. Syn. CAMELSPIN experiment. A 2-D NMR experiment similar to the 2-D NOESY experiment, except that the ROESY experiment employs a spin-lock using the B_1 field of the applied RF, thus skirting the problem of the cancellation of the NOE cross peak when correlation times become long enough to reduce the rate constant for the dipolar double-quantum spin flip.

will fail to make the transverse dipolar relaxation pathways efficient; thus, the ROESY experiment never suffers from the adverse effects noted in the discussion of the diminution of the double quantum relaxation pathway.

The downside of the ROESY experiment is its poorer sensitivity relative to that of the NOESY experiment for small molecules in nonviscous solutions. If we are given the choice between running the NOESY experiment and running the ROESY experiment, we should elect to run the NOESY experiment.

In the 2-D NOESY experiment, if the diagonal peaks are phased positively (and fully absorptive), NOE cross peaks for small molecules will be negative. Cross peaks arising from chemical exchange will, on the other hand, be positive, thus allowing the differentiation between the two. For larger molecules (e.g., proteins), the sign of the diagonal and all cross peaks will be the same. If chemical exchange is suspected as the cause of one or more of the cross peaks in the spectrum of a large molecule, we can collect a ROESY spectrum to identify those cross peaks that are due to chemical exchange (see Section 7.7.2).

For NOESY experiments with longer mixing times, spin diffusion can generate cross peaks between one spin and another spin more than 0.5 nm distant through an intermediate spin. Spin diffusion can be identified by collecting a series of 2-D NOESY spectra and examining the volume integrals of the cross peaks in question. By examining the NOE buildup, we can distinguish between the cross peaks from direct dipolar interaction (the NOE) and the cross peaks stemming from spin diffusion because the latter will build up more slowly as a function of increasing mixing time.

A typical 2-D NOESY spectrum collected from a 20 mM sample in a 5 mm inverse probe at 500 MHz will take one to two hours.

7.7.2 **ROESY**

We may recall that the TOCSY experiment (discussed in Chapter 6) also makes use of a spin lock for mixing. The main difference between the TOCSY spin lock and the ROESY spin lock lies in the frequency of the rotating frame used for the spin lock. For the TOCSY experiment, the frequency of the spin lock is typically placed in the middle of the spectral window (normally at the frequency of the transmitter), whereas for the ROESY experiment the spin lock frequency is placed far away from the spectral window. In most cases, the location of the spin lock frequency well away from the center of the spectral window will suppress the appearance of TOCSY cross peaks in the ROESY spectrum. The pulses used in the ROESY spin lock are typically about 90 μs, compared to about 30 μs for the TOCSY spin lock.

Just as the sign of NOESY cross peaks can provide information, so too can the sign of the ROESY cross peaks. ROESY cross peaks will be negative (relative to the phase of the diagonal) if they arise from direct dipolar interactions (the ROE), whereas those cross peaks stemming from spin diffusion (a three spin effect or a relayed ROE) will be positive. TOCSY cross peaks are also observed in many ROESY spectra, but the sign of these cross peaks is also positive—again allowing them to be readily distinguished from ROE cross peaks. Cross peaks arising from chemical exchange in the ROESY spectrum are also positive.

The range of molecular correlation times in the liquid state is not great enough to vary the sign of the cross peaks we observe in the ROESY spectrum because the time scale of the dipolar interactions is so much lower as a result of the use of the much weaker $\mathbf{B}_1$ field instead of $\mathbf{B}_0$.

When, in the ROESY experiment, the ROE and TOCSY (or ROE and relayed ROE, or ROE and chemical exchange) interactions both generate cross peak intensity at a particular location in the 2-D spectrum, we may observe that the resulting cross peak contains a mixed phase. In some cases, therefore, the volume integral of a cross peak may not serve as an accurate gauge of spin proximity. Careful analysis of a ROESY spectrum should always be accompanied by the TOCSY spectrum of the same molecule to allow the identification of cross peaks affected by the TOCSY interaction. The efficiency with which TOCSY cross peaks are generated in a ROESY spectrum can be lessened with careful placement of the frequency of the spin lock pulse train, but complete elimination of TOCSY-generated cross peaks is unlikely because of the relative strength of the TOCSY cross peak generating mechanism (J-coupling) versus the mechanism giving rise to the ROE (the dipolar interaction).

A reasonable ROESY spectrum on a 20 mM sample in a 5 mm inverse probe run at 500 MHz can be expected to take from two to four hours.

■ REFERENCE

[1] A. A. Bothner-By, R. L. Stephens, J.-M. Lee, C. D. Warren, R. W. Jeanloz, *J. Am. Chem. Soc.*, **106**, 811–813 (1984).

8

Molecular Dynamics

Molecular dynamics covering a wide range of time scales produce an array of effects in NMR spectroscopy. In large molecules, motion of different segments of a molecule may yield measurably distinct relaxation times, thus allowing us to differentiate between signals from different parts of a molecule. Conformational rearrangements can change the chemical shifts of NMR-active nuclei and the J-couplings observed between various spins. Rapid molecular motions average shifts and/or J-couplings, whereas slower motions may make discovering the underlying mechanistic motions difficult. In many cases, molecular motion and chemical exchange may give broad NMR lines devoid of coupling information.

Fortunately, most modern NMR spectrometers include variable temperature (VT) controlling equipment that allows the sample temperature to cover a wide range. Varying the sample temperature may allow us to observe signals that would be poorly suited to supplying desired information at ambient temperature.

Probes containing pulsed field gradient (PFG) coils, however, can often only tolerate a more limited range of temperatures compared to their PFG-coil-lacking counterparts; this reduced operating temperature range is attributable to the limitations associated with the materials used to construct these technologically sophisticated probes and the need to minimize thermal stress. Typical temperature ranges for a normal liquids NMR probe are from about $-100°C$ to $+120°C$, and PFG probes may only tolerate temperatures in the range of $-20°C$ to $+80°C$. Individual vendors list the temperature range recommended for each of their probes.

For the purposes of structural elucidation and resonance assignment, a cursory understanding of molecular dynamics and relaxation is

useful, but often not essential. Recognizing when a particular resonance is broadened as a result of exchange and knowing what step
or steps we might take to compensate for or to minimize the adverse
effects of a dynamically broadened resonance are useful skills to
possess. The information presented in this chapter will help us
develop these skills.

8.1 **RELAXATION**

Relaxation is the process by which a perturbed spin system returns
to equilibrium. In NMR spectroscopy, there are three principal measures of the relaxation rate observed for a given set of spins: T_1, T_2,
and $T_{1\rho}$.

T_1 relaxation is also called spin-lattice relaxation. It involves the
exchange of photons between the spins in question and the lattice
(the rest of the world). T_1 relaxation returns the net magnetization
vector to its equilibrium position along the +z-axis of the laboratory and also that of the rotating frame (recall that the two frames of
reference share the same z-axis).

T_2 relaxation is also called spin-spin relaxation. It involves the
exchange of photons between the spins in question and other nearby
spins. The T_2 relaxation mechanism is the means by which the component of the net magnetization vector in the **xy** plane decays to zero
(its equilibrium value).

$T_{1\rho}$ relaxation. The diminution of
the net magnetization vector in the
rotating frame of reference as the net
magnetization vector is subjected to
a **B_1** spin lock.

$T_{1\rho}$ relaxation involves the diminution of the net magnetization vector
in the rotating frame of reference as the net magnetization vector is
subjected to a **B_1** spin lock. Measurement of the $T_{1\rho}$ relaxation time
is accomplished by first tipping the net magnetization vector into the
xy plane with a 90° (or other) pulse, and then shifting the phase of
the applied RF so that the magnetic field component of the RF acts
as the magnetic field about which the net magnetization is forced to
precess *in the rotating frame*. Because the length of the net magnetization vector immediately following the initial 90° pulse is much larger
(due to **B_0**) than the net magnetization's equilibrium value in the
spin-locking condition (the **B_1** field is perhaps 20,000 times weaker
than the **B_0** field), the length of the net magnetization vector will
decay. This decay can be measured with an appropriately designed
NMR pulse sequence.

The T_1 and $T_{1\rho}$ relaxation rates will reach minimum values at a given
correlation time, τ_c (the minima will occur for two different τ_c's). The

T_2 relaxation rate, however, will continue to get shorter and shorter as τ_c increases.

In practice, relaxation times are rarely used to elucidate the structure of smaller molecules. Relaxation studies involving macromolecules (polymers) and other large molecules, however, are well known to yield important structural information.

8.2 RAPID CHEMICAL EXCHANGE

Rapid chemical exchange is often observed in ^1H spectra when our sample contains labile protons. Labile protons are most often those found on heteroatoms in hydroxyl, carboxyl, and amino groups. In special cases, other ^1H's may be observed to undergo rapid chemical exchange if there is a combination of several conditions that each contribute toward making a particular ^1H especially labile, e.g., if the ^1H is alpha to several carbonyls or if there is a strong propensity for the molecule to tautomerize.

Rapid chemical exchange. A chemical exchange process that occurs so rapidly that two or more resonances coalesce into a single resonance.

Rapid chemical exchange means that the exchange takes place on a time scale faster than any that can be resolved by using the instrument. As an aside, the time scale that can be observed with an NMR spectrometer is referred to as the NMR time scale; in fact, the NMR time scale may vary over many orders of magnitude, with the specific time scale depending on what experiment is being conducted.

NMR time scale. The time scale of dynamic processes that can be observed with an NMR spectrometer.

In the case of a simple multisite exchange of protons, the exchange can be said to be rapid if only one ^1H resonance is observed and if this resonance is a singlet and relatively narrow peak devoid of fine structure from J-coupling. The location along the chemical shift axis of the observed ^1H resonance from a proton exchanging between two or more sites is the average of the chemical shifts weighted by their relative populations. If a proton jumps from one site to another more rapidly than the time frame needed to observe the splitting of its resonance by J-coupling to another spin, then this proton will generate a resonance devoid of splitting.

8.3 SLOW CHEMICAL EXCHANGE

Slow chemical exchange can be more difficult to observe by NMR. For example, a ^1H may slowly exchange over time with deuterons in the solvent. Immediately after the solute is dissolved in the solvent, it may be possible to observe a resonance due to this slowly exchanging site, but over time, this resonance may disappear and be replaced by the shift of the ^{1}H on the solvent.

Slowly exchanging NMR-active nuclei or groups will still show what is considered normal behavior—they will show J-couplings and their chemical shifts will not be averaged—but over time these resonances may disappear or "exchange away" as a result of exchange with solvent or other chemical species present in solution.

8.4 INTERMEDIATE CHEMICAL EXCHANGE

Intermediate chemical exchange is the most difficult type of exchange to recognize because it often goes completely unnoticed. Intermediate exchange typically involves the extreme broadening of the resonance in question. In many cases, the broad peak may not be recognized for what it is, especially if automated baseline correction procedures are used to process the spectrum.

If there is the potential for chemical exchange, we should examine the frequency spectrum before we apply baseline correction. Increasing the vertical scale (how big the biggest peaks in the spectrum are relative to the maximum peak height that can be accommodated in the computer display) by several orders of magnitude can often reveal the presence of a broad peak.

When we prepare samples, we can take steps to minimize the extreme broadening of resonances susceptible to exchange broadening. We can use new and/or freshly distilled solvents (deuterated chloroform gets acidic after sitting on the shelf for six months), and we can also ensure that the pH of the sample is correct. When we observe the ^{1}H resonances of proteins and polypeptides in aqueous media, the rate of exchange of the labile backbone amide protons will be modulated by the pH of the solution. Typically, the optimal pH for minimizing this exchange is 4–5.

Intermediate chemical exchange is often readily amenable to study by variable temperature NMR, because the rate of exchange can be modulated by several factors of two by changing the temperature by tens of degrees Celsius. The rule of thumb taught in beginning chemistry courses that changing the temperature by ten degrees Celsius will halve or double the rate of a reaction (including exchange) shows that, in the case of intermediate exchange, there is often a readily accessible range of temperatures that should allow the elucidation of which resonances participate. Functionalized cyclohexane rings interconverting between the two chair conformations provide some of the best examples of intermediate-exchange-induced resonance broadening, but many other examples exist.

Whether two exchanging positions will show one or two NMR resonances (or something in between) is a function of the difference (in hertz) of their two chemical shifts. Because the hertz separation between two chemically distinct sites is a function of field strength (the shift difference in ppm is constant, but running the sample in a higher field strength instrument will result in a greater separation of chemical shifts when measured in hertz), the point at which two resonances merge and become one—the coalescence point—will occur at lower temperatures on higher frequency NMR instruments. If we wish to study chemical or conformational exchange by NMR and we have access to multiple NMR instruments (each with a different operating frequency), we can avoid excessive heating or cooling of our sample by choosing the optimal NMR frequency.

Mathematical fitting of observed line shapes can be used to extract the activation energy, E_a, for dynamic exchange processes by using an Arrhenius plot wherein the slope of the log K (K is the rate of exchange) versus inverse absolute temperature is proportional to activation barrier.

If we wish to assign the resonances to the atomic sites of a molecule, the indication that exchange is complicating our spectra is normally not welcome. Carrying out NMR studies at higher frequencies or at lower temperatures are two ways in which exchange broadening can be reduced.

It is important to understand that other phenomena may also introduce resonance broadening, such as a long molecular correlation time. Slow molecular tumbling (a long τ_c) makes the T_2 relaxation time short, so the net magnetization in the **xy** plane will decay very quickly, thus making it impossible to determine the frequency of the signal accurately (the resonances we observe in this case will be very broad). The remedy (increasing the temperature, thereby decreasing the line width) for a viscosity-broadened or similarly correlation-time-affected NMR resonance is the opposite of what to do to resolve multiple exchange-broadened resonances. It is important to keep this in mind when we examine our NMR data and are making decisions as to which experiment we should next carry out and/or how we should adjust our experimental parameters.

Coalescence point. The moment in time or the temperature at which two resonances merge to become one resonance. Mathematically, coalescence occurs when the curvature of the middle of the observed spectral feature changes sign from positive to negative.

Activation energy, E_a. The energy barrier that must be overcome to initiate a chemical process.

8.5 **TWO-DIMENSIONAL EXPERIMENTS THAT SHOW EXCHANGE**

Several NMR experiments can indicate the presence of chemical or conformational exchange. In some experiments, exchange produces cross peaks that are viewed as an annoyance. In other cases, the experiment may be carried out for the purpose of demonstrating the presence of exchange.

The TOCSY experiment can show cross peaks that arise from chemical exchange, usually between a protic solvent signal and a molecular site that has labile protons. In molecules with molecular weights over 1 kDa, the exchange-generated cross peaks in a TOCSY spectrum will be observed to have a sign opposite that of the cross peaks arising from J-couplings. Typically, TOCSY experiments are not used to explore chemical exchange; thus, the presence of signal from exchange is viewed as a complication rather than a beneficial result.

Carrying out the NOESY experiment for the express purpose of detecting exchange is termed the EXSY (for <u>ex</u>change <u>s</u>pectroscop<u>y</u>) experiment [1]. The EXSY experiment will show cross peaks between two resonances that undergo exchange during the mixing time of the experiment. When the rates of the forward and reverse reactions are not the same (i.e., if the system is not at equilibrium), the intensity of the two cross peaks will be unequal. The differential of the volume integrals of the two observed cross peak intensities will depend on the relaxation rates of the spins in the two sites and also on the rates of the forward and reverse reactions. For an irreversible reaction (where δ_r is the chemical shift of the reactant, and δ_p is the chemical shift of the product), the $(f_1 = \delta_r, f_2 = \delta_p)$ cross peak will be the only cross peak observed. The $(f_1 = \delta_p, f_2 = \delta_r)$ cross peak will not be observed. To observe a cross peak, sufficient exchange (reaction conversion) must take place during the mixing time of the EXSY experiment, and the T_2 relaxation times of the reactant and product cannot be too much shorter than the exchange mixing time—otherwise, all the signal will disappear before it can be detected.

■ REFERENCE

[1] J. Jeener, B. H. Meier, P. Bachmann, R. R. Ernst, *J. Chem. Phys.*, **71**, 4546–4553 (1979).

Chapter 9

Strategies for Assigning Resonances to Atoms Within a Molecule

The assignment of resonances to specific atoms in molecules can vary in difficulty from trivial to confounding. Some molecules lend themselves to resonance assignment readily with the application of a few simple rules. For other molecules, however, we make a series of preliminary assumptions or tentative assignments and then check our 2-D cross peaks in the gHMBC and/or gCOSY to determine whether we have a consistent set of assignments or (and this is more likely) a number of questionable, implausible, or far-fetched assignments that seriously call into question the validity of our tentative assignments.

Resonances we assign with certainty are called entry points because they establish a beachhead or toehold by which we can progressively work across the molecule, accounting for all expected resonances.

Entry point. The initial pairing of a readily recognizable spectral feature to the portion of the molecule responsible for the feature.

Different NMR experiments and even different types of information found in the same NMR data set (1-D or 2-D) provide sometimes conflicting implications regarding assignments. Entry points are typically those resonance assignments that are beyond reproach, those in which we place complete confidence.

Delving only a little way into the assigning the resonances of a complex molecule (with many overlapping resonances) will often immediately reveal conflicts. As a general goal, we will work to develop our ability to rank the significance and trustworthiness of each piece of spectral information. In the evaluation of the myriad conflicting pieces of NMR evidence, the most basic truth is:

Trust the information found in the 1-D NMR spectrum first.

For example, the J-couplings, multiplicities, and integrals found in the 1-D ^{1}H spectrum are to be trusted more than the relative intensities of some cross peaks in the 2-D ^{1}H-^{1}H TOCSY spectrum.

9.1 **PREDICTION OF CHEMICAL SHIFTS**

Chemical shifts are one of the most useful indicators we have of chemical environment. Inductive effects from atoms one or two bonds distant can often be readily recognized and put to good use. The additive nature of these inductive effects is also extant, thus allowing us to further refine our chemical shift intuition. Not only do inductive effects play a significant role in affecting chemical shifts, but conjugation, shielding, and through-space proximity may as well.

Consultation of tables containing chemical shifts of ^1H and ^{13}C atoms based on their chemical environment is something we do a lot of initially. However, as our assignment skills develop and mature we find that this practice is required less often. Many software packages that are commercially available at the time of this writing are able to predict ^1H and ^{13}C chemical shifts on the basis of a user-supplied chemical structure. However, these software packages are of only limited utility once we encounter greater molecular complexity.

An important caveat is that chemical shifts can often lead to incorrect assignments of resonances. Chemical shifts are influenced by many factors; e.g., chemical shifts reflect not only the electronegativity of nearby atoms but also bond hybridization as manifested through constraints imposed by molecular geometry, and proximity to aromatic and other electron-rich systems.

In carrying out the assignment of observed resonances to atoms in a molecule of known structure, we must balance the urge to use chemical shift arguments with a healthy skepticism of the many ways in which chemical shifts may be influenced by less-than-obvious factors. That is, avoid whenever possible using small differences in chemical shifts to make resonance assignments.

With that said, it should also be stated that chemical shifts are the single most accessible and readily useful aspect of the spectrum of a typical organic molecule. Identification of entry points is often done by using simple chemical shift arguments; and little if any corroborating information is expected, given a unique and well-isolated chemical shift.

For example, the ^1H resonance of a carboxylic acid proton or an aldehyde proton is typically in the range of 9–10 ppm, far downfield and well-separated from the other resonances in the ^1H spectrum. In the ^{13}C chemical shift range, carbonyls are similarly found well downfield (at 160–250 ppm) of the other ^{13}C resonances in the spectra of most organic compounds.

In many cases, the combination of chemical shift information with other data such as resonance integral/intensity or multiplicity will provide the means of identifying key resonances in a molecule.

9.2 PREDICTION OF INTEGRALS AND INTENSITIES

Prediction of the ratios we will observe in comparing ^{1}H integrals and ^{13}C intensities is easy. We simply count up the number of ^{1}H's on a given atom in the molecule and that is the normalized integral value we should expect if we take care to ensure that our ^{1}H signal is allowed to fully relax between successive scans. If two ^{1}H's of a methylene group are diastereotopic and are near a chiral center or occupy different environments as the axial and equatorial ^{1}H's do in a cyclohexane ring in the chair conformation, what we may have initially thought would be one resonance that would integrate to two ^{1}H's may in fact be observed as two resonances that integrate to one ^{1}H each. ^{1}H's on heteroatoms (mainly nitrogen and oxygen) will often appear broader. The observed integrals from the resonances of these ^{1}H's will usually be lower than the expected values.

There are several possible reasons to account for why we observe the low integral values for ^{1}H's bound to heteroatoms despite having a sufficiently long relaxation delay between scans. First, relaxation (T_2) may occur to a greater extent for those ^{1}H's whose signals are broad as a result of the time delay between the read pulse and the start of the digitization of the FID. Because a broad resonance in the frequency domain corresponds to a rapidly decaying signal amplitude in the time domain, broad resonances will often generate low integrals. A second possible reason for a low integral value is that baseline correction of the spectrum may wipe out the edges of broad resonances, thus subtracting intensity from the peak. A third possible reason for a low integral value of a ^{1}H on a heteroatom may result from partial chemical exchange of these ^{1}H's with deuterons (^{2}H's) in the solvent, especially if the solvent is deuterated water or methanol.

9.3 PREDICTION OF ^{1}H MULTIPLETS

We can predict how a resonance from a single atomic site will be split by J-coupling into a multiplet. We do this by considering what other NMR-active spins are two and three bonds away from the atom in question. That is, we use ^{2}J's and ^{3}J's. In special cases, we may have a molecule in which we expect to observe a ^{4}J as a result of an alignment of bonds in a planar or nearly planar conformation that looks like a letter W. We can use the methodology in Chapter 6 to predict multiplets, and record these predictions by using the abbreviations s for singlet, d for doublet, t for triplet, q for quartet, d^2 for a doublet of doublets, d^3 for the doublet of doublets of doublets, d^4 for a doublet of doublets of doublets of doublets, dq for a double of quartets, dt for a doublet of triplets, dq for a doublet of quartets, and so on.

Recall that ^{1}H's with a low pK_a value (e.g., ^{1}H's on heteroatoms) often will not show multiplicities because chemical exchange occurs too rapidly to allow the relatively small J-coupling to be resolved during the digitization of the FID. We must take care to consider that geminal ^{1}H's (e.g., those on a methylene group) may be diastereotopic and thus may have different chemical shifts, thus allowing them to couple with each other to give each an additional, and typically very large, ^{2}J.

Once multiplicities have been predicted for each ^{1}H resonance, we examine our list to look for unique multiplicities. We may be able to identify some of our ^{1}H resonances simply on the basis of the observed couplings in the ^{1}H 1-D spectrum.

9.4 GOOD BOOKKEEPING PRACTICES

A good starting point when we are given a molecule to assign is to tabulate all the ^{1}H and ^{13}C resonances we expect to see. We start with a drawing of our molecule using bond-line notation, with each atom except for the hydrogens assigned a number (it is okay to leave out the numbering on certain atoms that will not be appear in the ^{1}H or ^{13}C spectra). We should try to follow the IUPAC numbering scheme—the *CRC Handbook of Chemistry and Physics* has a good deal of information on this methodology, and *The Merck Index* has the correct (i.e., previously agreed upon by others) numbering written out explicitly for many molecules. We will also want or need to differentiate between diastereotopic ^{1}H's.

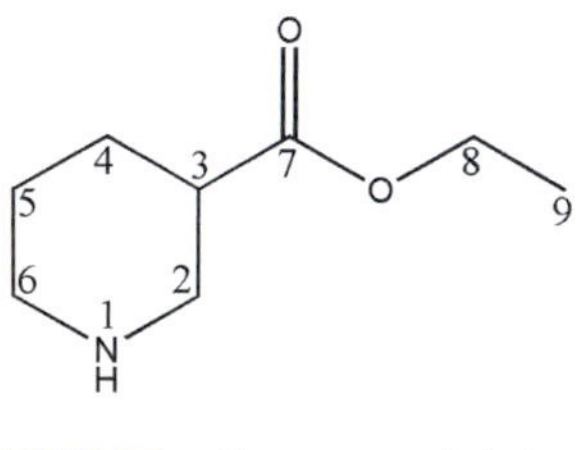

In general, it is a good practice to assume that a six-membered ring adopts a chair conformation; if this is the case, we will want to differentiate between axial and equatorial ^{1}H's. We build a model if we can—this model helps clarify the picture of the molecule we develop.

Consider the molecule ethyl nipecotate (Figure 9.1). After we draw the molecule and number the atoms whose resonances we will assign, we can make a table with seven columns for the ^{1}H NMR data and another table with five columns for the ^{13}C NMR data (Tables 9.1 and 9.2).

■ FIGURE 9.1 The structure of ethyl nipecotate, including the numbering of the relevant atoms for the assignment of the ^{1}H and ^{13}C NMR spectra.

Table 9.1 Format for table to contain predicted and observed ^1H NMR shifts (δ), integrals (int), and multiplicities (mult) for ethyl nipecotate. pred'd = predicted, obs'd = observed, d = doublet, q = quartet.

#	^1H δ (pred'd)	^1H δ (obs'd)	^1H int (pred'd)	^1H int (obs'd)	^1H mult (pred'd)	^1H mult (obs'd)
1	1–5		<1		Singlet (broad)	
2	2.4		2		$2 \times d^2$	
3	1.8		1		d^4	
4	1.5		2		$2 \times d^4$	
5	1.3		2		$2 \times d^5$	
6	2.2		2		$2 \times d^3$	
8	3.8		2		$2 \times q$	
9	1.1		3		t	

Table 9.2 Format for a table containing predicted and observed ^{13}C NMR shifts (δ) and intensities (int) for ethyl nipecotate. pred'd = predicted, obs'd = observed, s = strong, m = medium, w = weak.

#	^{13}C δ (pred'd)	^{13}C δ (obs'd)	^{13}C int (pred'd)	^{13}C int (obs'd)
2	48		s	
3	36		m	
4	30		s	
5	26		s	
6	39		s	
7	170		w	
8	57		s	
9	18		s	

For each column of predictions, we fill in as many guesses as we can. Making initial guesses is a good way to improve our predictive skills; later we can compare the correct answers with our predictions to see where we went wrong. Without looking at any spectra or consulting any tables, I have filled in as many of the boxes as I can in columns 2, 4, and 6 of Table 9.1 and columns 2 and 4 of Table 9.2.

9.5 ASSIGNING ¹H RESONANCES ON THE BASIS OF CHEMICAL SHIFTS

From simple chemical shift arguments, we can hope to readily identify the ¹H's on carbons 2, 6, and 8 because these ¹H's are on carbons bound to heteroatoms. To save time and space, we write the aforementioned ¹H's as H2's, H6's, and H8's. Because oxygen is more electronegative than nitrogen, we expect to find the H8's farther downfield than the H2's or H6's. We also expect to find the H2's slightly farther downfield relative to the H6's because the H2's are also adjacent to a methine group (position 3) rather than a methylene group (position 5).

The most important reason for participating in the exercise of predicting chemical shifts lies not in getting the correct value, but in ascertaining the order in which we will encounter the shifts as we move from one end of the spectrum to the other. An alternative method for predicting shifts might simply be to start by identifying the ¹H resonances we expect to find at the extremes of the spectra (farthest downfield and farthest upfield). We can then typically use the resonances found at the extremes of the spectral window as the entry points for subsequent assignment of the other resonances.

The order of the ¹H chemical shifts (from left to right, from greatest chemical shift to smallest) can be written as

$$H8 > H2 > H6 > H3 > H4 > H5 > H9$$

H1's resonance can be almost anywhere in the spectrum because H1 is on the nitrogen atom; the chemical shift of the resonance of a ¹H bound to a heteroatom defies accurate prediction. Hydrogen bonding may prevent the electronegative heteroatom from withdrawing electron density.

We should not rely exclusively on chemical shift arguments to distinguish between ¹H's in nearly the same chemical environment. In our consideration of ethyl nipecotate, the relative chemical shift ranking of H4 and H5 should be considered tentative; we must remain aware that our chemical shift predictions based on electronegativity alone should be viewed with a healthy amount of skepticism; we will use other methods to confirm or refute this tentative assignment. We need not agonize over the relative ranking of H4 and H5, because other unique spectral attributes will allow us to definitively identify the H4 and H5 resonances. We should bear in mind that ethyl nipecotate is anomalously ideal. In the real world, chemical shift arguments often lead to incorrect assignments.

9.6 ASSIGNING ¹H RESONANCES ON THE BASIS OF MULTIPLICITIES

Unique multiplicities also offer an excellent means of establishing a starting (entry) point from which to work around the molecule. The ethyl group attached to the oxygen (positions 8 and 9) provides us with distinctive multiplicities and ¹H integrals. As long as the molecule's chiral center is sufficiently far away (through space) from the methylene group (position 8), the H8 resonance will integrate to two ¹H's and display a diagnostic quartet multiplicity. Besides lying farthest upfield because of the electron donating character of the methyl group relative to that of the methylene and methine groups, the H9's will integrate to three ¹H's and will appear as a triplet.

Examination of the 1-D ¹H spectrum of ethyl nipecotate (Figure 9.2) allows us to immediately identify the resonances from the ethyl group (positions 8 and 9). The methyl group's resonance is observed at 1.04 ppm, integrates to three protons, and shows the 1:2:1 triplet splitting pattern clearly. The methylene group at position 8 produces the most downfield resonance at 3.92 ppm because of its proximity to the oxygen atom. The H8's integrate to two protons and show the 1:3:3:1 quartet splitting pattern.

Examination of the 1-D ¹H spectrum also allows us to identify the H1 resonance (the amino proton) because of its lack of fine structure (J-couplings) at 1.38 ppm. Again, this lack of fine structure is caused

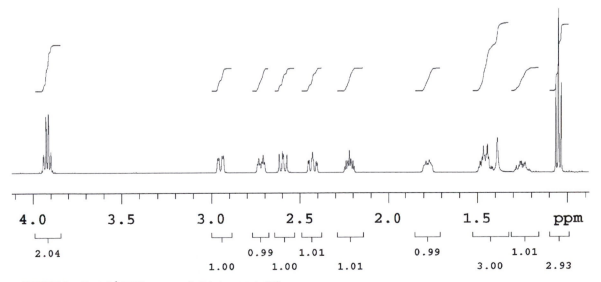

■ FIGURE 9.2   The 1-D ¹H NMR spectrum of ethyl nipecotate in CDCl₃.

by chemical exchange on a time scale too fast for observation of J-couplings during FID detection. As an aside: If a sample is so carefully prepared that all traces of acid or base are removed (including residual water and other protic impurities normally found in trace amounts), it may then be possible to observe fine structure (J-couplings) in ^{1}H's bound to heteroatoms such as nitrogen and oxygen.

Beyond the identification of the resonances from the ethyl group and the amino proton, the 1-D ^{1}H spectrum can appear daunting. Only through the examination of the multiplicity patterns of the resonances will we be able to make more progress in assigning the resonances, unless we resort to tedious matching of J-couplings or simplistic chemical shift arguments.

Rigorous analysis of multiplets is tedious, but often it can yield a great deal of information. Multiplet analysis also approaches the limit of the NMR interpretation skills of old-school chemists. Even though we will progress far beyond this level of sophistication, it is important that we understand this methodology because we may have to explain our assignments to an old-school chemist using this reasoning—even if we arrived at our assignments through the use of more modern methodologies. Multiplet analysis can also be used to corroborate (or refute) assignments made by other means.

We can analyze the multiplicity of each ^{1}H resonance to reveal how many ^{1}H's are two or three bonds distant. Put more simply: The splitting pattern of one ^{1}H shows how many other ^{1}H's are two and three bonds away. Let's start with the resonance at 2.96 ppm. We say that this resonance is a doublet of doublets (d^2) because it is composed of two pairs of partially overlapping peaks that together integrate to one ^{1}H. Because only the protons at position 2 are predicted to show the d^2 splitting pattern, we make the tentative assumption that the resonance at 2.96 ppm arises from the ^{1}H on C2 that is *gauche* to H3.

Recall that we initially assumed that our saturated six-membered ring adopts a chair conformation and that the bulky groups—in this case, the ethyl ester side chain—will be found in the equatorial and not the axial position. Thus, the H3 will be axial and its multiplet will show two large *trans* ^{3}J's to axial ^{1}H's at positions 2 and 4. H3 will also couple to the ^{1}H's that are equatorial at positions 2 and 4. In order for H3 to show a small coupling to one of the H2's, the dihedral angle must only be 60° (this is a *gauche* coupling), and therefore the H2 showing the small coupling must be equatorial. This concept is critical—build the model if this analysis is still unclear.

The axial proton at the 2 position is expected to be observed at about 2.56 ppm (0.4 ppm upfield from 2.96 ppm, the shift of the H2 equatorial proton, or $H2_{eq}$ for short). Because the number of protons two and three bonds distant for is the same for both $H2_{ax}$ and $H2_{eq}$, we expect to see a similar splitting pattern. But, because the axial H2 ($H2_{ax}$) will be *trans* (will have a 180° dihedral angle) with respect to $H3_{ax}$, we expect to observe a d^2 that resembles a triplet (this pattern is called a pseudotriplet and is written with the Greek letter psi preceding the letter t: Ψt). The resonance at 2.59 ppm fits this description exactly, and therefore this resonance must be $H2_{ax}$.

Note that $H2_{eq}$ is 0.37 ppm downfield from $H2_{ax}$ even though both are in the "same" bonding environment. This phenomenon is often seen whenever we compare the shifts of axial and equatorial methylene protons on six-membered rings in the chair conformation; because this 0.4 ppm shift offset is observed so often, it behooves us to commit this tidbit of information to memory for possible future use. The average of the two H2 shifts is 2.78 ppm, which is farther downfield than all protons except those on the methylene group of the ethoxy group (position 8).

The ¹H's at position 6 should be the next easiest to assign on the basis of their expected multiplicities. The H6's are expected to split each other through a geminal coupling and also be split by the two H5's through vicinal couplings. $H6_{ax}$ is expected to show two large couplings and a small coupling: a large ^2J, a large ^3J due to the 1,2 diaxial *trans* coupling to $H5_{ax}$, and a small ^3J due to the *gauche* coupling to $H5_{eq}$. In short, we are looking for two resonances with a d^3 splitting pattern. We expect the resonance from the equatorial ¹H on position 6 to show two small couplings (*gauche* ^3J's) and one large (*geminal* ^2J) coupling. Thus, $H6_{eq}$ will appear as two pseudotriplets side by side. The resonance at 2.72 ppm fits this description. The resonance from the axial ¹H on position 6 will, following the same line of logic, appear as a multiplet with two large couplings (*geminal* ^2J and *trans* ^3J) and one small (*gauche* ^3J) coupling to yield a pseudotriplet of small doublets. Because the *geminal* ^2J and the *trans* ^3J may differ slightly, the middle peak (or leg) of the pseudotriplet may receive intensity contributions that are slightly offset with respect to their frequencies, and thus the middle leg of the pseudotriplet may fail to overlap enough to give the expected 1:2:1 ratio of the heights of the legs of the multiplet. This spreading out of the middle leg of the pseudotriplet may instead generate extra lines in the multiplet. The fine structure of the resonance at 2.43 ppm shows that the *geminal* ^2J

Pseudotriplet, Ψt. A triplet-like splitting pattern caused by the identical coupling of the resonance of the observed spin to two other spins not related to each other by symmetry.

and the *trans* ^{3}J for H6$_{ax}$ are slightly different, thus making the center portion of the multiplet appear more like a rounded triplet than the doublets flanking it. Again, we see that the magnitudes of the ^{3}J's agree with the relative offset between the axial and equatorial ^{1}H's at position 6: The resonance from the equatorial proton (with its small ^{3}J) lies 0.29 ppm downfield from that of its axial counterpart.

The analysis of the remainder of the 1-D ^{1}H spectrum now becomes more difficult. We expect H3 to be a d^4 with two large (*trans* ^{3}J) couplings and two small (*gauche* ^{3}J) couplings to make an overall pattern a ΨtΨt (a pseudotriplet of pseudotriplets). The multiplet from H3 may end up looking like a 1:2:3:4:3:2:1 septet due to partial overlap of narrow triplets, or it may be even more complicated. Note that summing the leg intensity numbers for the septet above gives a total of 16, which is of course a power of 2. Put another way, because H3 is coupled to four ^{1}H's, it will be a d^4, and there should thus be 2^4 or 16 individual intensity contributions observed.

If we are unable to clearly discern 16 individual intensity contributions to the multiplet at 2.22 ppm, we are not alone. At some point, picking apart multiplets must be regarded as more of an art and less of a science. Once we are reduced to having to distinguish between resonances in a molecule that are all d^4, d^5, or higher predicted multiplicities, we can either resort to chemical shift arguments (this is a cop-out) or to the more sophisticated methodology discussed later in this chapter.

Aside: Many old-school chemists pride themselves on their ability to pick apart a multiplet to extract the coupling constant information contained therein. The advantage of this methodology is that every (homonuclear) coupling constant observed will appear twice in the spectrum (assuming no resonances are outside of the spectral window). Thus, we can piece together molecular connectivity by matching up particular J-couplings through analysis of the multiplets observed with the 1-D spectrum. The disadvantage is that this process is time consuming, fails when multiplicities become complex or when signals overlap, and is only truly needed when specific dihedral angles are required for detailed modeling.

9.7 ASSIGNING ^{1}H RESONANCES ON THE BASIS OF THE gCOSY SPECTRUM

The modern NMR instrument will have z-axis pulsed field gradient capabilities. This capability allows the collection of the absolute

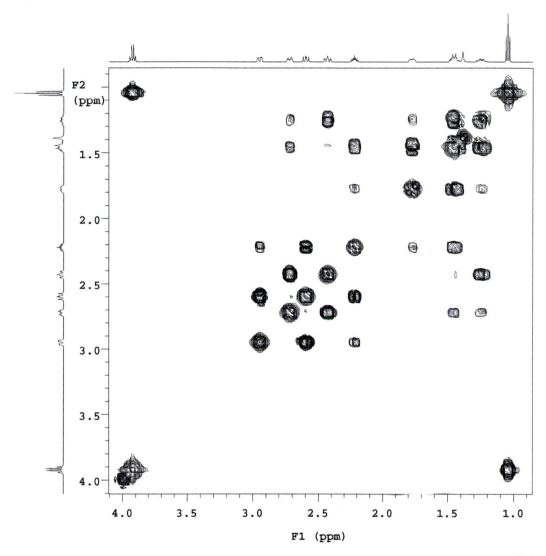

■ **FIGURE 9.3**   The 2-D ^1H-^1H gCOSY spectrum of ethyl nipecotate.

value ^1H-^1H 2-D gradient-selected COSY spectrum (gCOSY) in as little as 2–4 minutes given a reasonably concentrated sample. That is, the gCOSY spectrum can often be collected in less time than it takes to collect the 1-D ^{13}C spectrum! Collecting a gCOSY spectrum should be viewed as entirely normal and routine unless we are studying molecules that are so simple or so unusual that the information gained through the gCOSY is inconsequential.

Figure 9.3 shows the gCOSY spectrum of ethyl nipecotate. The gCOSY spectrum contains the 1-D spectrum along its diagonal and a number of off-diagonal peaks (cross peaks). The gCOSY cross peaks appear when one resonance is J-coupled to another. The larger the J-coupling, the larger the integrated volume of the cross peak. An important caveat lies in the last sentence: Broad peaks may appear to generate weaker cross peaks because they are spread out. We must account for the size of the footprint of a given cross peak when we are making an argument regarding cross peak intensity. This issue will come up later.

To gain a level of comfort and familiarity with the information content of the gCOSY spectrum, we begin by examining the ^{1}H resonances of ethyl nipecotate already assigned. Gratifyingly, the ethoxy group shows strong off-diagonal cross peaks between the methylene resonance (H8's) at 3.92 ppm and the methyl resonance (H9's) at 1.05 ppm. Another pair of resonances we know to share a common J-coupling are those from the geminal ^{1}H's at position 2 of the six-membered ring. The *geminal* ^{2}J coupling between the H2$_{ax}$ and H2$_{eq}$ resonances at 2.59 ppm and 2.96 ppm (respectively) generate the pair of cross peaks at $(f_1=2.59\,ppm, f_2=2.96\,ppm)$ and $(f_1=2.96\,ppm, f_2=2.59\,ppm)$.

Although gCOSY spectra are often symmetrized to improve their appearance (as has been done for the gCOSY spectrum in Figure 9.3), this mathematical operation can introduce spurious cross peaks and therefore should be applied with caution. This distortion occurs when two intense resonances generate what are called t_1 ridges (they should more properly be referred to as f_1 ridges), which are lines of noise that will, for the value(s) of f_2 corresponding to the intense resonances, give a ridge of noise that will cover the entire range of the f_1 spectral window. Because symmetrization will only preserve intensity if it is symmetrically distributed with respect to the diagonal, the presence of two t_1 ridges will give rise to two false cross peaks for the values of $(f_1 = x, f_2 = y)$ and $(f_1 = y, f_2 = x)$ if the two resonances with the t_1 ridges occur at $f_2 = x$ and $f_2 = y$. Prior to performing symmetrization of a 2-D data set, we must examine the unsymmetrized data set for the presence of multiple t_1 ridges. Symmetrization can still be performed if more than one t_1 ridge is present, but great care must be taken to avoid mistaking t_1-ridge-induced cross peaks for actual (J-coupling-induced) cross peaks.

Only homonuclear 2-D spectra can be symmetrized. Furthermore, there is the requirement that the data matrix to be symmetrized has to have the same number of rows and columns of data points (the matrix must be square). If we are going to measure J-couplings by using a DQF-COSY spectrum, we will typically transform the data set as a $1k \times 16k$ matrix to improve the digital resolution along the f_2 dimension, as this increase in the size of the matrix along f_2 reduces the uncertainty from lack of precision. Thus, the dimensions of the $1k \times 16k$ data matrix will not allow us to perform the symmetrization operation.

As expected, the H2 resonances at 2.59 ppm ($H2_{ax}$) and 2.96 ppm ($H2_{eq}$) also show cross peaks to the H3 resonance at 2.22 ppm. Notice that the ($H2_{ax}$, H3) cross peak is more intense than the ($H2_{eq}$, H3) cross peak, as expected because larger J's generate more intense cross peaks: The ^3J between $H2_{ax}$ and H3 (which is axial) is a *trans* coupling whereas the ^3J between $H2_{eq}$ and H3 is a *gauche* coupling. The same intensity differences are observed below the diagonal for the cross peaks arising from the (H3, $H2_{ax}$) and (H3, $H2_{eq}$) ^3J's.

9.8 THE BEST WAY TO READ A gCOSY SPECTRUM

Plots containing a 2-D spectrum normally also include the appropriate 1-D spectrum along the top and to the left of the 2-D spectrum. Whenever possible, the 1-D spectrum is used in lieu of the actual projection of the 2-D spectrum because the 1-D spectrum's digital resolution is smaller (the number of points per unit frequency is greater). We start on the left side of the plot and first consider the 1-D ^1H spectrum that serves as a projection of the 2-D spectrum. On this 1-D spectrum, we locate a resonance of interest, for example the $H2_{eq}$ resonance at 2.96 ppm, and then move horizontally (parallel to the f_1 frequency axis) until we encounter the peak on the diagonal (the line that connects the lower left corner to the upper right corner of a homonuclear 2-D spectrum). From this diagonal peak, we can then move either horizontally or vertically to see what other resonances show cross peaks with the resonance in question. Upon encountering a cross peak when moving vertically, we then move to the left to the 1-D projection to determine what other resonance participates in generating the cross peak just encountered. By analogy, when moving horizontally off of the diagonal and encountering a cross peak, move vertically to the 1-D projection at the top of the plot to determine what resonance is coupled to the resonance from which we originally departed horizontally (on the diagonal). If our 2-D spectrum is not symmetrized, we will typically find that the resolution in the f_2 dimension is better than that in the f_1 dimension. In cases where resonance overlap is present, we may wish to limit our search for cross peaks to horizontal movement from the diagonal (assuming f_2 is the vertical axis—Bruker NMR data is typically plotted with f_2 as the horizontal axis so when examining a 2-D spectrum collected using a Bruker instrument we will want to move vertically from the diagonal to find cross peaks).

Now that we have used the gCOSY spectrum to identify H3, we can continue around the ring to identify the ^{1}H resonances at position 2. Starting on the diagonal at the point where $f_1 = 2.22$ ppm and $f_2 = 2.22$ ppm, we move upwards and see that H3 shares two cross peaks with two additional resonances. These two cross peaks must correspond to the H4's; as expected, the H4$_{eq}$ shows a weaker cross peak than does H4$_{ax}$ to the lone H3 (which is axial). Again, recall that we expect the downfield H4 to be equatorial and the upfield H4 to be axial. Looking to the left from the first cross peak encountered as we move up from the H3 diagonal peak, we see that the resonance corresponding to H4$_{eq}$ has a chemical shift of 1.78 ppm. If we look to the left from the second cross peak above the H3 position on the diagonal, we arrive at the 1-D projection in the vicinity of 1.41–1.49 ppm. If we consult the integrals on the 1-D ^{1}H spectrum, we find that this region integrates to two protons, thus indicating that not only does H4$_{ax}$ resonate at this position, but a second ^{1}H does as well. If we return to the cross peak on the gCOSY spectrum between H3 and H4$_{ax}$, we can see another cross peak to the left whose center is slightly lower (at a higher ppm value) than the center of the cross peak between H3 and H4$_{ax}$. Thus, we can differentiate between the two resonances in the 1.41–1.49 ppm range. The center of the (H3, H4$_{ax}$) cross peak shows us that H4$_{ax}$ is centered at 1.46 ppm, while the center of the as-yet-unassigned resonance overlapping with H4$_{ax}$ is at 1.47 ppm.

We can summarize what we have just discovered: the 2-D gCOSY spectrum allows us to determine more precisely the chemical shifts of resonances that overlap in the 1-D ^{1}H NMR spectrum.

Because of the overlap problems with H4, we will attempt to get at the H5 resonances from position 6. Recall that, on the basis of chemical shifts and multiplicities, we assigned the two resonances at 2.43 and 2.72 ppm as those corresponding to H6$_{ax}$ and H6$_{eq}$, respectively. Encouragingly, we see a strong cross peak in the gCOSY spectrum between resonances at 2.43 and 2.72 ppm; this cross peak's strength is consistent with the large ^{2}J coupling we expect between *geminal* ^{1}H's.

If we now start at 2.43 ppm on the diagonal, we can move upward to find the two cross peaks that correspond to H5$_{eq}$ and H5$_{ax}$. Given our expectation that H5$_{eq}$ will resonate downfield from H5$_{ax}$, we expect to encounter the cross peak with the chemical shift of the resonance of H5$_{eq}$ first; we expect this cross peak to be weaker than the cross peak between the H5$_{ax}$ and H4$_{ax}$ resonances. In fact, the cross peak between the H5$_{eq}$ and H4$_{ax}$ resonances is almost unobserved in Figure 9.3. In contrast, the cross peak between the H5$_{ax}$ and H4$_{ax}$ resonances is strong, as expected.

Moving up from the position of the H4$_{eq}$ resonance on the diagonal at 1.78 ppm, we encounter two cross peaks of roughly equal intensity, indicating that the two *gauche* ^3J's from H4$_{eq}$ to H5$_{ax}$ and H5$_{eq}$ are nearly the same. Tracing the center of the cross peaks to the 1-D spectrum on the left side of the plot, we see that H5$_{eq}$ resonates at 1.47 ppm and H5$_{ax}$ resonates at 1.25 ppm. Thus, H5$_{eq}$ is responsible for the resonance that overlaps with that of H4$_{ax}$.

We have now assigned all of the ^{1}H resonances of ethyl nipecotate, and we leave as an exercise the identification of the cross peaks between the resonances from the H4's and H5's.

9.9 ASSIGNING ^{13}C RESONANCES ON THE BASIS OF CHEMICAL SHIFTS

In the ^{13}C 1-D spectrum, we can use chemical shifts arguments based on the electronegativity of nearby atoms to identify the carbonyl ^{13}C resonance (carbon 7, or C7 for short). That is, because C7 is doubly bonded to an oxygen atom, we expect the chemical shift of its resonance to be the most downfield (highest ppm value or δ) of all the ^{13}C resonances we observe from ethyl nipecotate. Moving from left to right in the ^{13}C 1-D spectrum, we then expect to encounter C8, then C2, and then C6. Because oxygen is more electronegative than nitrogen, the ^{13}C next to the oxygen (C8, the methylene carbon of the ethyl group) is expected to lie farther downfield than C2 or C6. Because C2 is also adjacent to the methine carbon C3 (which is slightly more electron-density starved than C6 because C6 is adjacent to a less electron-density-starved methylene group, C5), we expect C2 to lie farther downfield from but nonetheless very close to C6.

Just as we did for the ^{1}H resonances, the order of the ^{13}C chemical shifts (left to right, highest ppm to lowest) can be written

$$C7 > C8 > C2 > C6 > C3 > C4 > C5 > C9$$

As we did for the ^{1}H resonances, we can make a table (Table 9.2) listing the shifts and intensities we predict for the ^{13}C resonances. We could, of course, consult published tables.

Prediction of ^{13}C resonance intensities is a useful exercise. Nonprotonated carbons tend to relax slowly because they do not have the strong magnetic dipole moment of a nearby proton to provide them with an efficient spin-lattice (T_1) relaxation mechanism. Under normal experimental conditions for the collection of 1-D ^{13}C NMR spectra, the relaxation delay between scans will be less than the T_1 relaxation time constant of the nonprotonated ^{13}C's in a molecule. Because the nonprotonated ^{13}C's lack sufficient time to relax before the next scan takes place, their resonances tend to be less intense than the resonances from protonated ^{13}C's. By the same reasoning, methine ^{13}C's tend to relax less efficiently than do methylene and methyl ^{13}C's, thus methine ^{13}C resonances sometimes exhibit intermediate intensity. In cases where a *tert*-butyl group is present, the resonances from the methyl ^{13}C's of this group will be very intense, because more than one carbon site of the molecule contributes to the methyl ^{13}C resonance (the methyl carbon atoms of the *tert*-butyl group are homotopic). Isopropyl groups have prochiral methyl groups; but, in the absence of a chiral center in the molecule (or a chiral solvent), or perhaps just by coincidence, these methyl ^{13}C's may produce a doubly strong signal.

Now we are ready to examine the actual 1-D ^{13}C spectrum of ethyl nipecotate (Figure 9.4). On the basis of its downfield position and

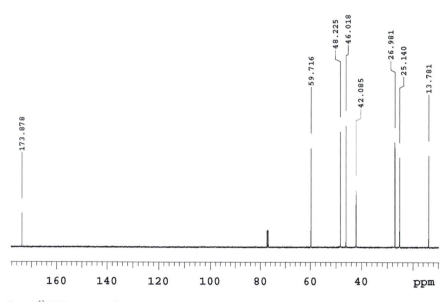

■ **FIGURE 9.4**   The 1-D ^{13}C NMR spectrum of ethyl nipecotate in CDCl$_3$.

its low relative intensity, we assign the resonance at 173.9 ppm to the carbonyl ^{13}C (C7). (We limit our reporting of ^{13}C chemical shifts to ±0.1 ppm unless particular resonances are separated by less than 0.1 ppm.)

The methylene adjacent to the oxygen atom (C8) is likely responsible for the resonance at 59.7 ppm, and the two resonances at 48.2 and 46.0 ppm are likely from C2 and C6, respectively. (Note: chemical shift arguments that differentiate between resonances that are this close should be viewed with a great deal of skepticism; we confirm this type of assignment by other means, if possible.) The less intense resonance at 42.1 ppm is expected to be from the only methine carbon C3. (Recall that the methine resonance was expected to possibly be less intense than the methylene and methyl ^{13}C resonances.)

Looking all the way upfield (to the right), we see that the resonance at 13.8 ppm can be assigned to the methyl carbon (C9). Notice that the methyl resonance is not the most intense resonance in the ^{13}C spectrum, indicating that ^{13}C's having additional bound protons do not always generate the most intense resonances.

If pressed on the subject, we would argue that the last two resonances to be assigned at 27.0 and 25.1 ppm should be attributed to C4 and C5, respectively. Again, the small difference between the two resonances calls strongly for the use of other methods to confirm this tentative assignment.

9.10 PAIRING ¹H AND ¹³C SHIFTS BY USING THE HSQC/HMQC SPECTRUM

A method exists to (in most cases) unambiguously pair protonated ^{13}C resonances to ^{1}H resonances through the $^{1}J_{CH}$ coupling. The heteronuclear single quantum correlation (HSQC) experiment generates a 2-D spectrum with the ^{13}C chemical shift scale on one axis (normally the horizontal axis on a Varian instrument) and the ^{1}H chemical shift scale on the other (normally vertical) axis. Cross peaks appear when there is a ^{1}J of 125–155 Hz between a ^{13}C and a ^{1}H in the molecule. Although 98.9% of the proton signal must be discarded through a process called phase cycling (because 98.9% of the carbon atoms at any molecular site are ^{12}C's), it is still more efficient to detect the ^{1}H signal and subtract the large $^{1}H\text{-}^{12}C$ signal from the overall signal to leave only the $^{1}H\text{-}^{13}C$ signal. Collecting heteronuclear correlation (HETCOR) information through direct detection of the ^{13}C signal is almost never required, even if we are using a conventionally configured NMR probe (with X coil closer to sample and the ^{1}H coil on the outside). Many old-school chemists persist in using the HETCOR experiment, when what they should more properly be using is the HSQC or HMQC experiment, which requires far less instrument time to generate the same quality of data. One conceivable instance in which the HETCOR (direct ^{13}C detection) experiment would be preferable to the HSQC experiment occurs when two protonated ^{13}C's generate resonances so close to each other on the chemical shift axis that collecting a larger number of ^{13}C data points to resolve the slight chemical shift difference between the two ^{13}C resonances is afforded more readily by extending the number of points in the FID of the HETCOR, rather than by increasing the number of HSQC FIDs collected. That is, using the HETCOR experiment, ^{13}C resolution (along the f_2 frequency axis) may be improved more readily by collecting more t_2 data points instead of using the HSQC experiment and collecting more t_1 points to provide better f_1 resolution in the indirect (^{13}C) dimension.

The HSQC spectrum shows cross peaks between the resonances of ^{1}H's and the resonances of the ^{13}C's to which the ^{1}H's are attached. Resonances from terminal alkyne ^{1}H's may fail to show a cross peak to the alkynic ^{13}C resonance due to a ^{1}J$_{CH}$ of ~220 Hz. If we have a molecule with an unusual ^{1}J$_{CH}$, we can adjust a delay in the HSQC or HMQC pulse sequence to make the experiment particularly sensitive to a given coupling. If the 1-D ^{1}H spectrum shows the resonance from a ^{1}H in question to be well resolved from other ^{1}H resonances, we can directly measure the spacing of the ^{13}C satellite peaks (the intensity of each satellite peak is 0.55% compared to the intensity of the center peak) and thus determine the value of the ^{1}J$_{CH}$ coupling directly. Then, armed with this information, the delay in the HSQC/HMQC parameters can be adjusted, the spectrum can be recollected, and the resulting data set will show only those cross peaks with ^{1}J$_{CH}$'s in the vicinity (say ±30 Hz) of the target ^{1}J$_{CH}$.

Returning to ethyl nipecotate, we see its 2-D ^{1}H-^{13}C HMQC spectrum in Figure 9.5. Note that there is no true diagonal in this spectrum, because it has for its axes two different chemical shift scales (it is a heteronuclear correlation, after all). Nonetheless, there exists what is called a pseudodiagonal. The cross peaks are all roughly scattered in a relatively narrow strip that extends from the lower left of the spectrum to the upper right. Deviations from the pseudodiagonal often indicate a large electronic shielding gradient, due possibly to the

Pseudodiagonal. The line connecting the upper-right corner to the lower-left corner of a heteronuclear 2-D spectrum, especially a ^{1}H-^{13}C HMQC or HSQC 2-D spectrum.

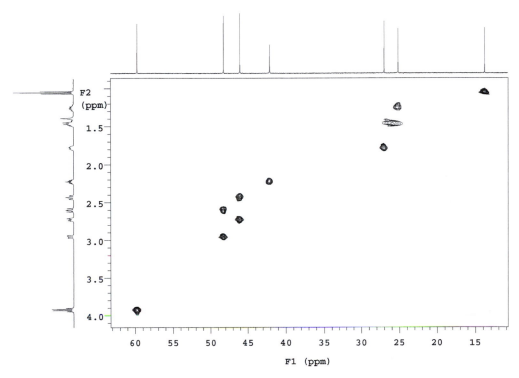

■ **FIGURE 9.5**   The 2-D ¹H-¹³C HMQC NMR spectrum of ethyl nipecotate in CDCl₃.

proximity of an atom with a high atomic number (like bromine or iodine) or an aromatic π system.

The left side of Figure 9.5 shows the 1-D ¹H spectrum, and the top of the spectrum shows the 1-D ¹³C spectrum. The two cross peaks arising from the ethyl group of ethyl nipecotate are the easiest to identify; they are also the first cross peaks we encounter if we start at either end of the pseudodiagonal. In the lower left of the HMQC spectrum, we see a cross peak between the two isochronous methylene ¹H's (see the integral of the resonance at 3.92 ppm in the ¹H 1-D spectrum in Figure 9.2) whose resonance is the quartet at 3.92 ppm (H8's) and the ¹³C at 59.7 ppm (C8). In the upper right of the HMQC spectrum, we see a cross peak correlating the resonance of the three methyl ¹H's making up the triplet at 1.05 ppm (H9's) with the ¹³C resonance at 13.8 ppm (C9). Although these two HMQC cross peaks do not provide us with any new information, they do confirm our earlier assignments based on chemical shifts, multiplicities, and integrals/intensities.

Examination of the remaining cross peaks in the spectrum reveal the astounding utility of the HMQC experiment. The resonances from each remaining methylene group (positions 2, 4, 5, and 6) generate two cross peaks (per methylene group) in the HMQC spectrum. The resonance of every ^{13}C bearing two ^{1}H's that produce anisochronous resonances will show two cross peaks in the HMQC spectrum. We can locate these cross peaks by descending from the ^{13}C methylene resonances in the 1-D ^{13}C spectrum at the top of the plot. Now our elaborate examination of the multiplicities of the ^{1}H's on positions 2 and 6 seems largely superfluous—instead we could have skipped ahead to the HMQC and seen that the four ^{1}H's on positions 2 and 6 are staggered such that, in going from high δ to low δ, we encounter $H2_{eq}$, then $H6_{eq}$, then $H2_{ax}$, and then $H6_{ax}$. (Admittedly, we still need to compare the widths of the two H2 multiplets to determine which has the 1,2-diaxial splitting to H3.)

By the process of elimination, we can readily identify the H3-C3 cross peak in the HMQC spectrum: The C3 resonance correlates only with a ^{1}H resonance whose integral indicates just one ^{1}H. That is, even though C8 and C9 correlate to single resonances in the 1-D ^{1}H spectrum, the integrals of the H8 and H9 regions of the 1-D ^{1}H spectrum show that more than one ^{1}H resonates there.

Finally we come to the assignment of the shifts of the ^{1}H's and ^{13}C's at positions 4 and 5. Clearly the isolated ^{1}H at 1.25 ppm is bound to the ^{13}C at 25.1 ppm, and the isolated ^{1}H at 1.78 ppm is bonded to the ^{13}C at 27.0 ppm. What is less clear is how to split up the two ^{1}H's in the 1.4–1.5 ppm range. What we see on the spectrum is two partially overlapping cross peaks. One cross peak lies to the upper right, and the second lies to the lower left (relative to the center of the overlapping mess).

Phaseable. The ability to effectively control the relative absorptive versus dispersive character of an NMR spectrum through partitioning of the displayed data between orthogonal (real and imaginary) subsets.

Had this HMQC spectrum been collected with more than 32 phase-sensitive points in the t_1 time dimension (64 FIDs were collected, but each t_1 evolution delay had two FIDs to make the f_1 dimension phase sensitive or "phaseable"), the resolution of these cross peaks would likely have been achieved. This HMQC spectrum is useful in that it shows us that a spectrum can be collected in a very short amount of time (8 scans/FID $\times$ 64 FIDs $\times$ 1 s/scan = 512 s or just under 9 minutes). However, it also shows us that with short 2-D collection times, the f_1 resolution may be poor (if the number of FIDs collected is small, i.e., is less than 128 or 256), or the signal-to-noise ratio may suffer if the number of scans/FID is reduced. The HMQC spectrum in Figure 9.5 was collected with 8 scans/FID—probably 4 or even 2 could have been possible with this relatively concentrated sample, ~50 mM.

Tracing the centers of what we must suppose are two overlapping ellipsoidal cross peaks horizontally to the left, we can determine that one cross peak is centered at 1.46 ppm (the ^1H bound to the ^{13}C at 27.4 ppm), and the other is centered at 1.47 ppm (the ^1H bound to the ^{13}C at 25.6 ppm). Thus, we see that the 2-D cross peaks can be used to measure chemical shifts when overlap in the 1-D spectrum prevents it.

If we return briefly to the gCOSY spectrum in Figure 9.3, we can see that the ^1H resonance at 1.25 ppm shows cross peaks to the H6's at 2.43 and 2.72 ppm, while the ^1H resonance at 1.78 ppm shows a cross peak to H3 at 2.22 ppm. This additional piece of information allows us to determine that the ^1H resonance at 1.47 ppm can be assigned to position 5 (along with the ^1H resonance at 1.25 ppm and the ^{13}C resonance at 25.1 ppm), and the ^1H resonance observed at 1.46 ppm can be assigned to position 4 (along with the ^1H resonance at 1.78 ppm and the ^{13}C resonance at 27.0 ppm). That is, we expect H3 to couple more strongly to the H4's than to the H5's. Likewise, we expect the H6's to couple more strongly with the H5's than with the H4's. By using the HMQC spectrum to identify well-resolved ^1H resonances that are paired with resonances that overlap in the 1H 1-D spectrum, we can then use the gCOSY spectrum to determine how to assign the overlapping resonances centered at 1.46 and 1.47 ppm.

We have therefore confirmed the expectation (based on chemical shift arguments) that the C4/H4 resonances are observed farther downfield relative to those of C5/H5.

On the basis of our earlier examination of multiplets in uncluttered regions of the 1-D ^1H spectrum, we expect the equatorial ^1H's to resonate downfield (higher ppm) relative to their axial counterparts. So, we expect that H4$_{eq}$ lies at 1.78 ppm and H5$_{ax}$ lies at 1.25 ppm. If we return to the 1-D ^1H spectrum, we can see that the ^1H resonance at 1.78 ppm appears to show only one large coupling (the *geminal* ^2J coupling between H4$_{eq}$ and H4$_{ax}$). It is also gratifying to note that the ^1H resonance at 1.25 ppm can, with hindsight, be seen to be composed of an approximate pseudoquartet due to the three large couplings expected to be experienced by H4$_{ax}$ (the *geminal* ^2J to H4$_{eq}$ and the two 1,2-diaxial *trans* ^3J couplings to H5$_{ax}$ and H3, the axial methine proton).

Pseudoquartet, Ψq. A quartet-like splitting pattern caused by the identical coupling of the resonance of the observed spin to three other spins not related by symmetry.

9.11 ASSIGNMENT OF NONPROTONATED ^{13}C'S ON THE BASIS OF THE HMBC SPECTRUM

Ethyl nipecotate contains only one nonprotonated carbon site. Because of this lack of multiple nonprotonated carbon sites, it fails to serve as a useful example for illustrating the power of the 2-D ^{1}H-^{13}C heteronuclear multiple bond correlation (HMBC) experiment in assigning the resonances from nonprotonated ^{13}C's. As a general rule, a molecule with few nonprotonated carbons will rarely require data from the HMBC experiment.

6-Ethyl-3-formylchromone (Figure 9.6), on the other hand, has a number of nonprotonated carbon atoms. Five of its twelve carbons are sp^2-hybridized, nonprotonated carbons. The 1-D ^{13}C NMR spectrum of 6-ethyl-3-formylchromone is shown in Figure 9.7. The five non-protonated ^{13}C resonances are easily spotted due to their lower intensities; their chemical shifts are 176.3, 154.8, 143.4, 125.2, and 120.4 ppm.

A chemical shift argument is sufficient to identify the ketone carbonyl ^{13}C resonance (C4) at 176.3 ppm. (Note that the aldehyde ^{13}C resonance at 189.0 ppm, C11, is more intense as a result of protonation.)

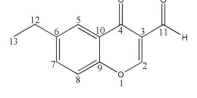

■ **FIGURE 9.6**   The numbered structure of 6-ethyl-3-formylchromone.

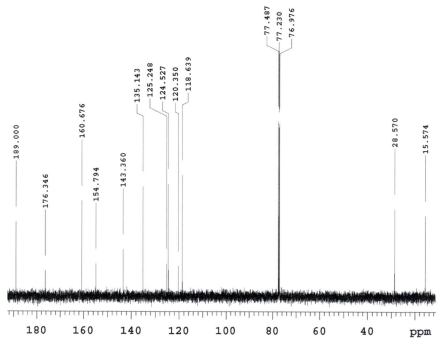

■ **FIGURE 9.7**   The 1-D ^{13}C NMR spectrum of 6-ethyl-3-formylchromone in CDCl$_3$.

Of the four remaining nonprotonated ^{13}C resonances (C3, C6, C9, and C10), we can predict that the C9 resonance will be most downfield because it is alpha to an oxygen atom. The C6 and C10 resonances should be observed at about the same chemical shift because of resonance considerations; but C10 we expect will resonate slightly farther downfield because of the withdrawal of electron density caused by the oxygen on C4. C3's position relative to C6 and C10 may be difficult to predict because C3 lies outside of a clearly defined aromatic electron system.

Although we might be tempted to indulge in further speculation by invoking more chemical shift arguments dealing with electronegativity and resonance, there is a sounder method for unequivocally assigning the remaining nonprotonated ^{13}C resonances in the molecule to their corresponding molecular sites. Recall that the ^{3}J Karplus diagram predicts that a maximum coupling constant will occur when the dihedral angle defined by the two sets of three adjacent atoms in a four-atom set is 180°. Put another way, when the two bonds on either side of a common bond are *trans*, the ^{3}J-coupling is at a maximum. When two bonds separated by a third, common bond are *cis*, the ^{3}J-coupling will be smaller than for the *trans* configuration, but will still likely be observed.

The simplest method for distinguishing between the nonprotonated ^{13}C resonances in 6-ethyl-3-formylchromone is to examine the structure (Figure 9.6) and identify the carbon sites that are expected to show *trans* ^{3}J$_{CH}$'s. C3 is not expected to show any *trans* ^{3}J's to ^{1}H's. C6 is *trans* to H8. C9 is *trans* to H2, H5, and H7. C10 is *trans* to H8. On the basis of these geometries, we therefore expect C9 to show three cross peaks in the 2-D ^{1}H-^{13}C HMBC spectrum to ^{1}H's on sp^2-hybridized carbon atoms. Figure 9.8 shows the 2-D ^{1}H-^{13}C HMBC spectrum of 6-ethyl-3-formylchromone.

Before proceeding further we will quickly assign a portion of the 1-D ^{1}H spectrum, which appears along the left side of Figure 9.8. The ^{1}H spectrum is relatively easy to assign by using our knowledge of chemical shifts and J-couplings. H11 (the aldehyde proton) resonates farthest downfield at 10.39 ppm, and the four remaining ^{1}H's on sp^2-hybridized carbons resonate at 8.53 ppm, 8.09 ppm, 7.58 ppm, and 7.46 ppm (we disregard the solvent resonance from the small amount of protonated chloroform at 7.27 ppm). Clearly, we expect H7 and H8 to be strongly coupled by a ^{3}J$_{HH}$, and indeed the two ^{1}H resonances at 7.46 ppm and 7.58 ppm show a large splitting. Additionally, we expect to observe a slight splitting from the W-coupling (^{4}J) between H7 and H5. The resonance at 8.09 ppm has a much lower

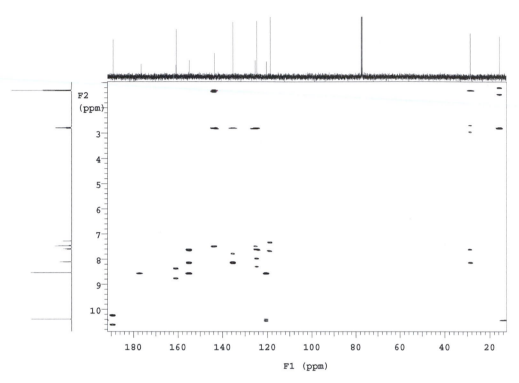

■ **FIGURE 9.8** The 2-D ^{1}H-^{13}C HMBC NMR spectrum of 6-ethyl-3-formylchromone in CDCl$_3$.

height intensity (but presumably still has the same integral) relative to the resonance at 8.53 ppm. Therefore, we must assume that the resonance at 8.09 ppm is that of H5 and not H2, because the W-coupling will broaden the H5 resonance without affecting its integral.

Likewise, the argument that slight broadening allows us to distinguish the resonances of H5 from that of H2 also allows us to identify the resonance of H7 (as opposed to that of H8) as being that observed at 7.58 ppm. That is, H7's resonance is a multiplet that shows not only a strong ^{3}J to H8, but H7's resonance also split by a ^{4}J to H5. Thus, the overall height of the H7 resonance is reduced by this additional splitting relative to the height of the H8 resonance.

Examination of the 2-D ^{1}H-^{13}C HMBC NMR spectrum of 6-ethyl-3-formylchromone shows that of the five nonprotonated ^{13}C resonances (identified by the low peak heights in the 1-D ^{13}C spectrum at the top of the figure), one shows more cross peaks to ^{1}H resonances on sp^2-hybridized carbons than the others. The ^{13}C resonance at 154.8 ppm shows three strong cross peaks to H2, H5, and H7 (at 8.53 ppm,

8.09 ppm, and 7.58 ppm, respectively). The ^{13}C resonance at 154.8 ppm therefore must be that of C9, because C9 is *trans* to H2, H5, and H7. Note that even though the ^3J from H2 to C9 passes through a hetero-atom (an oxygen atom), the HMBC cross peak is still formidable.

The ^1H at position 8 is *trans* to both C6 and C10, so we cannot easily differentiate between the C6 and C10 ^{13}C resonances using HMBC cross peaks to H8 alone. That is, the H8 resonance at 7.46 ppm shows a strong cross peak to a ^{13}C signal at 143.4 ppm and a weak cross peak to the ^{13}C signal at 125.2 ppm. If we examine the upper portion of the HMBC spectrum, however, we see that the ^1H reso-nances from the ethyl group show cross peaks to only one nonpro-tonated ^{13}C resonance at 143.4 ppm (the cross peak from the H12's show a cross peak to the protonated ^{13}C resonance at 124.5 ppm, not to the nonprotonated ^{13}C resonance at 125.2 ppm). Thus, we can conclude that the C6 resonance is that at 143.4 ppm, and there-fore that the C10 resonance is the one found at 125.2 ppm. Why the H8-C10 cross peak is so much weaker than we expect is a matter for some debate. Perhaps this discrepancy is caused by the proximity of the heteroatom at position 1 of the bicyclic ring system.

The carbon at position 3 must generate the resonance at 120.4 ppm. This resonance shows two HMBC cross peaks, one to the reso-nance from H2 and the other to that of H11. Interestingly, both of these cross peaks are due to ^2J-couplings; these cross peaks are not expected because C2 and C11 are both sp^2-hybridized (recall that ^2J's are expected to be small when the bond angle is near 120°). One of the reasons that these HMBC cross peaks may appear on the plot is that both H2 and H11 are "tall" resonances; that is, both H2 and H11 are not broadened by coupling to other ^1H's, so the cross peaks that they do generate in the HMBC have all their intensity packed into a relatively small area, thus making these cross peaks more prominent relative to cross peaks of comparable intensity involving ^1H's that are more spread out as a result of homonuclear J-coupling. Put another way, even though the volume of these ^2J-spawned HMBC cross peaks is very likely much smaller than that of the cross peaks arising from *trans* ^3J's, the narrow width of the H2 and H11 resonances allows these weaker cross peaks to rise above the minimum threshold in the plot. A more accurate means of assessing relative cross peak intensity is to obtain volume integrals of the cross peaks in the HMBC spec-trum, but this practice is rare.

Strategies for Elucidating Unknown Molecular Structures

Determining the correct molecular structure of an unknown, given only its NMR spectra, can be daunting. We start by drawing the molecular fragments that correspond to the entry points we recognize in the spectra. Onto these fragments we continue to add atoms either individually or as functional groups on the basis of our interpretation of the remaining resonances we observe. This process proceeds until we account for every observed resonance and simultaneously piece together the unknown molecular structure.

Sometimes multiple interpretations are possible. When we cobble together a complete molecule from a dizzyingly complex set of spectra, we try to develop in parallel only a limited number of fragments. Only one plausible molecule should ultimately be generated that accounts for every observed resonance.

The interpretation that is best (i.e., correct) is often the simplest. This simplicity is perhaps the most important guiding precept for molecular structure elucidation. The best course is to initially explore all possibilities and then reduce the manifold options to the simplest explanation for all evidence. This admonition is a restatement of Occam's razor, which is

Entities should not be multiplied unnecessarily.

That is, we will keep it simple. Until a new piece of evidence rules out the simplest theory, we will keep to that explanation. The other key introductory point is to practice, practice, practice. Knowledge of the molecular weight of the molecule through mass spectrometric methods may also be available with unit mass resolution. Knowing the molecular weight helps us a great deal in our quest to determine an unknown molecular structure. Having the empirical formula is, of course, even better—if we have access to a high-resolution mass spectrometer, we can often determine the mass of the parent ion to six or seven significant digits, thereby allowing us to calculate the empirical formula if the molecular weight is below about $700\,g\,mol^{-1}$.

To illustrate the utility of the methodology presented in this chapter, a single problem will be presented throughout the course of the chapter. Figure 10.1 shows the entire 1-D ^1H spectrum of Compound X dissolved in 99% D_2O. First, we must consider what solvent is present; the solvent may contribute resonances to the ^1H and/or ^{13}C spectra. Also be aware that some solvents are protic (or, more properly, deuterotic); protic solvents will exchange away (replace with $^2H's$) the $^1H's$ in our molecule at sites with low $pK_a's$, and in some instances may also protonate (deuteronate) Lewis base sites in our molecule. In this example, the solvent (D_2O) will exchange away all labile protons, meaning that amino, hydroxyl, and carboxyl protons will not be observed. Also note the prominent ^1H resonance at approximately 4.65 ppm, arising from the HOD present in the sample. This peak is always seen when D_2O is the solvent, and if the bottle from which the solvent came is old (or has been left open to a moist atmosphere for any length of time), the peak can be very intense relative to the solute resonances, especially if the solute concentration is low.

Figure 10.2 shows the 1-D ^{13}C spectrum obtained from the same sample of Compound X. There is no solvent peak in this spectrum because the solvent lacks $^{13}C's$. Also note that this 1-D spectrum shows a low signal-to-noise ratio, as is often the case when the amount of sample, its solubility, and/or the amount of NMR instrument time available is minimal. Keep in mind that some (most likely nonprotonated) ^{13}C resonances in Compound X may not have a sufficiently large signal-to-noise ratio to be clearly revealed.

10.1 INITIAL INSPECTION OF THE ONE-DIMENSIONAL SPECTRA

Our first step is to inspect the 1-D spectra (^1H and ^{13}C) for obvious attributes, as this may help speed analysis. Put another way, we should look at the 1-D spectra for clues to the nature of the molecule. Conveniently, integration of the ^1H resonances is supplied with the 1-D ^1H spectrum in Figure 10.1. Summing the integrals by rounding every nonintegral value (except for the solvent resonance at about 4.65 ppm), we observe twelve ^1H resonances. Notice that we round the integral of the resonance at 10.20 ppm from 0.44 up to 1,

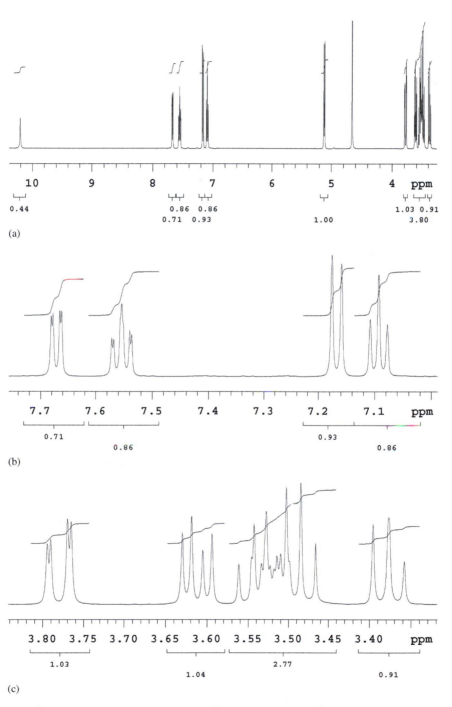

(a)

(b)

(c)

■ **FIGURE 10.1** The 1-D ^{1}H spectrum of Compound X in D$_2$O. (a) The entire ^{1}H spectrum; (b) the aromatic ^{1}H region; (c) a portion of the aliphatic ^{1}H region.

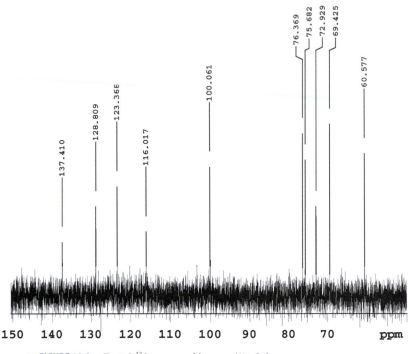

■ FIGURE 10.2   The 1-D ^{13}C spectrum of Compound X in D_2O.

since rounding this integral down to 0 implies that this resonance does not exist. We might be tempted to double every integral so the ^1H resonance at 10.20 ppm achieves the more-satisfying integral of 0.88. In keeping with Occam's razor, however, we proceed with the assumption that doubling the integrals is not yet called for. Because we are observing a ^1H resonance at >10 ppm, we can conclude that this resonance is either from a carboxylic acid group, an aldehyde group, or possibly an aromatic hydroxyl group. The solvent is 99% D_2O, so the resonance at 10.20 ppm must arise from an aldehyde ^1H— otherwise the solvent deuterons would exchange with this resonance and we would not observe the resonance. The low value of the integral associated with the resonance at 10.20 ppm is also consistent with the diagnosis of an aldehyde ^1H, because aldehyde ^1H's typically have much longer T_1 relaxation times than do aliphatic ^1H's. Thus, we have already accounted in another way for the low ^1H integral of 0.44 at 10.20 ppm (we didn't have to double the integrals after all).

Moving upfield to the middle of the ^1H 1-D spectrum, we next encounter four ^1H's between 7.0 and 7.7 ppm (see Figure 10.1b). This chemical shift location is in the range expected for ^1H's on an aromatic ring. Because these resonances correspond to four aromatic ^{1}H's, we assume that we have a single aromatic ring that is disubstituted. We will establish the ring's substitution pattern later in Section 10.4.

Continuing our survey of the ^{1}H 1-D spectrum, we proceed to the right into the olefinic and aliphatic chemical shift region. Here we observe the resonance from one ^{1}H at 5.11 ppm that is split into a doublet, then we find the solvent at 4.65 ppm, and finally we encounter the resonances of six more ^{1}H's in the range from 3.3 to 3.8 ppm (see Figure 10.1c; although the spectral window ends at 3.3 ppm, give the spectroscopist who collected and plotted the data credit for having observed and included all resonances from the 1-D spectrum). Because no ^{1}H resonances have chemical shifts lower than 3.3 ppm, we must assume that this molecule lacks simple alkyl moieties like ethyl or isopropyl groups. Furthermore, because those ^{1}H resonances in the aliphatic region are observed on the left side of aliphatic chemical shift range (see Section 4.2), we must assume that a number of heteroatoms are present to cause this downfield shift.

The 1-D ^{13}C spectrum (Figure 10.2) shows 10 resonances ranging from 137.4 to 60.6 ppm. We may reasonably assume that an aldehyde ^{13}C resonance lies further downfield (around 200 ppm, we will check this later). Just as we observed the resonances of four aromatic ^{1}H's in the ^{1}H 1-D spectrum, here in the ^{13}C 1-D spectrum we also see four ^{13}C resonances (137.4, 128.8, 123.4, and 116.0 ppm) in the aromatic and/or sp^2-hybridized ^{13}C chemical shift range. This number of ^{13}C resonances is consistent with the idea that we have a single aromatic ring in the molecule; because it is likely that the ring is disubstituted (we know this from the 1-D ^{1}H spectrum), we may not be observing the two nonprotonated ^{13}C's. The ^{13}C resonance at 100.1 ppm is a little too far upfield to be comfortably lumped into the aromatic ^{13}C chemical shift range (recall that benzene has a ^{13}C chemical shift of 128 ppm, so we don't want to deviate too far from this shift, especially upfield). Thus, there are six ^{13}C resonances from 100.1 to 60.6 ppm that we will match up with the aliphatic ^{1}H resonances.

10.2 GOOD ACCOUNTING PRACTICES

Keeping track of the data in tabular form is a good practice, as we saw in the assignment of the resonances of a known structure. We have no way of knowing how to number the resonances in our unknown, so we can either assign a letter to each resonance in the spectrum as we proceed from left to right; or we can simply tabulate the resonances by placing each resonance's shift into the table in

Table 10.1 ^{1}H resonances of Compound X.

δ (ppm)	Multiplicity	Integral	Comments
10.20	s	0.44	aldehyde
7.68	d^2	0.71	big J, little J, aromatic
7.55	d^3 or dΨt	0.86	2 big J's, one little J, aromatic
7.17	d	0.93	big J, aromatic
7.09	Ψt	0.86	2 big J's, aromatic
5.11	d	1.00	aliphatic, medium J
3.79	d^2	1.03	aliphatic, big J, little J
3.61	d^2	1.04	aliphatic, big J, medium J
3.44–3.56	(overlapping)	2.77	aliphatic
3.37	d^2	0.91	aliphatic, 2 medium J's

Table 10.2 ^{13}C resonances of Compound X.

δ (ppm)	Intensity	Comments
137.4	medium	aromatic, protonated
128.8	medium	aromatic, protonated
123.4	medium	aromatic, protonated
116.0	medium	aromatic, protonated
100.1	strong	aliphatic
76.4	strong	aliphatic
75.7	strong	aliphatic
72.9	medium	aliphatic
69.4	strong	aliphatic
60.6	strong	aliphatic

the order in which the resonances occur in the spectrum. Table 10.1 shows the resonances in order from left to right for the 1-D ^{1}H spectrum. Table 10.2 shows the resonances for the 1-D ^{13}C spectrum.

The next step is to examine the 2-D ^{1}H-^{13}C HMQC spectrum (Figure 10.3). This spectrum allows us to pair the aldehyde ^{1}H

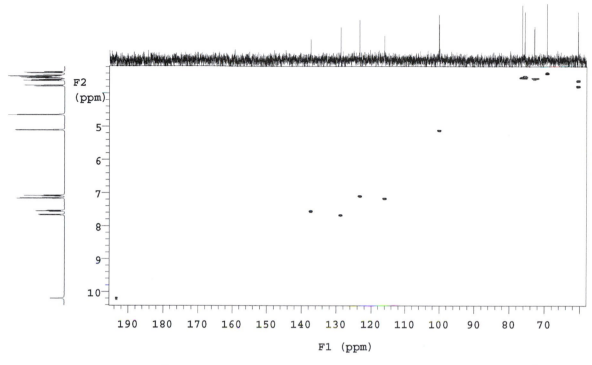

■ **FIGURE 10.3** The 2-D ^{1}H-^{13}C HMQC spectrum of Compound X in D$_2$O.

resonance to the aldehyde ^{13}C resonance through the cross peak at (δ_H=10.20 ppm, δ_C=193.3 ppm). Although we do not observe the aldehyde ^{13}C resonance directly in the 1-D ^{13}C spectrum, we can simply trace down from the center of the HMQC cross peak to the ^{13}C chemical shift axis to determine the approximate shift of the aldehyde carbonyl ^{13}C resonance. The four protonated aromatic ^{13}C resonances can be paired with the four aromatic ^{1}H resonances by reading the positions of the four cross peaks in the middle of the HMQC spectrum: (δ_H = 7.68 ppm, δ_C = 128.8 ppm), (δ_H = 7.55 ppm, δ_C = 137.4 ppm), (δ_H = 7.17 ppm, δ_C = 116.0 ppm), and (δ_H = 7.09 ppm, δ_C = 123.4 ppm). The most downfield of the aliphatic resonances is a methine group with cross peak coordinates of (δ_H = 5.11 ppm, δ_C = 100.1 ppm).

In the 1-D ^{1}H spectrum, the three ^{1}H's found in the range from 3.44 to 3.56 ppm overlap to such an extent that direct observation of their chemical shifts and J-couplings is difficult, if not impossible (see Figure 10.1c). However, the expansion of the 2-D ^{1}H-^{13}C HMQC spectrum in Figure 10.4 allows us to determine the shifts of the three ^{1}H's in this region by drawing a line horizontally (use a straightedge)

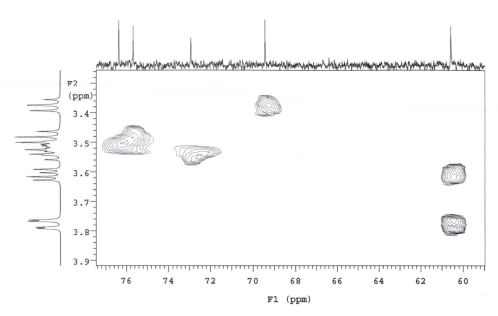

■ **FIGURE 10.4** An expansion of the 2-D
^{1}H-^{13}C HMQC spectrum of Compound X in D$_2$O.

to the ^{1}H chemical shift axis (to the left) from the center of each cross peak. Even though two of the cross peaks overlap, it is still possible to see that these two cross peaks are staggered. One cross peak is at the lower left of the overlapping cross peaks and the other is at the upper right. Although some intensity is present that fills the lower right-hand corner of the overlapping, staggered cross peaks, it must be that the two cross peaks are oriented with one on the lower left and the other on the upper right—no other relative orientation can better account for the observed spectral features.

Although the two resonances at 76.4 and 75.7 ppm are clearly resolved in the ^{13}C 1-D spectrum (above the HMQC spectrum expansion in Figure 10.4), the 2-D HMQC spectrum itself lacks the same resolution in the ^{13}C (f_1) dimension. F_1 resolution can be improved by adjusting upward the number of t_1 time increments (and therefore the number of digitized FIDs). Unfortunately, this adjustment lengthens the time of the experiment, so in practice we often barely resolve resonances in the ^{13}C chemical shift (f_1) domain of the 2-D spectrum that are clearly resolved in the ^{13}C 1-D spectrum. The HMQC cross peaks from the resonances of three overlapping (in the ^{1}H chemical shift domain) methine groups with ^{1}H chemical shifts in the 3.44–3.56 ppm range can be measured in this way. These cross peaks occur at (δ_H = 3.54 ppm, δ_C = 72.9 ppm), (δ_H = 3.51 ppm, δ_C = 76.4 ppm), and (δ_H = 3.47 ppm, δ_C = 75.7 ppm). A well-resolved methine cross peak is seen at (δ_H = 3.37 ppm, δ_C = 69.4 ppm).

Finally, two cross peaks arising from a single methylene (CH_2) group with two inequivalent (diastereotopic) protons are evident at ($\delta_H = 3.79$ ppm, $\delta_C = 60.6$ ppm) and ($\delta_H = 3.61$ ppm, $\delta_C = 60.6$ ppm). The two cross peaks stacked vertically from the methylene group must be from the resonances of two inequivalent ^1H's bound to a methylene ^{13}C. Two ^1H's attached to one ^{13}C must be from a methylene group— there is no other way to account for these cross peaks unless we make the assumption that the ^{13}C resonance at 60.6 ppm is due to two and not just one ^{13}C in the molecule. Remember Occam's razor: we proceed with the simplest explanation until we have a compelling reason to introduce a more complicated explanation.

10.3 IDENTIFICATION OF ENTRY POINTS

Entry points are often well-isolated resonances that suggest the presence of unique chemical bonding in the molecule. Other entry points include distinctive alkyl groups like ethyl groups. Determining how to start piecing together a molecule on the basis of obvious spectral features is a critical step. As we have already discovered, Compound X probably contains an aldehyde group, an aromatic ring (we only observe four of six ^{13}C's in the ^{13}C 1-D spectrum), and six sp^3-hybridized carbons, five of which are methine groups, and one of which is a methylene group. The downfield chemical shifts of the resonances of the aliphatic ^1H's and ^{13}C's suggest that other heteroatoms are also present in the molecule.

10.4 COMPLETION OF ASSIGNMENTS

Establishing the connectivity between the methyl, methylene, methine, and other functional groups is done with the gCOSY and HMBC spectra. Figure 10.5 shows the 2-D ^1H-^1H gCOSY spectrum of Compound X.

Assuming we have a disubstituted aromatic ring, we can now examine both the splitting pattern of the aromatic ^1H resonances and the gCOSY connectivity between these resonances to determine the substitution pattern of the ring. The splitting patterns of the aromatic ^1H resonances in Compound X show that two of the ^1H's experience one large coupling and two experience two large couplings. Because the resonances of adjacent ^1H's on an aromatic ring show a reasonably large *cis* ^3J, we can therefore deduce that the pseudotriplets (Ψt's) are due to aromatic ^1H's with two adjacent ^1H's (^1H's three bonds away on neighboring carbon atoms). The pseudotriplets at 7.56 and 7.09 ppm must therefore both be bracketed by two aromatic methines. The

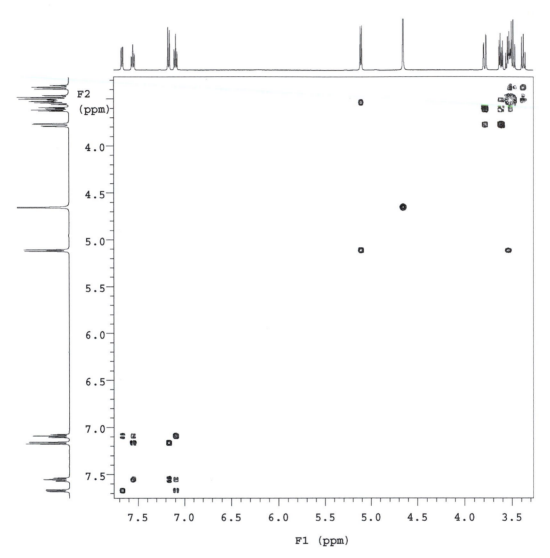

gCOSY spectrum further shows that the resonances that are pseudo-triplets correlate with each other (are coupled). The two doublets (one is actually a doublet of doublets, but this fact can be glossed over for now) each show a gCOSY cross peak to both a doublet and a pseudo-triplet, a pattern indicating that these methine groups bracket the two pseudo-triplets in the molecule.

An alternative way of arriving at an understanding of the coupling of the ^{1}H resonances in the aromatic chemical shift region involves starting with one of the doublets and counting the number of other resonances to which it is coupled. Because a doublet implies that there is only one nearby spin to which it is coupled, it is heartening to observe that there is only one pair of cross peaks (symmetrically placed with respect to the diagonal) for each doublet. Tracing from the doublet at 7.68 ppm, we see that this aromatic ^{1}H is coupled to the pseudotriplet at 7.09 ppm. The pseudotriplet at 7.09 ppm is, of course, coupled to the doublet at 7.68 ppm, but also to the pseudo-triplet at 7.56 ppm. The pseudotriplet at 7.56 ppm is coupled not only to the pseudotriplet at 7.09 ppm, but also to the doublet at 7.17 ppm.

The four aromatic ^{1}H resonances are thus all grouped together to form an isolated spin system. Interestingly, the two most downfield ^{1}H resonances (at 7.68 and 7.56 ppm) also show a small W (^{4}J) coupling to one another, while the two most upfield resonances (at 7.17 and 7.09 ppm) do not show the analogous ^{4}J. The ^{4}J we observe in the 1-D ^{1}H spectrum (Figure 10.1) is not sufficient to generate an observable cross peak on the gCOSY spectrum, although lowering the plot threshold would likely reveal the presence of this weaker cross peak.

Thus the ring is *ortho* (1,2) substituted. If the ring were *meta* (1,3) substituted, one of the ^{1}H's would be more isolated and would just possibly show smaller ^{4}J's to the other aromatic ^{1}H's. If the ring were *para* (1,4) substituted, only two ^{1}H resonances would be observed since rotation of the aromatic ring would, in a molecule of this small size, relate the ring ^{1}H's pairwise by symmetry (a C_2 rotation, i.e., a 180° ring flip) and give us only two resonances, each with an integral corresponding to two ^{1}H's.

Additional information can be gained by examining the chemical shifts of the four aromatic ^{1}H's. Recall that benzene has a ^{1}H shift of 7.16 ppm. Because the aromatic ^{1}H resonances in Compound X show shifts both above and below that of benzene, we can infer that both electron-withdrawing groups (EWGs) and electron-donating groups (EDGs) are present in the molecule. That is, through resonance, an EWG shifts two of the resonances well downfield from their expected position, while an EDG shifts one of the aromatic ^{1}H resonances upfield as well. It therefore makes sense to conclude that there is one EDG and one EWG attached to the aromatic ring.

Proceeding to the aliphatic side of the ^{1}H spectrum in the gCOSY spectrum, we encounter the resonance of a ^{1}H at 5.11 ppm that is split into a doublet and shows a cross peak to the ^{1}H resonance at 3.54 ppm. The ^{1}H resonance at 3.54 ppm, whose multiplicity we cannot easily determine because of overlap in the 1-D ^{1}H spectrum, shows what appears to be a coupling to the ^{1}H resonance at 3.47 ppm. The ^{1}H resonance at 3.47 ppm appears to show a coupling to a ^{1}H resonance at 3.37 ppm. The ^{1}H resonance at 3.37 ppm, clearly a pseudotriplet, shows a cross peak to the ^{1}H resonance at 3.51 ppm. The ^{1}H resonance at 3.51 ppm shows a cross peak to the ^{1}H resonance at 3.61 ppm. The ^{1}H resonance at 3.61 ppm, a doublet of doublets that we know is part of a methylene group, also shows a coupling to its geminal neighbor resonating at 3.79 ppm.

Because we know (1) the ^{1}H's from the methylene group resonating at 3.61 and 3.79 ppm must show a strong geminal (^{2}J) coupling to each other, and (2) the ^{1}H resonances both are doublets of doublets (see the 1-D ^{1}H spectrum in Figure 10.1 to confirm this), each of these ^{1}H's must couple with one other ^{1}H that is three bonds (or possibly four) bonds removed. Again, we will assume that the next coupling partner to the methylene ^{1}H's is not four bonds distant because of Occam's razor.

Despite the overlap in the 3.44–3.56 ppm region of the spectrum, there is still sufficient resolution in the 2-D ^{1}H-^{1}H gCOSY spectrum to differentiate the three ^{1}H's in this region. Figure 10.6 shows an expansion of the relevant portion of the gCOSY spectrum.

Sighting from the projection toward the diagonal, we can see that there are three distinct areas where cross peaks occur in this crowded spectral region. What may not be immediately apparent is that there exists a small box whose upper left and lower right corners are defined by the cross peak between the two resonances at 3.47 and 3.54 ppm. That is, instead of just intensity on the diagonal, the two cross peaks between the resonances at 3.47 and 3.54 ppm complete the upper left and lower right corners of the box (the lower left and upper right corners of this box are defined by the on-diagonal peaks from the resonances at 3.47 and 3.54 ppm).

We are still missing the shifts for the two nonprotonated ^{13}C resonances on the aromatic ring we previously noted as missing from the ^{13}C 1-D spectrum due to a low signal-to-noise ratio, as well as the shift of the carbonyl ^{13}C resonance of the aldehyde group. The 2-D ^{1}H-^{13}C HMBC spectrum (Figure 10.7) provides the shifts of the missing ^{13}C resonances. We see a number of cross peaks in the HMBC spectrum with ^{13}C chemical shifts of 193.4 ppm, 159.2 ppm, and 125.1 ppm; these ^{13}C chemical shifts are new to us. The most downfield shift must be from the aldehyde ^{13}C (which is itself an EWG), and the 159.2 ppm shift is likely due to the nonprotonated aromatic ^{13}C bonded to the

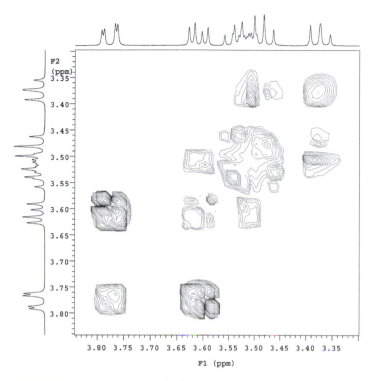

■ **FIGURE 10.6**   An expansion of the 2-D ^{1}H-^{1}H gCOSY spectrum of Compound X in D$_2$O.

electronegative EDG. The other shift must be from the nonprotonated aromatic ^{13}C bonded to the EWG (aldehyde group).

Putting all of our fragments together, we have an aldehyde group, an *ortho*-substituted aromatic ring, and five methines strung end to end with a methylene at the terminus. The need for an electron-withdrawing group on the ring suggests that the aldehyde group should be placed on the aromatic ring. The electron-donating group on the ring must contain an atom with a lone pair or be an alkyl group. Because the two nonprotonated aromatic ^{13}C resonances are both too far downfield for the electron donating group to be an alkyl group, we must assume that the electron-donating group is either an oxygen, a nitrogen, or a halogen atom.

Clearly we cannot satisfy the octet rule without adding more atoms to our molecule. Adding more atoms is also called for because we need something electronegative to pull the aliphatic ^{1}H resonances downfield from where we would normally expect to observe them.

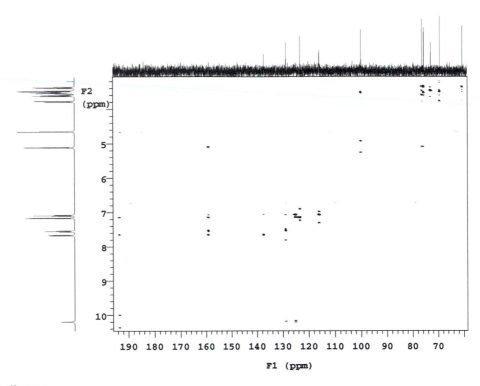

■ **FIGURE 10.7** The 2-D ^{1}H-^{13}C HMBC spectrum of Compound X in D$_2$O.

Anomeric methine group. The carbon-hydrogen group in a cyclic sugar that is bound to two oxygen atoms.

The downfield shift of the aliphatic methine group at (δ_H = 5.11 ppm, δ_C = 100.1 ppm) suggests the presence of two oxygen atoms, because one oxygen atom would not suffice to induce such a large downfield shift. In fact, the ^{13}C chemical shift of approximately 100 ppm is diagnostic and well known to sugar chemists, as this shift is characteristic of the anomeric ^{13}C (the anomeric methine group in a cyclic sugar is the one that is alpha to two oxygens). Without having two oxygens on this methine group, it is impossible to obtain the shifts we observe. Linear sugars do not show this behavior, only cyclic sugars.

The 2-D ^{1}H-^{13}C HMBC spectrum in Figure 10.7 also reveals how the sugar ring is attached to the aromatic ring through an oxygen atom. The question is, of course, which oxygen atom of the sugar ring is the attachment point? The answer is revealed by the cross peak at (δ_H = 5.11 ppm, δ_C = 159.4 ppm), which indicates that the anomeric methine group is also the attachment point to the aromatic ring. At this juncture, we can draw the structure of the molecule (Figure 10.8).

What remains to be done is to determine the stereochemistry of the sugar ring. Assuming that the ring adopts the chair conformation is

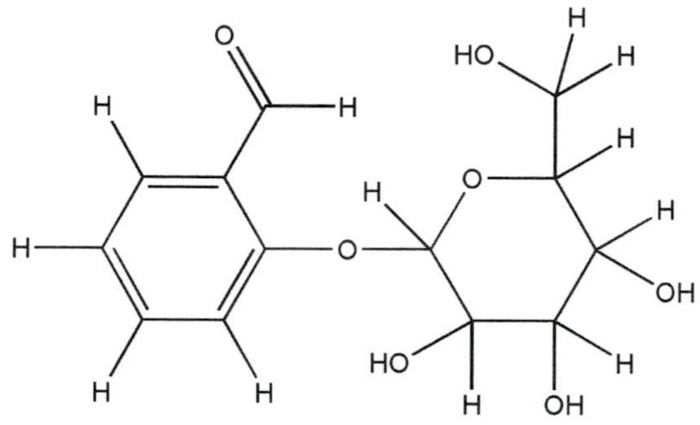

■ **FIGURE 10.8**   Structure of Compound X without stereochemistry.

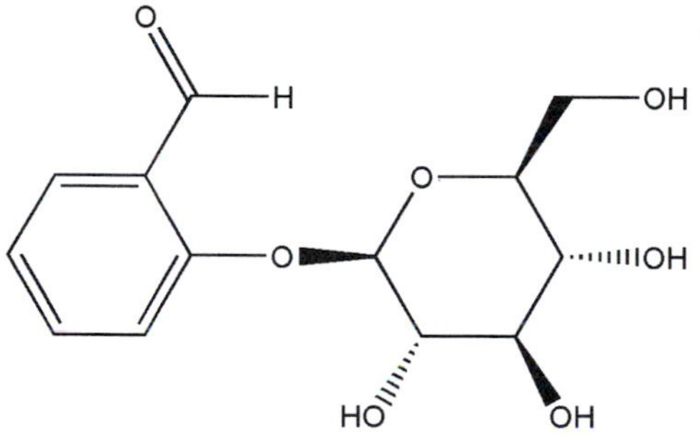

■ **FIGURE 10.9**   Structure of Compound X (helicin) with stereochemistry.

the starting point. What then remains is to look for 1,2-diaxial couplings (*trans* ^3J's). From these, we can determine which methine ^{1}H's
are axial. The overlap in the 1-D ^{1}H spectrum makes this task rather
difficult, but each of the first four multiplets starting from the anomeric methine shows only large (9–10 Hz) couplings. The ^{1}H resonances at 3.47 and 3.54 ppm are both triplets, only the middle ^{1}H
resonance at 3.51 ppm has a multiplicity that is difficult to extract
because it is coupled to three nearby ^{1}H's. Because all the couplings
are large, the sugar ring ^{1}H's must all be axial. Thus, the molecule's
structure can be redrawn as that shown in Figure 10.9.

Simple Assignment Problems

This chapter contains (relatively) simple NMR resonance assignment problems. The structure of the molecule is given, and it is up to the reader to assign the ^{1}H and ^{13}C resonances to the hydrogen and carbon sites in the molecular structure.

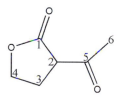

PROBLEM 11.1   **2-ACETYLBUTYROLACTONE IN CDCl$_3$ (SAMPLE 26)**

■ **FIGURE 11.1.struc**   The structure of 2-acetylbutyrolactone (with numbering).

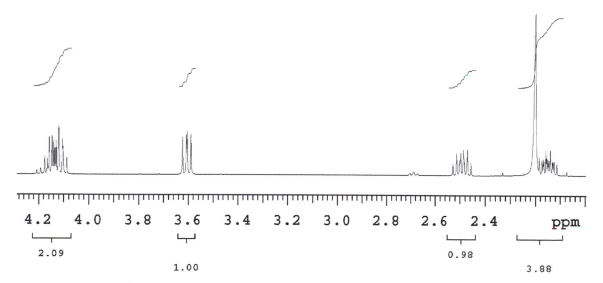

■ **FIGURE 11.1.h**   The 1-D ^{1}H NMR spectrum of 2-acetylbutyrolactone in CDCl$_3$.

■ **FIGURE 11.1.*c*** The 1-D ^{13}C NMR spectrum of 2–acetylbutyrolactone in CDCl$_3$.

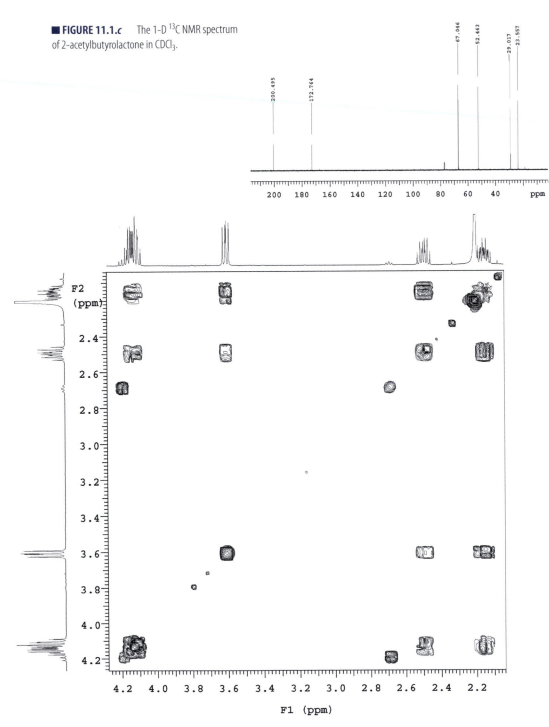

■ **FIGURE 11.1.*gcosy*** The 2-D ^{1}H-^{1}H gCOSY NMR spectrum of 2-acetylbutyrolactone in CDCl$_3$.

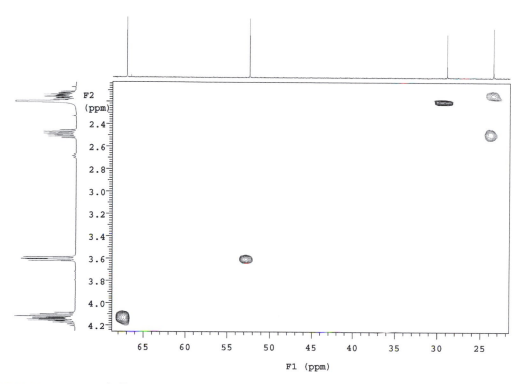

■ **FIGURE 11.1.***hmqc*   The 2-D ¹H-¹³C HMQC NMR spectrum of 2-acetylbutyrolactone in CDCl₃.

PROBLEM 11.2   α-TERPINENE IN CDCl₃ (SAMPLE 28)

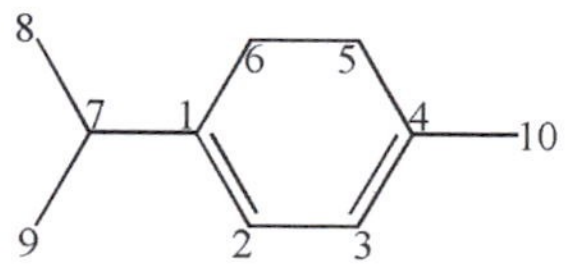

■ **FIGURE 11.2.***struc*   The structure of α-terpinene (with numbering).

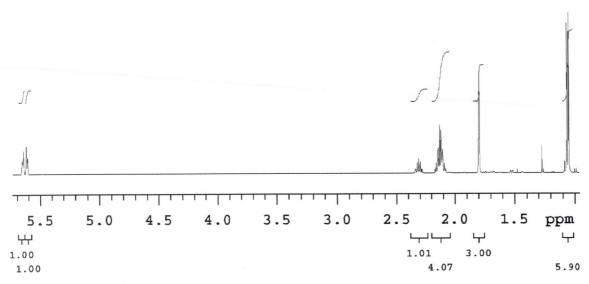

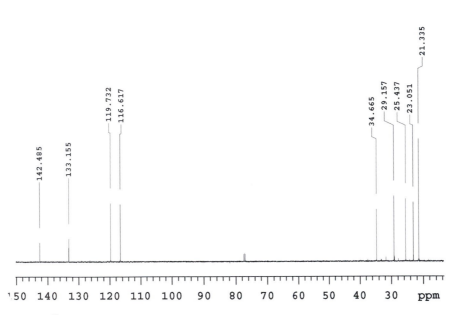

■ **FIGURE 11.2.h** The 1-D ^{1}H NMR spectrum of α-terpinene in CDCl$_3$.

■ **FIGURE 11.2.c** The 1-D ^{13}C NMR spectrum of α-terpinene in CDCl$_3$.

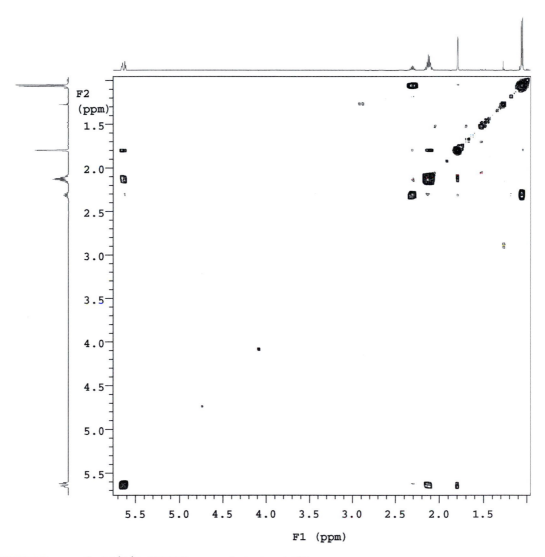

■ **FIGURE 11.2.*gcosy***   The 2-D ¹H-¹H gCOSY NMR spectrum of α-terpinene in CDCl₃.

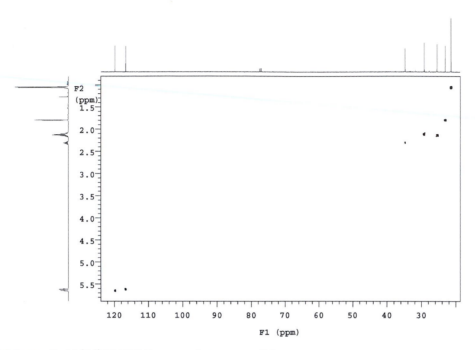

■ **FIGURE 11.2.*hmqc*** The 2-D ¹H-¹³C HMQC NMR spectrum of α-terpinene in CDCl₃.

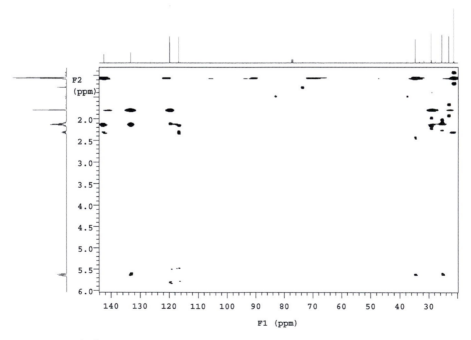

■ **FIGURE 11.2.*hmbc*** The 2-D ¹H-¹³C HMBC NMR spectrum of α-terpinene in CDCl₃.

PROBLEM 11.3   **(1*R*)-*ENDO*-(+)-FENCHYL ALCOHOL IN CDCl₃ (SAMPLE 30)**

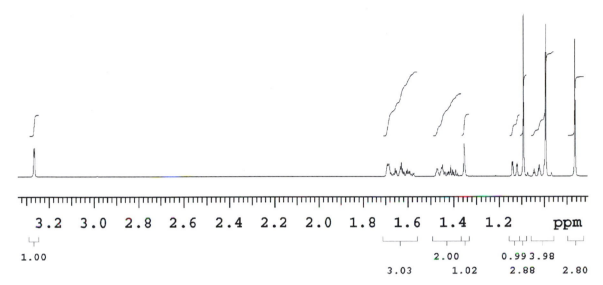

■ **FIGURE 11.3.*h*** The 1-D ¹H NMR spectrum of (1*R*)-*endo*-(+)-fenchyl alcohol in CDCl₃.

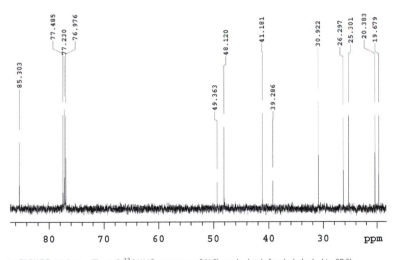

■ **FIGURE 11.3.*c*** The 1-D ¹³C NMR spectrum of (1*R*)-*endo*-(+)-fenchyl alcohol in CDCl₃.

■ **FIGURE 11.3.*struc*** The structure of (1*R*)-*endo*-(+)-fenchyl alcohol (with numbering).

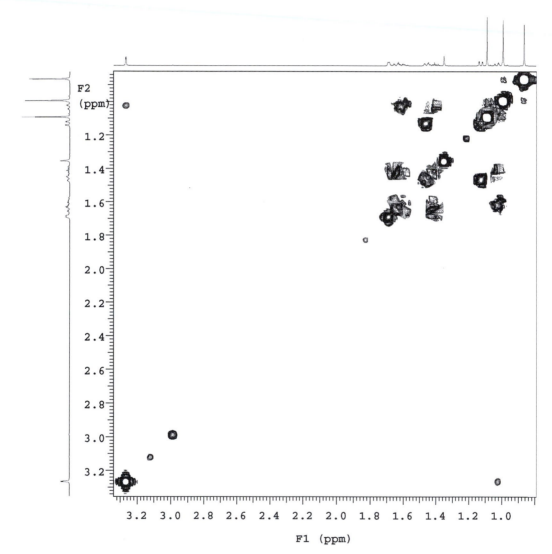

■ **FIGURE 11.3.***gcosy* The 2-D ¹H-¹H gCOSY NMR spectrum of (1*R*)-*endo*-(+)-fenchyl alcohol in CDCl₃.

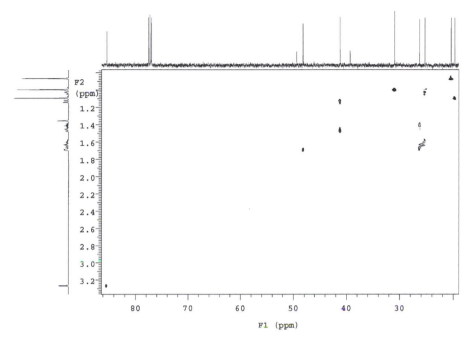

■ **FIGURE 11.3.***hmqc*   The 2-D ¹H–¹³C HMQC NMR spectrum of (1*R*)-*endo*-(+)-fenchyl alcohol in CDCl₃.

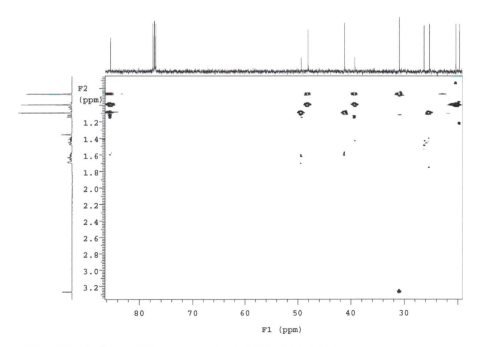

■ **FIGURE 11.3.***hmbc*   The 2-D ¹H–¹³C HMBC NMR spectrum of (1*R*)-*endo*-(+)-fenchyl alcohol in CDCl₃.

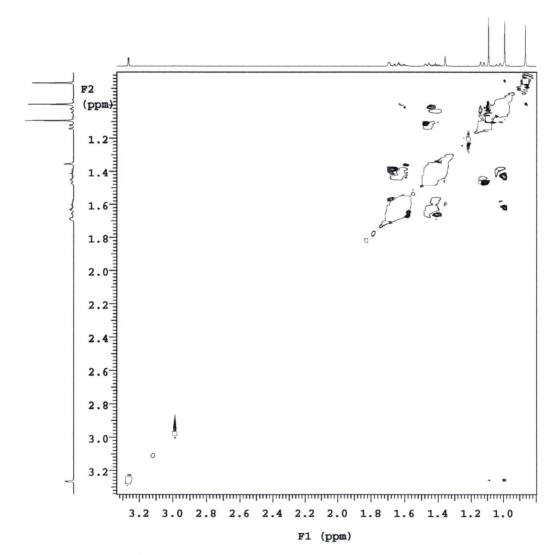

■ FIGURE 11.3.roesy The 2-D ^{1}H-^{1}H ROESY (200 ms mix) NMR spectrum of (1*R*)-*endo*-(+)-fenchyl alcohol in CDCl$_3$.

PROBLEM 11.4   **(−)-BORNYL ACETATE IN CDCl$_3$
(SAMPLE 31)**

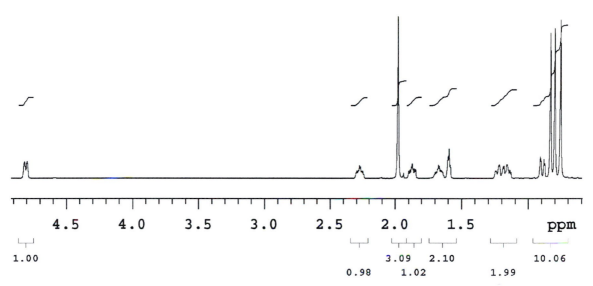

■ FIGURE 11.4.*h*   The 1-D ^{1}H NMR spectrum of (−)-bornyl acetate in CDCl$_3$.

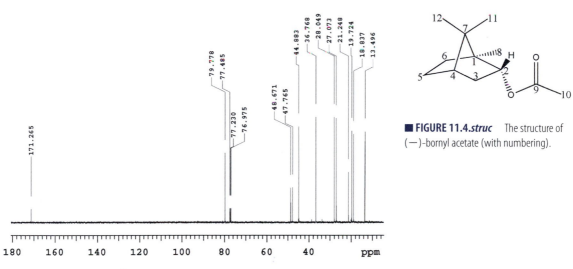

■ FIGURE 11.4.*struc*   The structure of (−)-bornyl acetate (with numbering).

■ FIGURE 11.4.*c*   The 1-D ^{13}C NMR spectrum of (−)-bornyl acetate in CDCl$_3$.

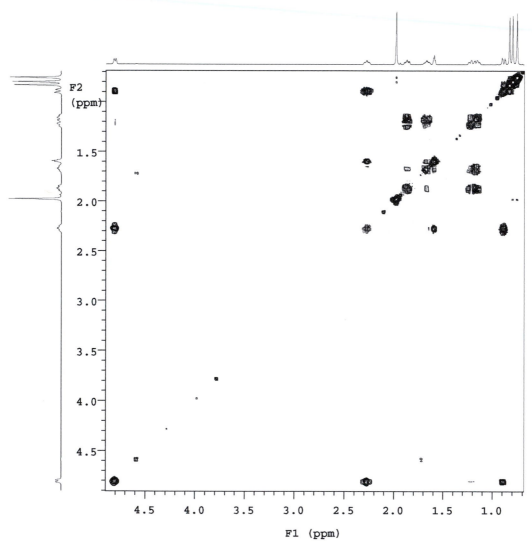

■ **FIGURE 11.4.*gcosy***   The 2-D ^{1}H-^{1}H gCOSY NMR spectrum (—)-bornyl acetate in CDCl$_3$.

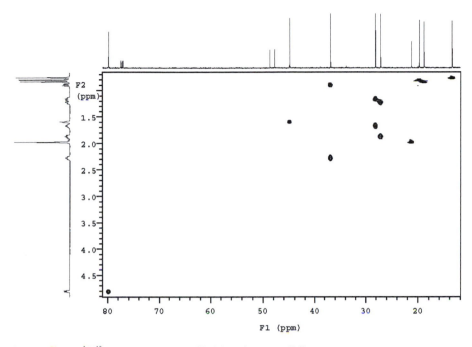

■ **FIGURE 11.4.*hmqc*** The 2-D ¹H-¹³C HMQC NMR spectrum of (−)-bornyl acetate in CDCl₃.

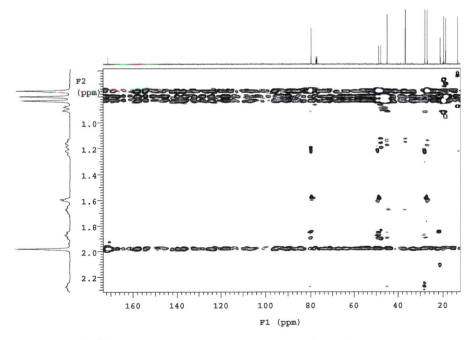

■ **FIGURE 11.4.*hmbc1*** The 2-D ¹H-¹³C HMBC NMR spectrum of (−)-bornyl acetate in CDCl₃ (region 1).

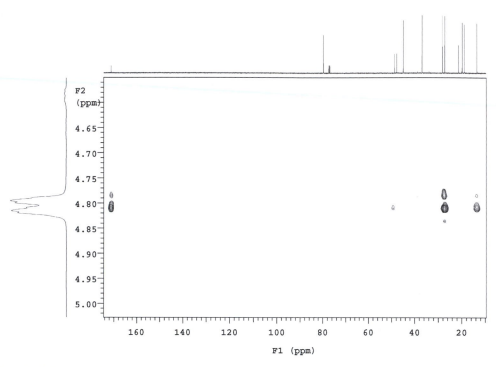

■ **FIGURE 11.4.*hmbc2*** The 2-D ^{1}H-^{13}C HMBC NMR spectrum of (—)-bornyl acetate in CDCl$_3$ (region 2).

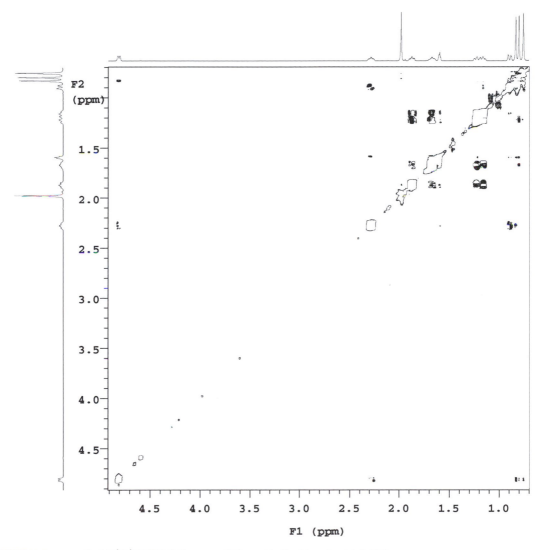

■ **FIGURE 11.4.*roesy*** The 2-D ¹H-¹H ROESY NMR spectrum (200 ms mix) of (—)-bornyl acetate in CDCl₃.

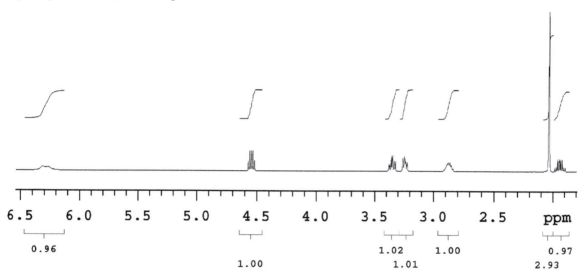

PROBLEM 11.5 **N-ACETYLHOMOCYSTEINE THIOLACTONE IN CDCl$_3$ (SAMPLE 35)**

■ **FIGURE 11.5.struc** The structure of N-acetylhomocysteine thiolactone (with numbering).

■ **FIGURE 11.5.h** The 1-D ^{1}H NMR spectrum of N-acetylhomocysteine thiolactone in CDCl$_3$.

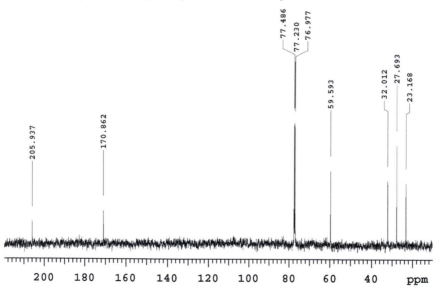

■ **FIGURE 11.5.c** The 1-D ^{13}C NMR spectrum of N-acetylhomocysteine thiolactone in CDCl$_3$.

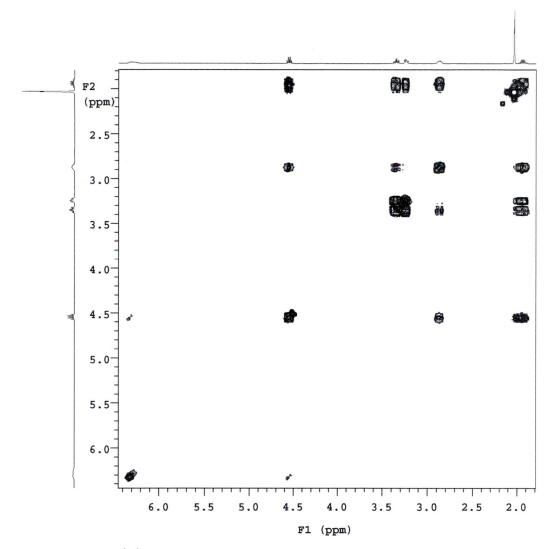

■ **FIGURE 11.5.***gcosy* The 2-D ^{1}H-^{1}H gCOSY NMR spectrum of *N*-acetylhomocysteine thiolactone in CDCl$_3$.

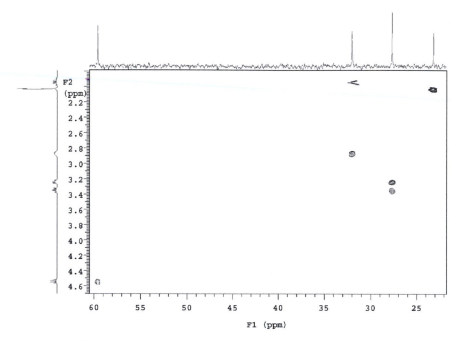

■ **FIGURE 11.5.*hmqc*** The 2-D ^{1}H-^{13}C HMQC NMR spectrum of *N*-acetylhomocysteine thiolactone in CDCl$_3$.

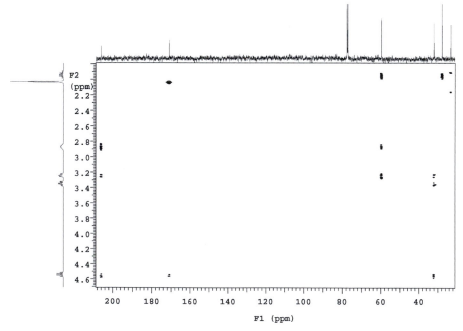

■ **FIGURE 11.5.*hmbc*** The 2-D ^{1}H-^{13}C HMBC NMR spectrum of *N*-acetylhomocysteine thiolactone in CDCl$_3$.

PROBLEM 11.6 **GUAIAZULENE IN CDCl$_3$ (SAMPLE 52)**

One of the ^{1}H resonances is very broad, perhaps due to exchange.

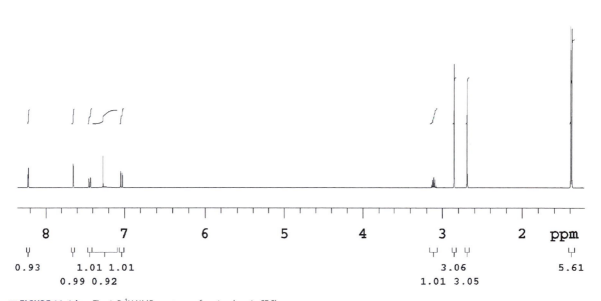

■ **FIGURE 11.6.struc** The structure of guaiazulene (with numbering).

■ **FIGURE 11.6.h** The 1-D ^{1}H NMR spectrum of guaiazulene in CDCl$_3$.

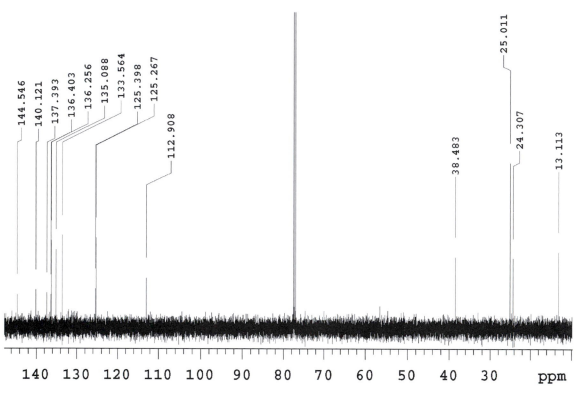

■ **FIGURE 11.6.c** The 1-D ^{13}C NMR spectrum of guaiazulene in CDCl$_3$.

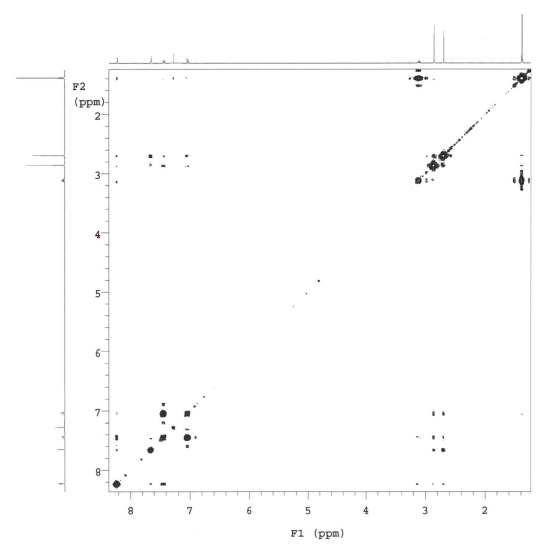

■ **FIGURE 11.6.*gcosy***   The 2-D ¹H-¹H gCOSY NMR spectrum of guaiazulene in CDCl₃.

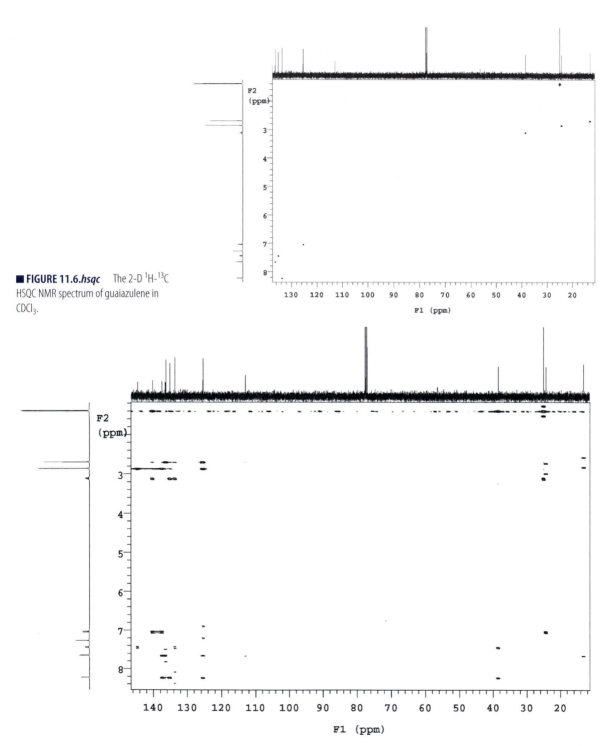

■ **FIGURE 11.6.***hsqc*   The 2-D ¹H-¹³C HSQC NMR spectrum of guaiazulene in CDCl₃.

■ **FIGURE 11.6.***ghmbc*   The 2-D ¹H-¹³C gHMB C NMR spectrum of guaiazulene in CDCl₃.

PROBLEM 11.7 **2-HYDROXY-3-PINANONE
IN CDCl₃ (SAMPLE 76)**

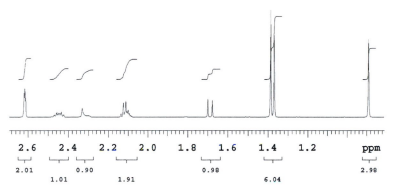

■ **FIGURE 11.7.h**   The 1-D ¹H NMR spectrum of 2-hydroxy-3-pinanone in CDCl₃.

■ **FIGURE 11.7.struc**   The structure of
2-hydroxy-3-pinanone (with numbering).

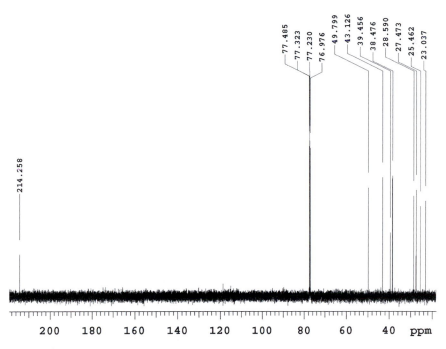

■ **FIGURE 11.7.c**   The 1-D ¹³C NMR spectrum of 2-hydroxy-3-pinanone in CDCl₃.

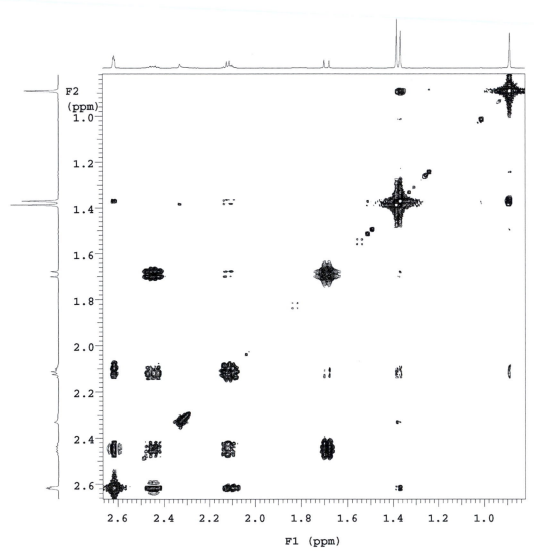

■ **FIGURE 11.7.***gcosy* The 2-D ^{1}H-^{1}H gCOSY NMR spectrum of 2–hydroxy-3–pinanone in CDCl$_3$.

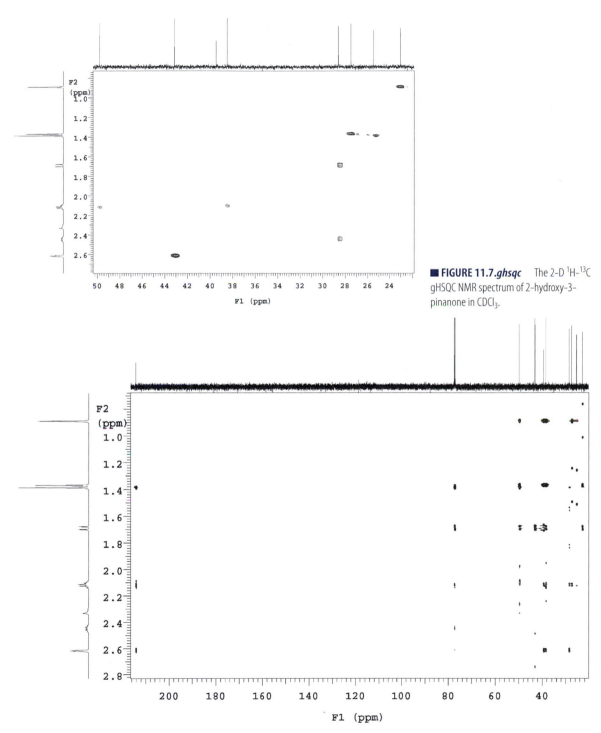

■ **FIGURE 11.7.***ghsqc*   The 2-D ^{1}H-^{13}C gHSQC NMR spectrum of 2-hydroxy-3-pinanone in CDCl$_3$.

■ **FIGURE 11.7.***ghmbc*   The 2-D ^{1}H-^{13}C gHMBC NMR spectrum of 2-hydroxy-3-pinanone in CDCl$_3$.

PROBLEM 11.8   **(R)-(+)-PERILLYL ALCOHOL IN CDCl₃ (SAMPLE 81)**

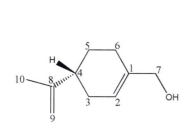

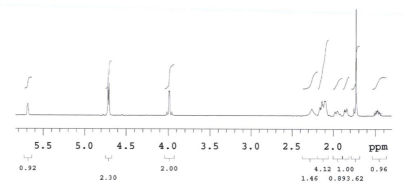

■ **FIGURE 11.8.struc**   The structure of (R)-(+)-perillyl alcohol (with numbering).

■ **FIGURE 11.8.h**   The 1-D ¹H NMR spectrum of (R)-(+)-perillyl alcohol in CDCl₃.

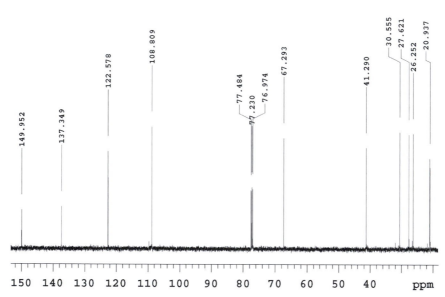

■ **FIGURE 11.8.c**   The 1-D ¹³C NMR spectrum of (R)-(+)-perillyl alcohol in CDCl₃.

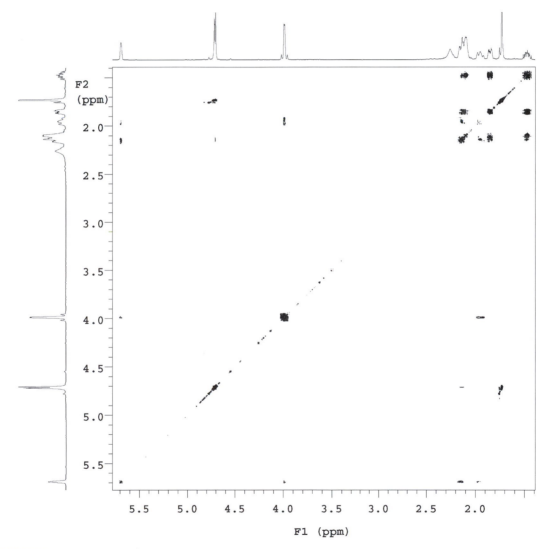

■ **FIGURE 11.8.***gcosy* The 2-D ^{1}H–^{1}H gCOSY NMR spectrum of (*R*)-(+)-perillyl alcohol in CDCl$_3$.

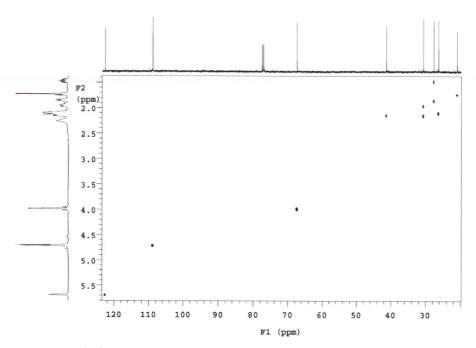

■ **FIGURE 11.8.***ghsqc*   The 2-D ^1H-^{13}C gHSQC NMR spectrum of (R)-(+)-perillyl alcohol in CDCl$_3$.

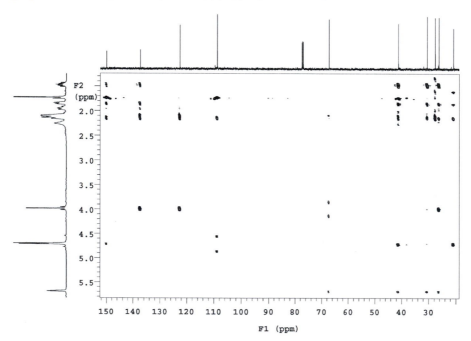

■ **FIGURE 11.8.***ghmbc*   The 2-D ^1H-^{13}C gHMBC NMR spectrum of (R)-(+)-perillyl alcohol in CDCl$_3$.

PROBLEM 11.9   **7-METHOXY-4-METHYLCOUMARIN IN CDCl₃ (SAMPLE 90)**

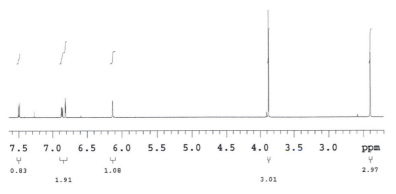

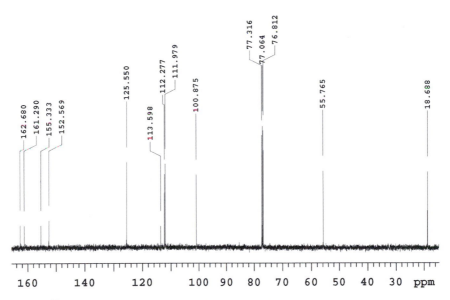

■ **FIGURE 11.9.*struc***   The structure of 7-methoxy -4-methylcoumarin (with numbering).

■ **FIGURE 11.9.*h***   The 1-D ¹H NMR spectrum of 7-methoxy-4-methylcoumarin in CDCl₃.

■ **FIGURE 11.9.*c***   The 1-D ¹³C NMR spectrum of 7-methoxy-4-methylcoumarin in CDCl₃.

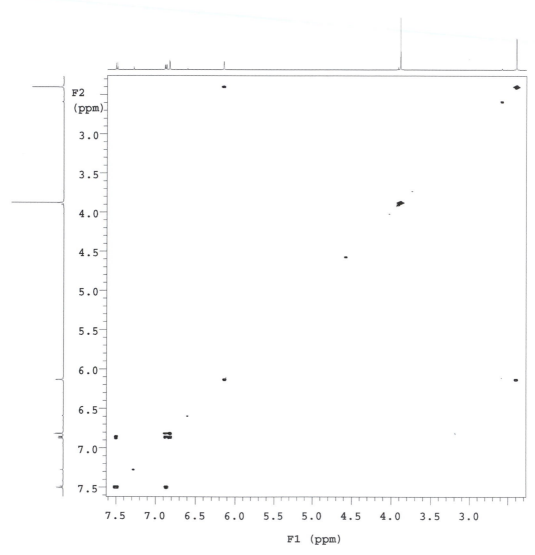

■ **FIGURE 11.9.***gcosy*   The 2-D ^{1}H–^{1}H gCOSY NMR spectrum of 7-methoxy-4-methylcoumarin in CDCl$_3$.

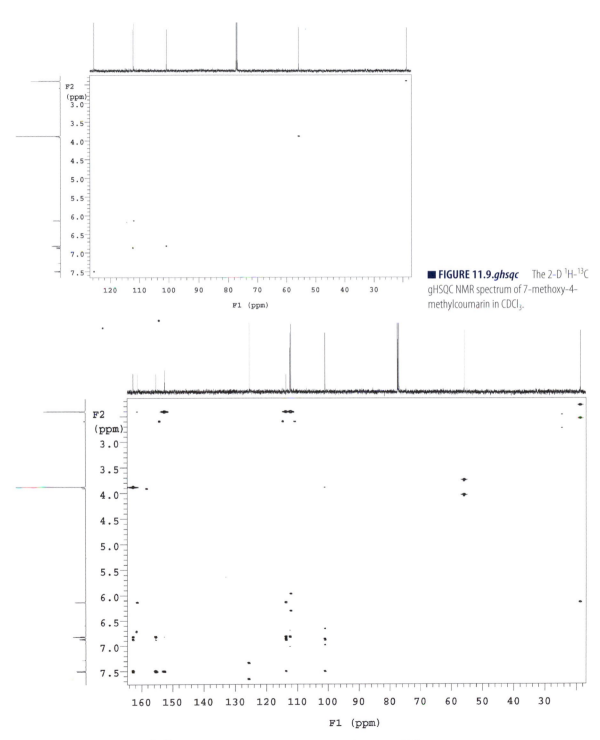

■ **FIGURE 11.9.*ghsqc*** The 2-D ^{1}H-^{13}C gHSQC NMR spectrum of 7-methoxy-4-methylcoumarin in CDCl$_3$.

■ **FIGURE 11.9.*ghmbc*** The 2-D ^{1}H-^{13}C gHMBC NMR spectrum of 7-methoxy-4-methylcoumarin in CDCl$_3$.

PROBLEM 11.10 **SUCROSE IN D₂O (SAMPLE 21)**

Recall that using a protic solvent such as D_2O will exchange away any ionizable ^1H's in the solute molecule.

■ FIGURE 11.10.struc The structure of sucrose (with numbering).

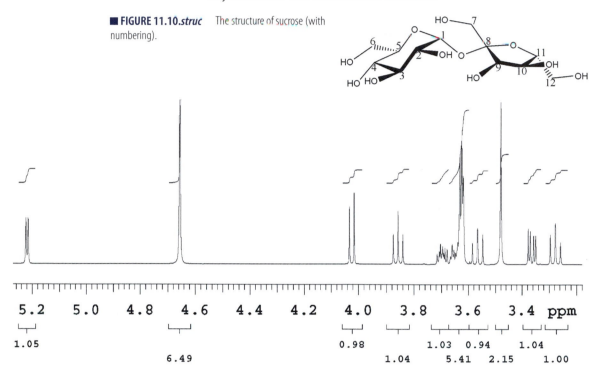

■ FIGURE 11.10.h The 1-D ^1H NMR spectrum of sucrose in D_2O.

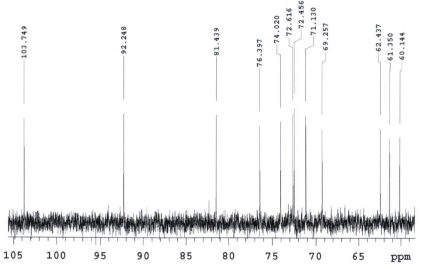

■ FIGURE 11.10.c The 1-D ^{13}C NMR spectrum of sucrose in D_2O.

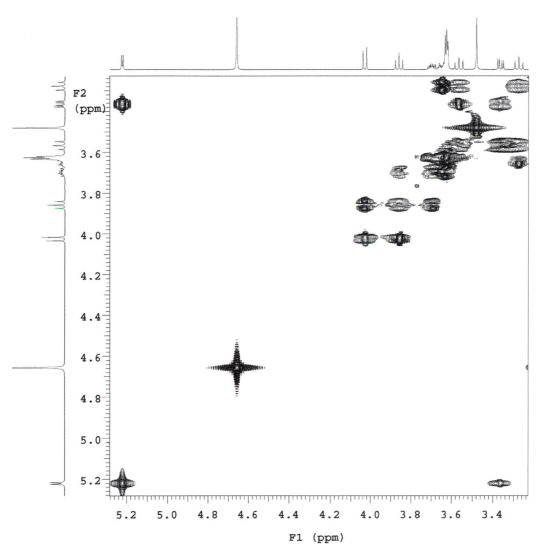

■ **FIGURE 11.10.***gcosy*   The 2-D ¹H-¹H gCOSY NMR spectrum of sucrose in D₂O.

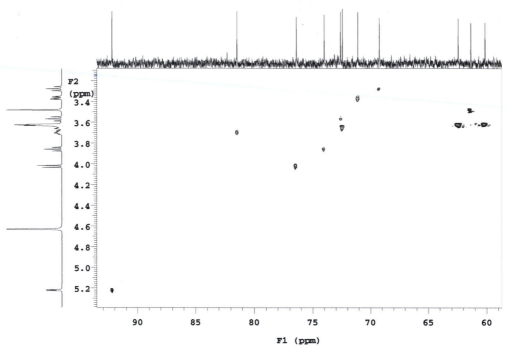

■ **FIGURE 11.10.***hmqc*   The 2-D ¹H-¹³C HMQC NMR spectrum of sucrose in D₂O.

Complex Assignment Problems

This chapter contains more difficult NMR resonance assignment problems. The structure of the molecule is given, and it is up to the reader to assign the ^1H and ^{13}C resonances.

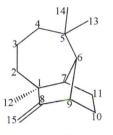

PROBLEM 12.1 **LONGIFOLENE IN CDCl₃ (SAMPLE 48)**

■ **FIGURE 12.1.*struc*** The structure of Longifolene (with numbering).

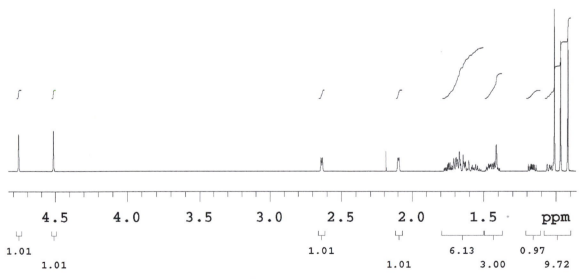

■ **FIGURE 12.1.*h*** The 1-D ^1H NMR spectrum of longifolene in CDCl₃.

233

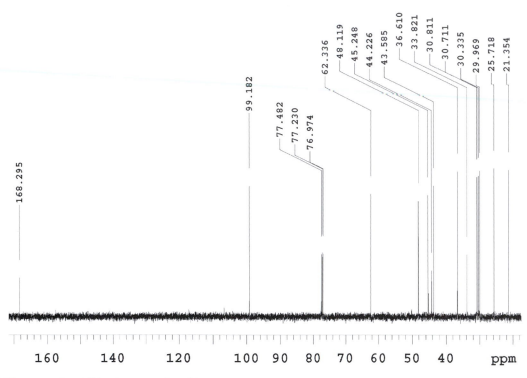

■ **FIGURE 12.1.c** The 1-D ^{13}C NMR spectrum of longifolene in CDCl$_3$.

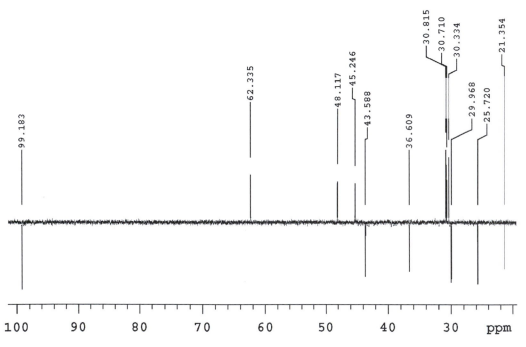

■ **FIGURE 12.1.dept135** The 1-D ^{13}C DEPT135 NMR spectrum of longifolene in CDCl$_3$.

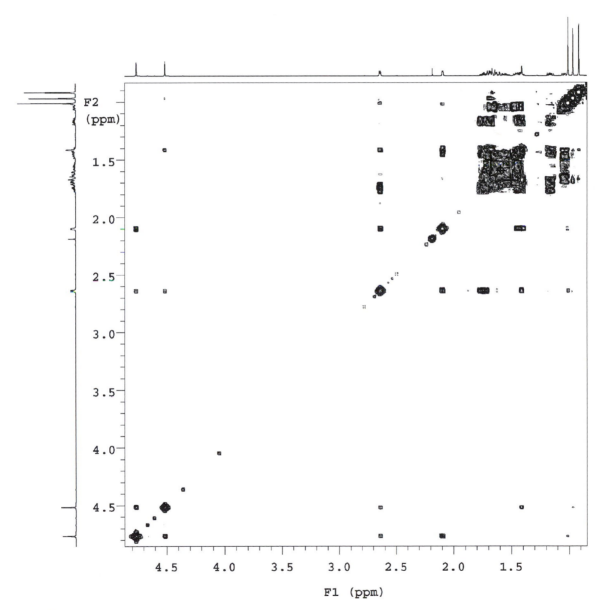

■ **FIGURE 12.1.*gcosy*** The 2-D ¹H–¹H gCOSY NMR spectrum of longifolene in CDCl₃.

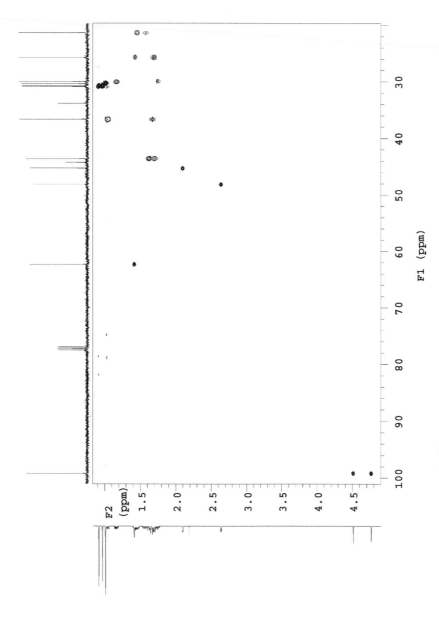

■ **FIGURE 12.1.hmqc** The 2-D ^{1}H–^{13}C HMQC NMR spectrum of longifolene in CDCl$_3$.

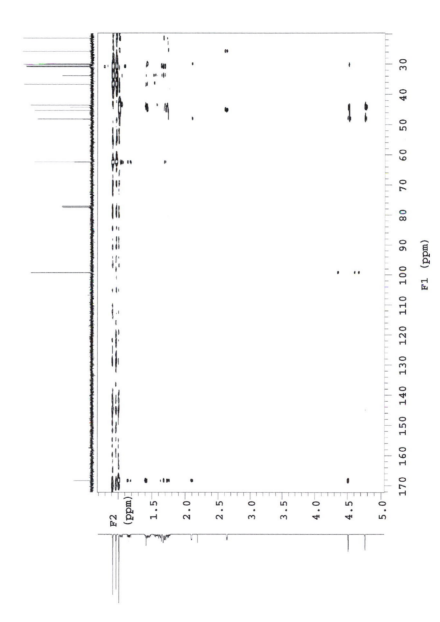

■ **FIGURE 12.1.hmbc** The 2-D ¹H–¹³C HMBC NMR spectrum of longifolene in CDCl₃.

PROBLEM 12.2 **(+)-LIMONENE IN CDCl$_3$ (SAMPLE 49)**

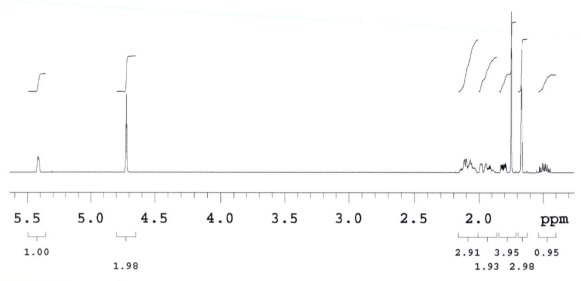

■ **FIGURE 12.2.h** The 1-D ^{1}H NMR spectrum of (+)-limonene in CDCl$_3$.

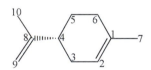

■ **FIGURE 12.2.struc** The structure of (+)-limonene (with numbering).

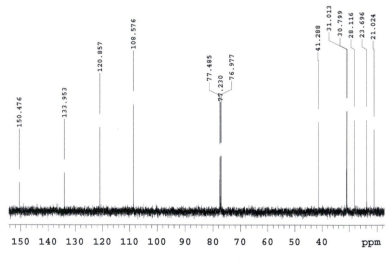

■ **FIGURE 12.2.c** The 1-D ^{13}C NMR spectrum of (+)-limonene in CDCl$_3$.

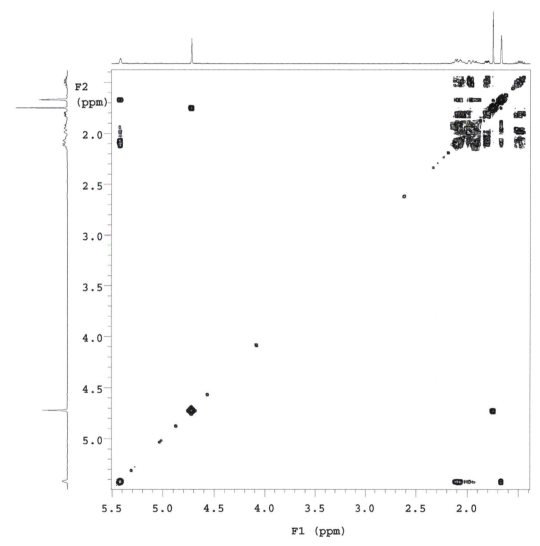

■ **FIGURE 12.2.*gcosy*** The 2-D ¹H-¹H gCOSY NMR spectrum of (+)-limonene in CDCl₃.

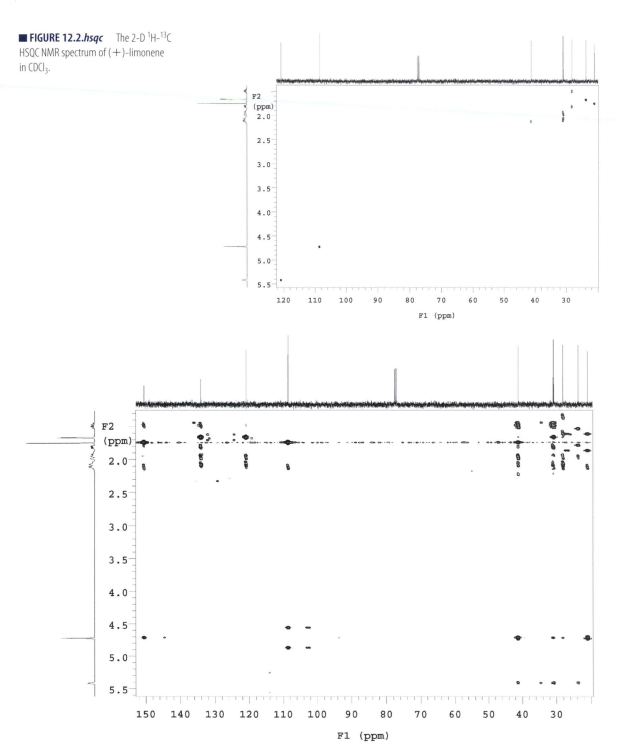

■ **FIGURE 12.2.***hsqc* The 2-D ^{1}H-^{13}C HSQC NMR spectrum of (+)-limonene in CDCl$_3$.

■ **FIGURE 12.2.***ghmbc* The 2-D ^{1}H-^{13}C gHMBC NMR spectrum of (+)-limonene in CDCl$_3$.

PROBLEM 12.3   **L-CINCHONIDINE IN CDCl₃ (SAMPLE 53)**

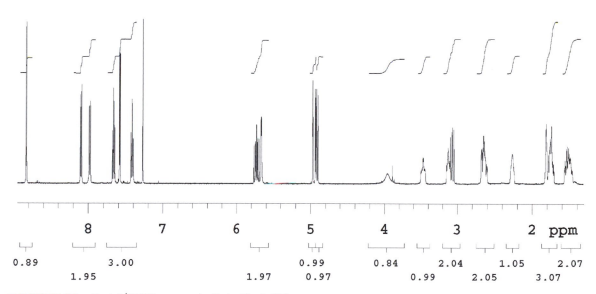

■ **FIGURE 12.3.*struc***   The structure of L-Cinchonidine (with numbering).

■ **FIGURE 12.3.*h***   The 1-D ¹H NMR spectrum of L-Cinchonidine in CDCl₃.

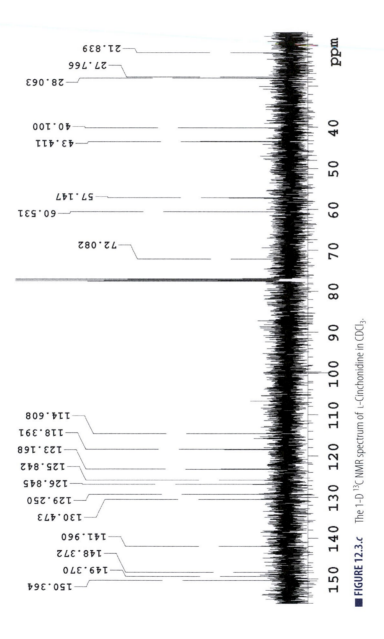

FIGURE 12.3.c   The 1-D ^{13}C NMR spectrum of L–Cinchonidine in CDCl$_3$.

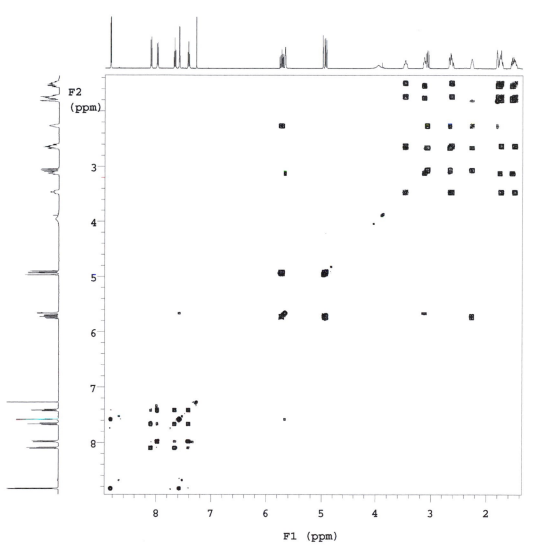

■ **FIGURE 12.3.*gcosy*** The 2-D ¹H–¹H gCOSY NMR spectrum of ʟ-Cinchonidine in CDCl₃.

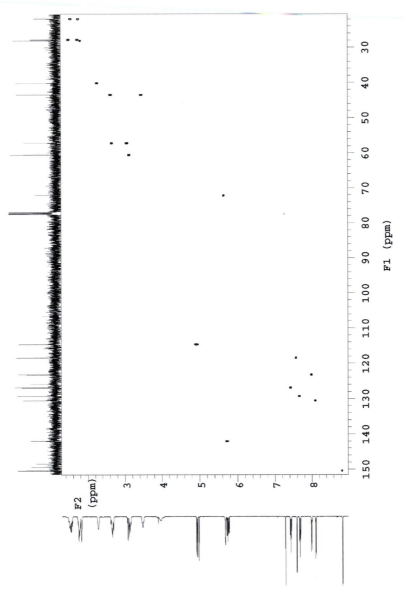

■ **FIGURE 12.3.***hsqc*   The 2–D ^{1}H–^{13}C HSQC NMR spectrum of L–Cinchonidine in CDCl$_3$.

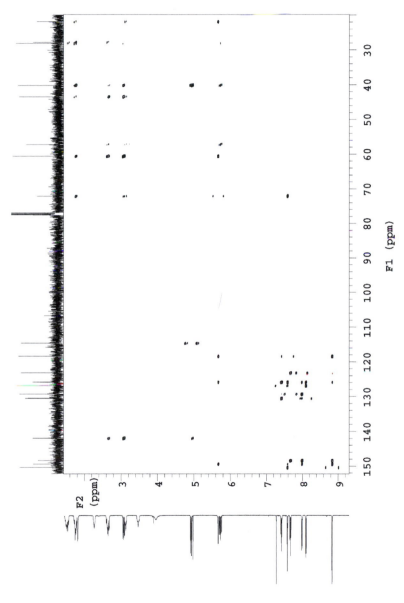

■ **FIGURE 12.3.ghmbc**   The 2-D ¹H-¹³C gHMBC NMR spectrum of L-Cinchonidine in CDCl₃.

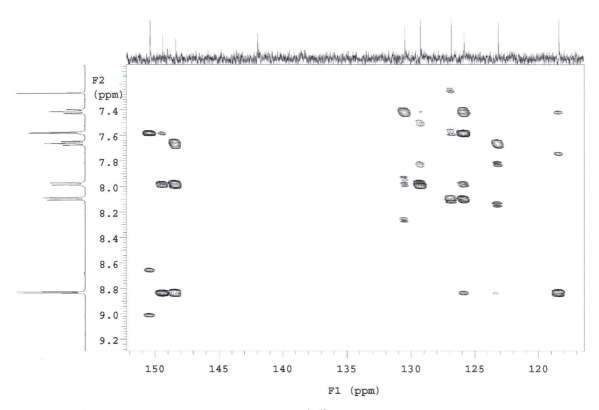

■ **FIGURE 12.3.*ghmbc_expansion*** An expansion of a portion of the 2-D ^{1}H-^{13}C gHMBC NMR spectrum of L-Cinchonidine in CDCl$_3$.

PROBLEM 12.4 **(3a*R*)-(+)-SCLAREOLIDE IN CDCl$_3$ (SAMPLE 54)**

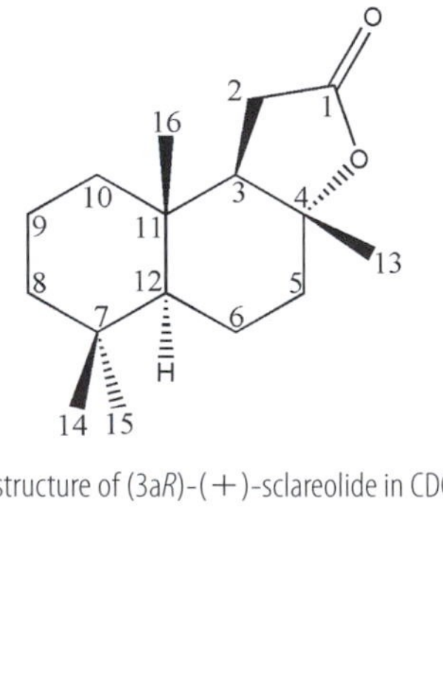

■ **FIGURE 12.4.*struc*** The structure of (3a*R*)-(+)-sclareolide in CDCl$_3$.

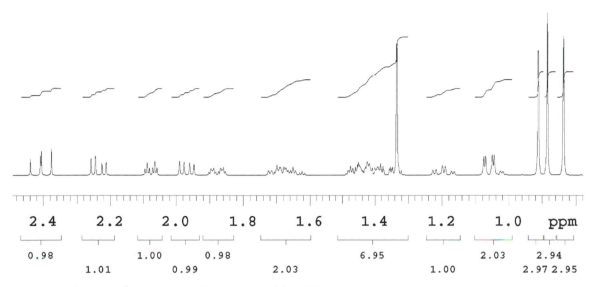

■ FIGURE 12.4.*h* The 1-D ¹H NMR spectrum of (3a*R*)-(+)-sclareolide in CDCl₃.

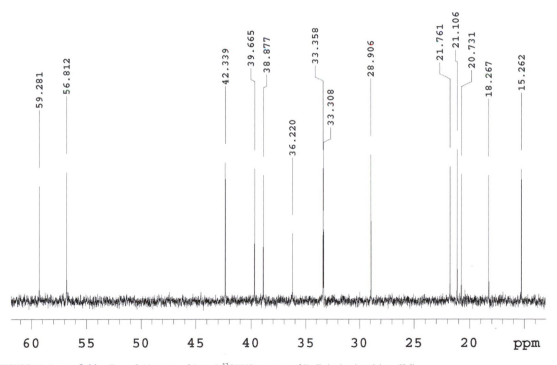

■ FIGURE 12.4.*c_upfield* The upfield portion of the 1-D ¹³C NMR spectrum of (3a*R*)-(+)-sclareolide in CDCl₃.

■ **FIGURE 12.4.*c_downfield*** The downfield portion of
the 1-D ^{13}C NMR spectrum of (3a*R*)-(+)-sclareolide in CDCl$_3$.

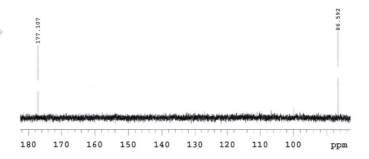

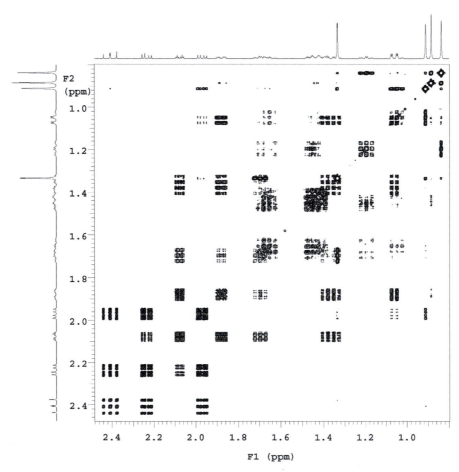

■ **FIGURE 12.4.*gcosy*** The 2-D ^{1}H-^{1}H gCOSY NMR spectrum of (3a*R*)-(+)-sclareolide in CDCl$_3$.

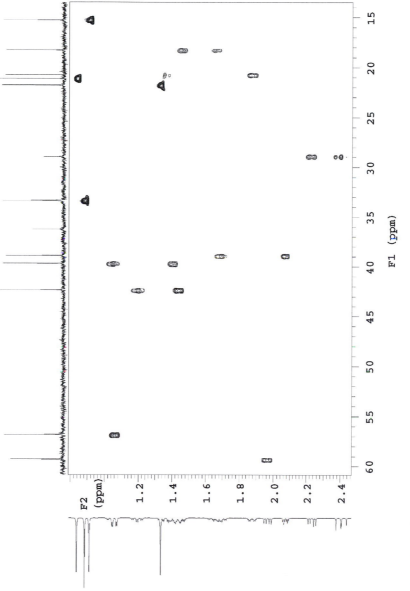

■ **FIGURE 12.4.*hsqc*** The 2-D ¹H–¹³C HSQC NMR spectrum of (3a*R*)-(+)-sclareolide in CDCl₃.

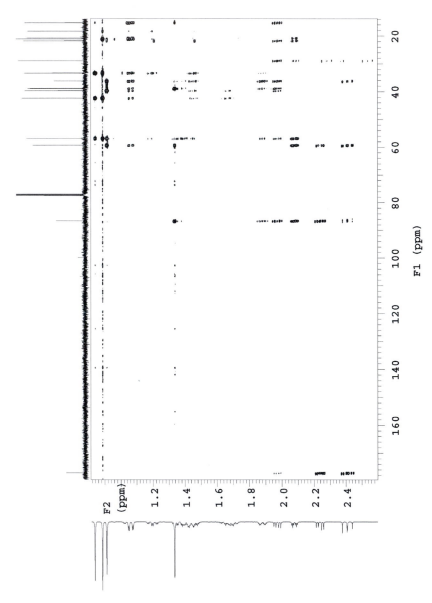

■ **FIGURE 12.4.ghmbc** The 2-D ^{1}H–^{13}C gHMBC NMR spectrum of (3aR)–(+)-sclareolide in CDCl$_3$.

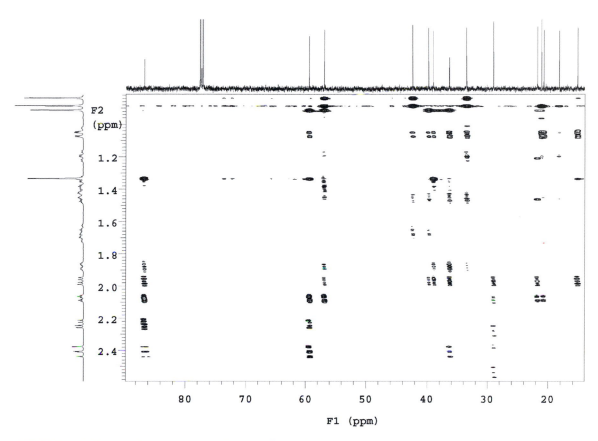

■ **FIGURE 12.4.*ghmbc_expansion*** An expansion of a portion of the 2-D ^{1}H-^{13}C gHMBC NMR spectrum of (3aR)-(+)-sclareolide in CDCl$_3$.

PROBLEM 12.5 **(−)-EPICATECHIN IN ACETONE-*d₆***
(SAMPLE 55)

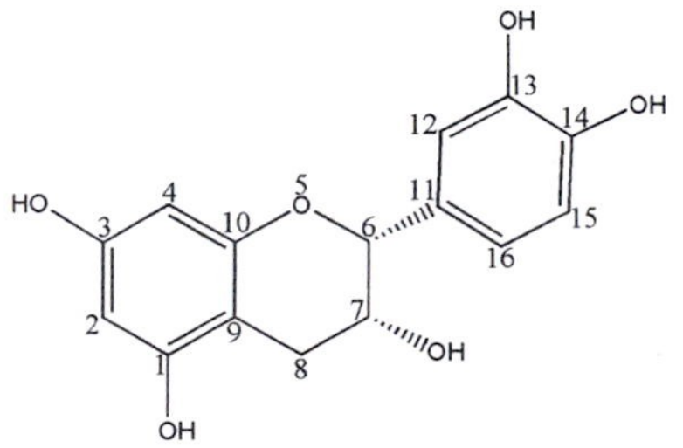

■ **FIGURE 12.5.*struc*** The structure of (−)-epicatechin in acetone-d_6 (with numbering).

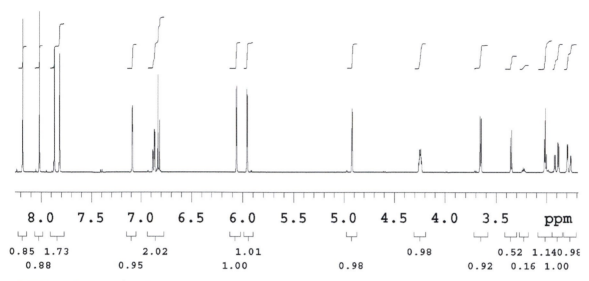

■ FIGURE 12.5.h   The 1-D ^{1}H NMR spectrum of (—)-epicatechin in acetone-d_6.

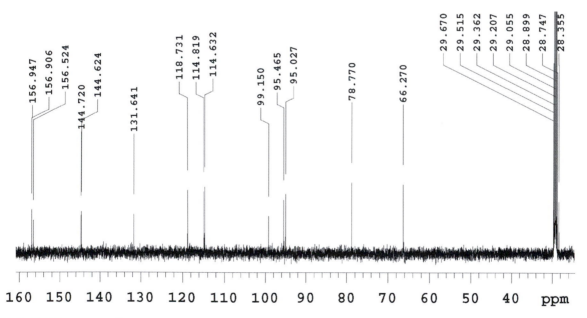

■ FIGURE 12.5.c   The 1-D ^{13}C NMR spectrum of (—)-epicatechin in acetone-d_6.

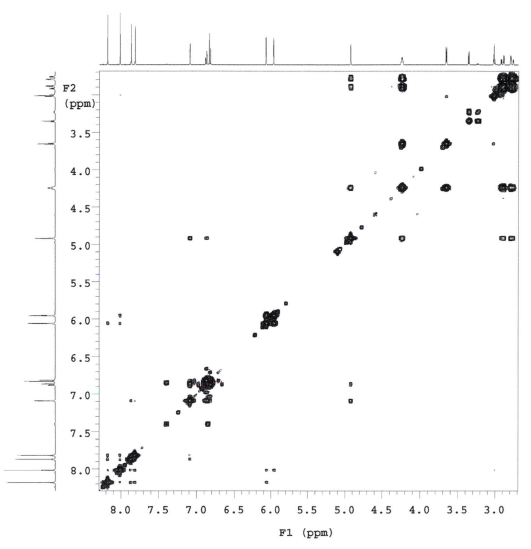

■ **FIGURE 12.5.gcosy** The 2-D ^{1}H-^{1}H gCOSY NMR spectrum of (−)-epicatechin in acetone-d_6.

■ **FIGURE 12.5.*hsqc*** The 2-D ^{1}H-^{13}C HSQC NMR spectrum of (−)-epicatechin in acetone-d_6.

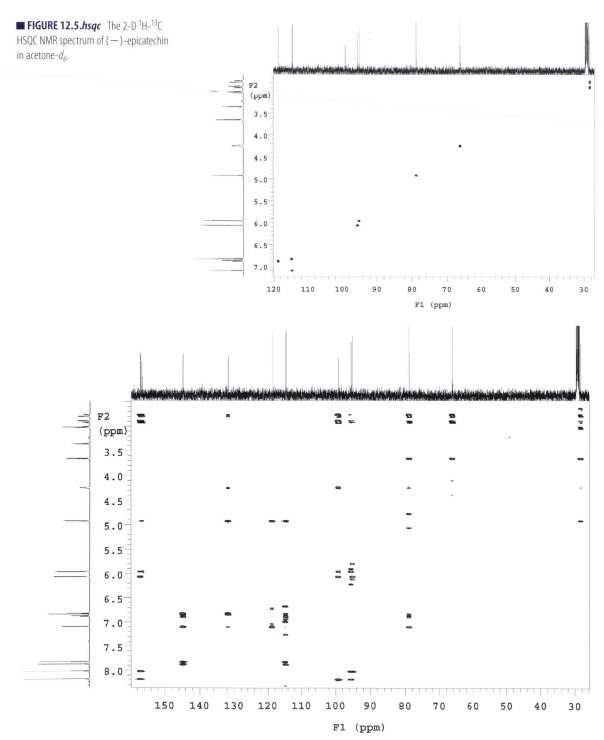

■ **FIGURE 12.5.*ghmbc*** The 2-D ^{1}H-^{13}C gHMBC NMR spectrum of (−)-epicatechin in acetone-d_6.

PROBLEM 12.6   **(−)-EBURNAMONINE IN CDCl₃ (SAMPLE 71)**

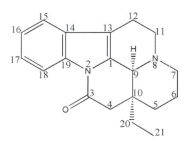

■ **FIGURE 12.6.struc** The structure of (—)-eburnamonine (with numbering).

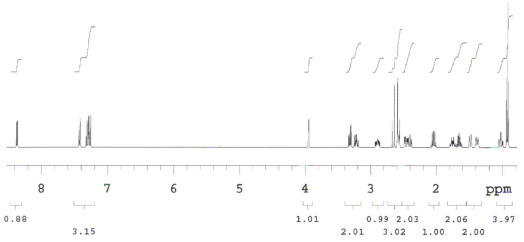

■ **FIGURE 12.6.h** The 1-D ¹H NMR spectrum of (—) -eburnamonine in CDCl₃.

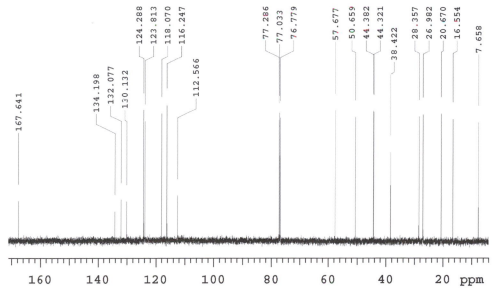

■ **FIGURE 12.6.c** The 1-D ¹³C NMR spectrum of (—)-eburnamonine in CDCl₃.

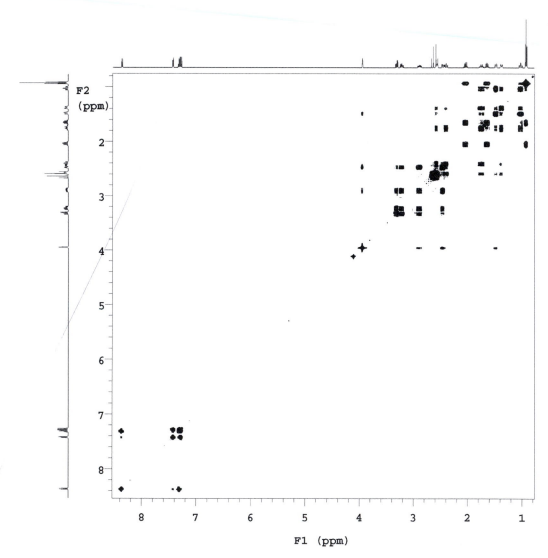

■ **FIGURE 12.6.***gcosy* The 2-D ¹H–¹H gCOSY NMR spectrum of (—)-eburnamonine in CDCl₃.

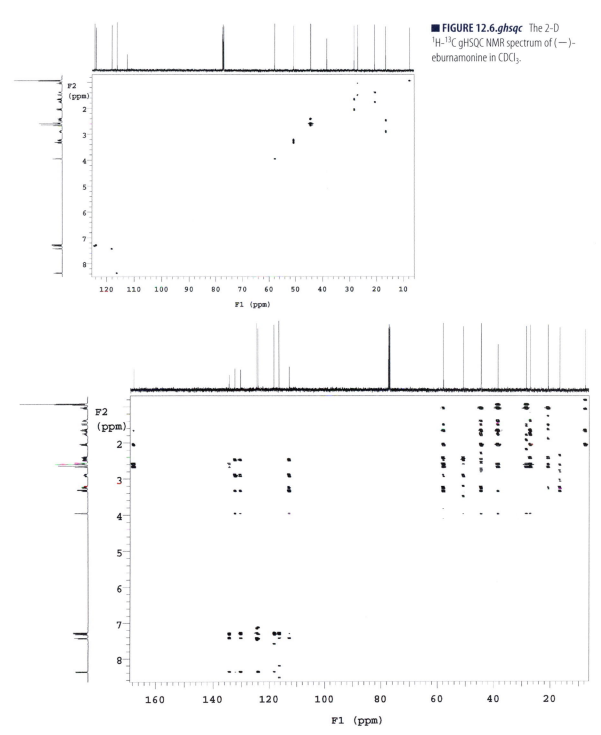

■ **FIGURE 12.6.*ghsqc*** The 2-D ¹H-¹³C gHSQC NMR spectrum of (—)-eburnamonine in CDCl₃.

■ **FIGURE 12.6.*ghmbc*** The 2-D ¹H-¹³C gHMBC NMR spectrum of (—)-eburnamonine in CDCl₃.

PROBLEM 12.7   ***TRANS*-MYRTANOL IN CDCl$_3$**
(SAMPLE 72/78)

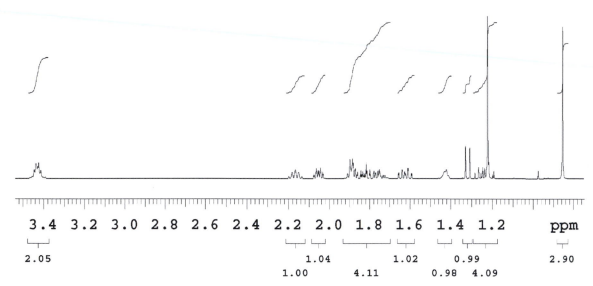

■ **FIGURE 12.7.*h*** The 1-D ^{1}H NMR spectrum of *trans*-myrtanol in CDCl$_3$.

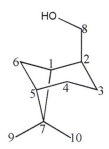

■ **FIGURE 12.7.*struc*** The structure of *trans*-myrtanol (with numbering).

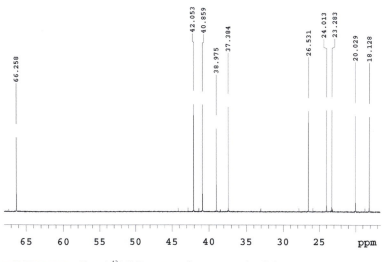

■ **FIGURE 12.7.*c*** The 1-D ^{13}C NMR spectrum of *trans*-myrtanol in CDCl$_3$.

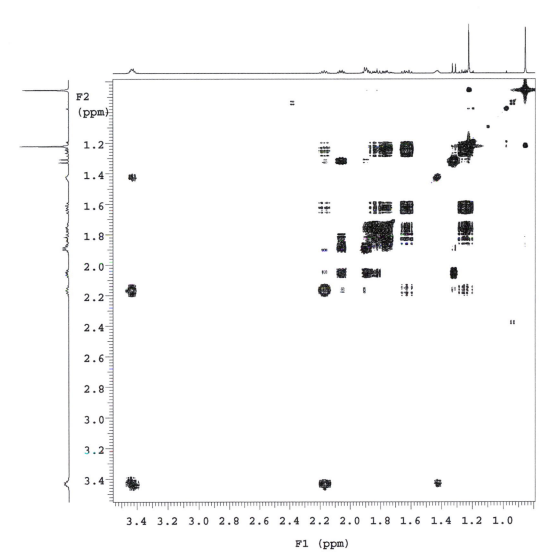

■ **FIGURE 12.7.*gcosy*** The 2-D ¹H-¹H gCOSY NMR spectrum of *trans*-myrtanol in CDCl₃.

■ **FIGURE 12.7.*ghsqc*** The 2-D ^1H-^{13}C gHSQC NMR spectrum of *trans*-myrtanol in CDCl$_3$.

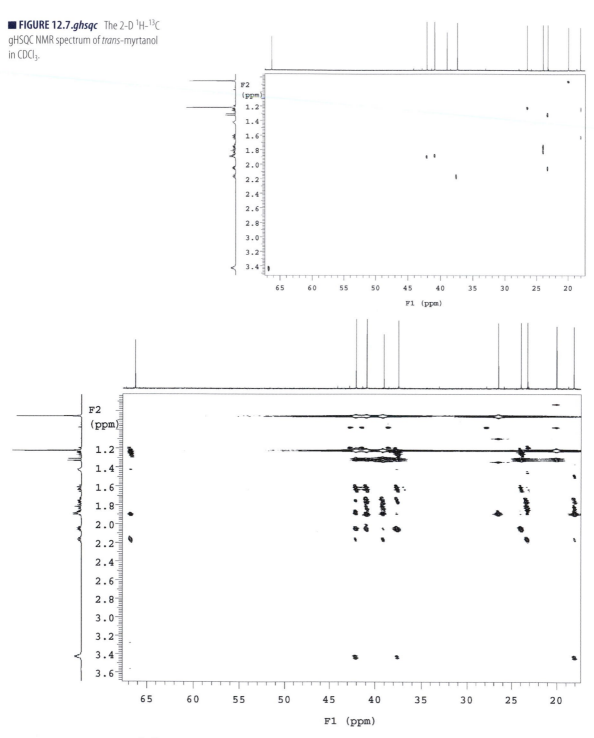

■ **FIGURE 12.7.*ghmbc*** The 2-D ^1H-^{13}C gHMBC NMR spectrum of *trans*-myrtanol in CDCl$_3$. Note the most intense cross peaks appear as "snow-capped" mountains.

PROBLEM 12.8   ***CIS*-MYRTANOL IN CDCl₃
(SAMPLE 73/77)**

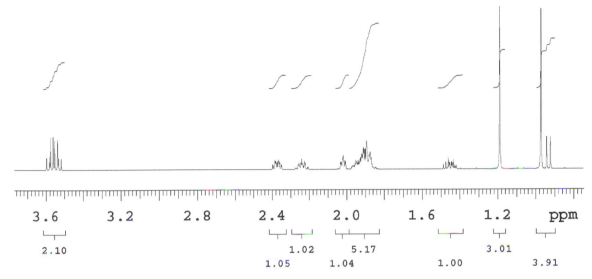

■ **FIGURE 12.8.*h*** The 1-D ¹H NMR spectrum of *cis*-myrtanol in CDCl₃.

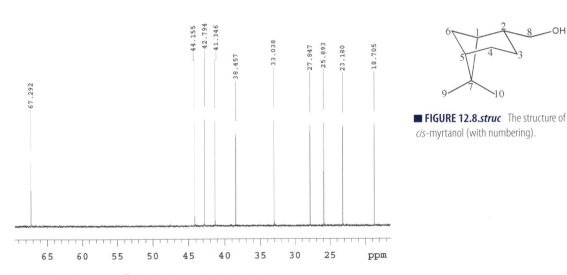

■ **FIGURE 12.8.*struc*** The structure of *cis*-myrtanol (with numbering).

■ **FIGURE 12.8.*c*** The 1-D ¹³C NMR spectrum of *cis*-myrtanol in CDCl₃.

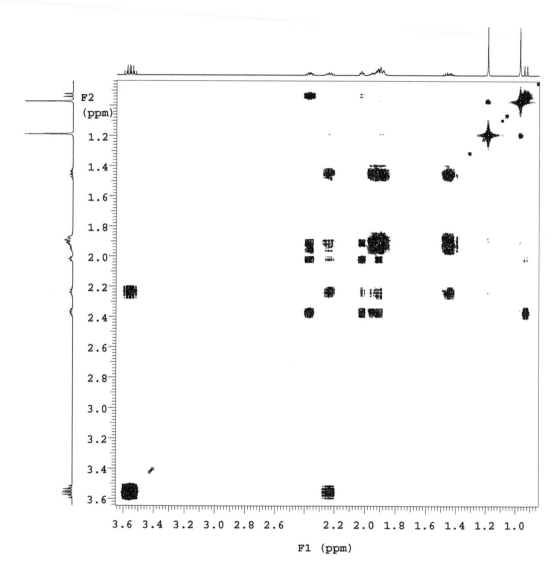

■ **FIGURE12.8.*gcosy*** The 2-D ¹H–¹H gCOSY NMR spectrum of *cis*-myrtanol in CDCl₃.

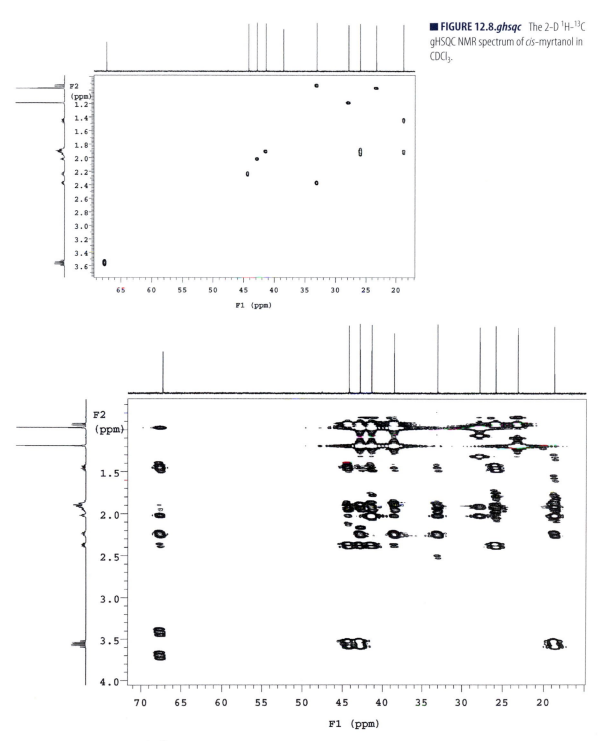

■ **FIGURE 12.8.***ghsqc* The 2-D ^{1}H-^{13}C gHSQC NMR spectrum of *cis*-myrtanol in CDCl₃.

■ **FIGURE 12.8.***ghmbc* The 2-D ^{1}H-^{13}C gHMBC NMR spectrum of *cis*-myrtanol in CDCl₃.

PROBLEM 12.9 **NARINGENIN IN ACETONE-d_6 (SAMPLE 89)**

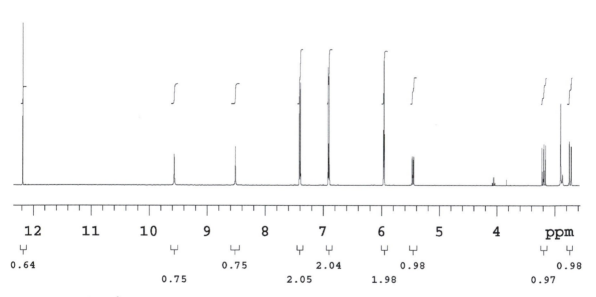

■ **FIGURE 12.9.struc** The structure of naringenin (with numbering).

■ **FIGURE 12.9.h** The 1-D ^{1}H NMR spectrum of naringenin in acetone-d_6.

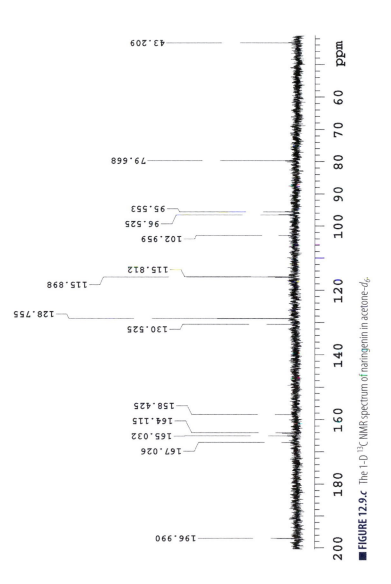

■ **FIGURE 12.9.c** The 1-D ¹³C NMR spectrum of naringenin in acetone-d_6.

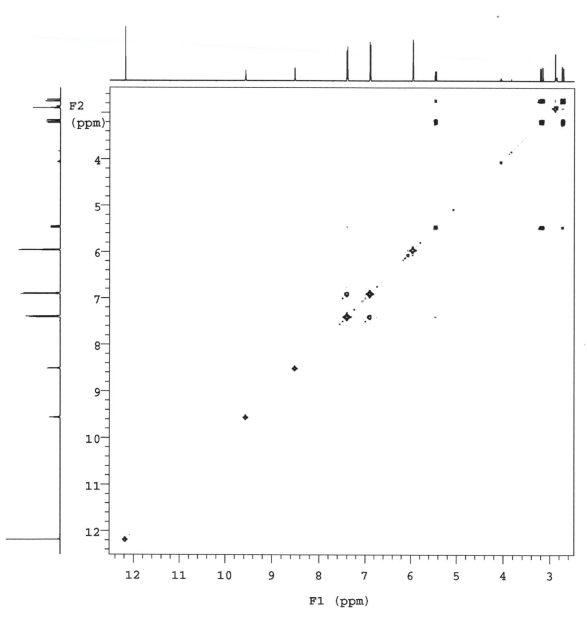

■ **FIGURE 12.9.*gcosy*** The 2-D ^{1}H-^{1}H gCOSY NMR spectrum of naringenin in acetone-d_6.

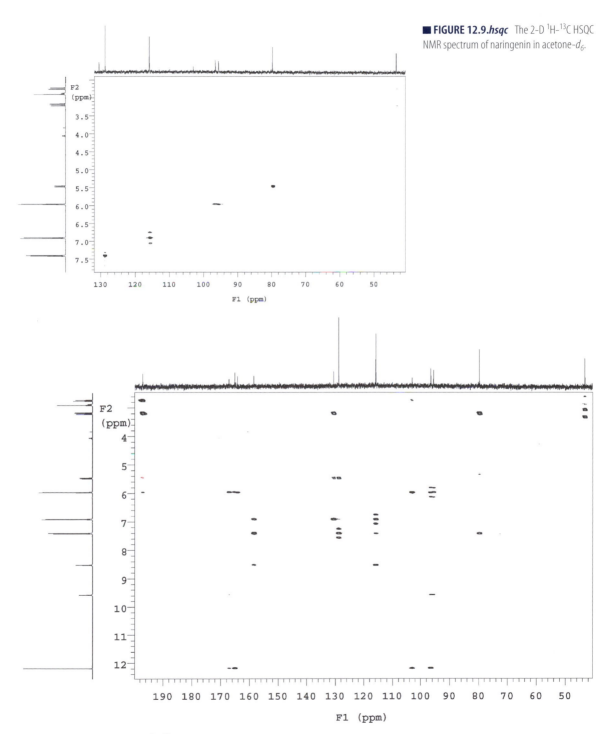

■ **FIGURE 12.9.*hsqc*** The 2-D ^{1}H-^{13}C HSQC NMR spectrum of naringenin in acetone-d$_6$.

■ **FIGURE 12.9.*ghmbc*** The 2-D ^{1}H-^{13}C gHMBC NMR spectrum of naringenin in acetone-d$_6$.

PROBLEM 12.10 **(−)-AMBROXIDE IN CDCl₃ (SAMPLE AMBROXIDE)**

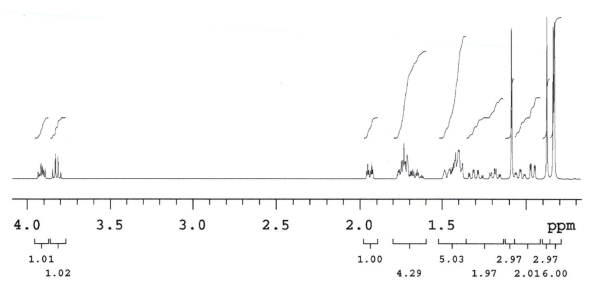

■ **FIGURE 12.10.*h*** The 1-D ¹H NMR spectrum of (−)-ambroxide in CDCl₃.

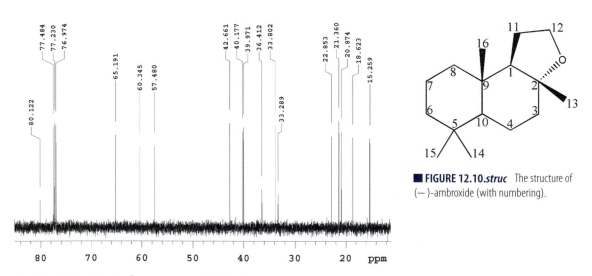

■ **FIGURE 12.10.*c*** The 1-D ¹³C NMR spectrum of (−)-ambroxide in CDCl₃.

■ **FIGURE 12.10.*struc*** The structure of (−)-ambroxide (with numbering).

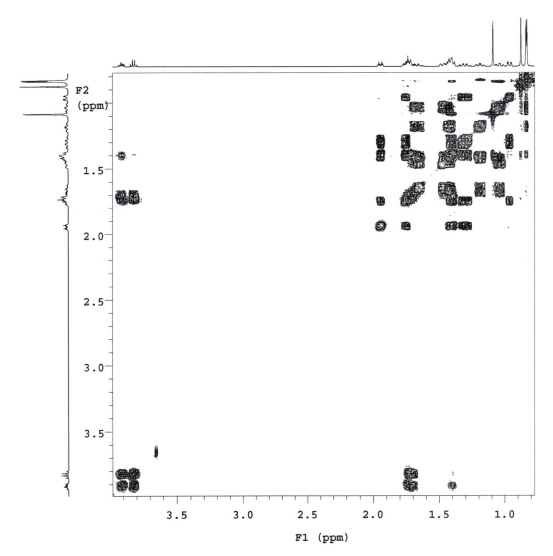

■ **FIGURE 12.10.***gcosy* The 2-D ¹H-¹H gCOSY NMR spectrum of (—)-ambroxide in CDCl₃.

■ **FIGURE 12.10.***hsqc* The 2-D ^{1}H-^{13}C HSQC NMR spectrum of (—)-ambroxide in CDCl$_3$.

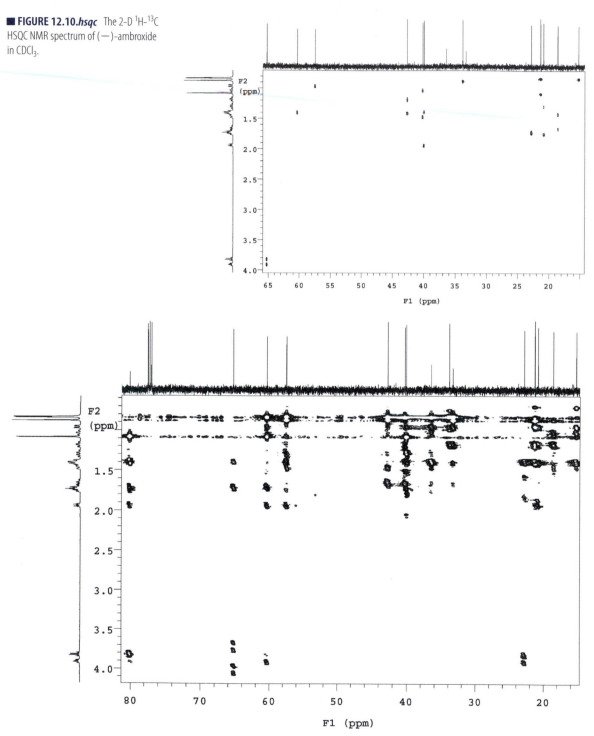

■ **FIGURE 12.10.***ghmbc* The 2-D ^{1}H-^{13}C gHMBC NMR spectrum of (—)-ambroxide in CDCl$_3$.

Chapter 13

Simple Unknown Problems

This chapter contains relatively simple NMR unknown problems. The molecular weight or empirical formula of the molecule is sometimes given, and it is up to the reader to elucidate the structure of the unknown. If you are an instructor and wish to confirm that you have the correct structure, you may contact the author via e-mail at jsimpson@mit.edu.

PROBLEM 13.1 **UNKNOWN 13.1 IN CDCl$_3$ (SAMPLE 20)**

The ^{13}C spectrum is missing a resonance at 157.0 ppm. Unknown 13.1 has a MW of 184.62.

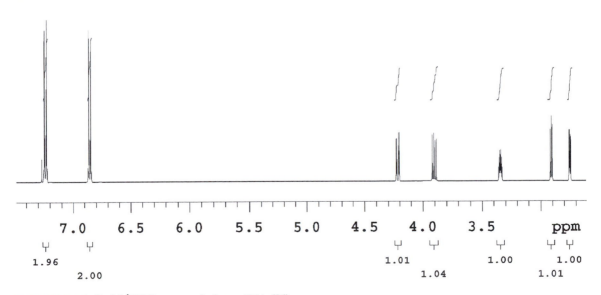

■ **FIGURE 13.1.h** The 1-D ^{1}H NMR spectrum of unknown 13.1 in CDCl$_3$.

■ **FIGURE 13.1.*c*** The 1-D ¹³C NMR spectrum of unknown 13.1 in CDCl₃.

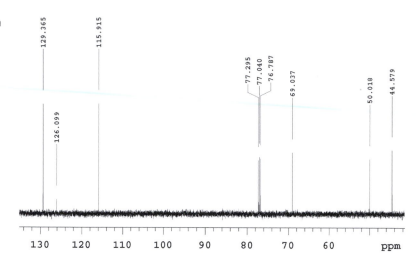

■ **FIGURE 13.1.*gcosy*** The 2-D ¹H-¹H gCOSY NMR spectrum of unknown 13.1 in CDCl₃.

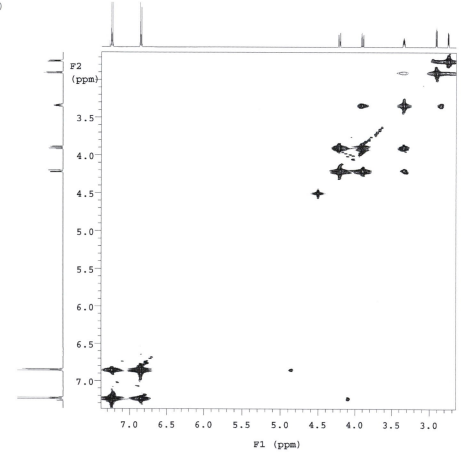

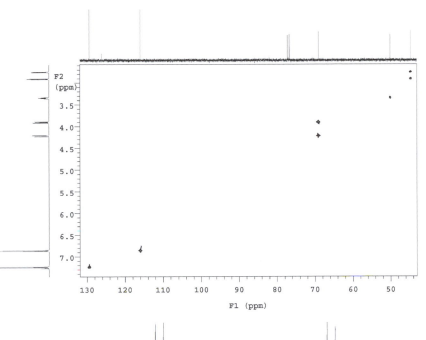

■ **FIGURE 13.1.***hmqc* The 2-D ¹H-¹³C HMQC NMR spectrum of unknown 13.1 in CDCl₃.

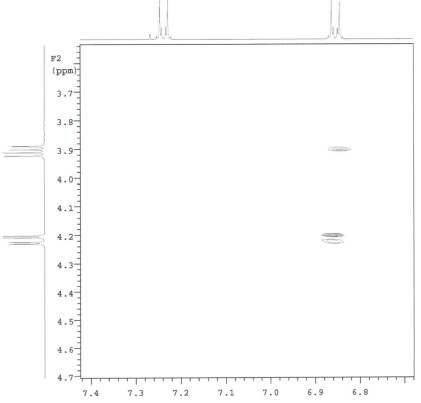

■ **FIGURE 13.1.***noesy* A portion of the 2-D ¹H-¹H NOESY NMR spectrum of unknown 13.1 in CDCl₃.

PROBLEM 13.2 **UNKNOWN 13.2 IN CDCl$_3$ (SAMPLE 41)**

The empirical formula of unknown 13.2 is C$_{15}$H$_{12}$O.

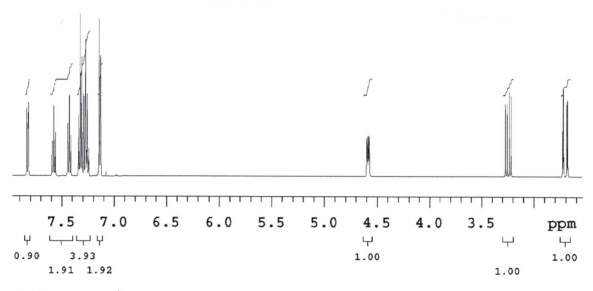

■ FIGURE 13.2.h The 1-D ^{1}H NMR spectrum of unknown 13.2 in CDCl$_3$.

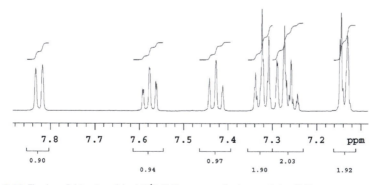

■ FIGURE 13.2.h_downfield The downfield region of the 1-D ^{1}H NMR spectrum of unknown 13.2 in CDCl$_3$.

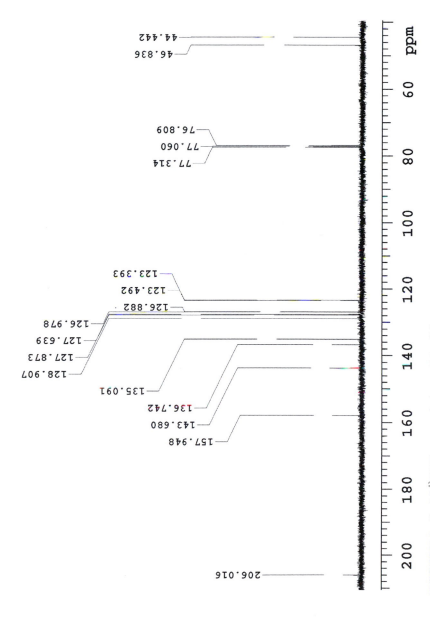

■ **FIGURE 13.2.c** The 1-D ¹³C NMR spectrum of unknown 13.2 in CDCl₃.

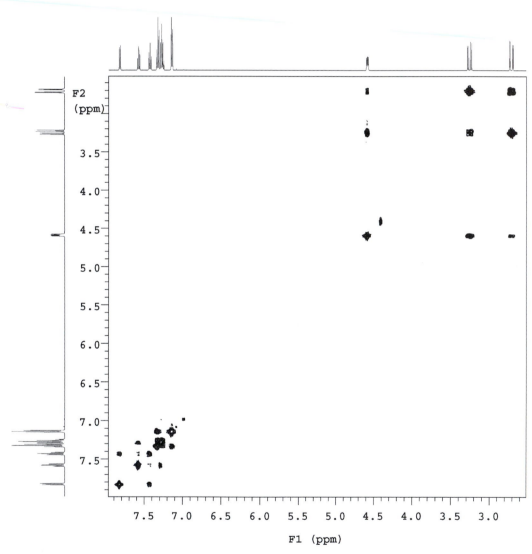

■ **FIGURE 13.2.***gcosy* The 2-D ^{1}H–^{1}H gCOSY NMR spectrum of unknown 13.2 in CDCl$_3$.

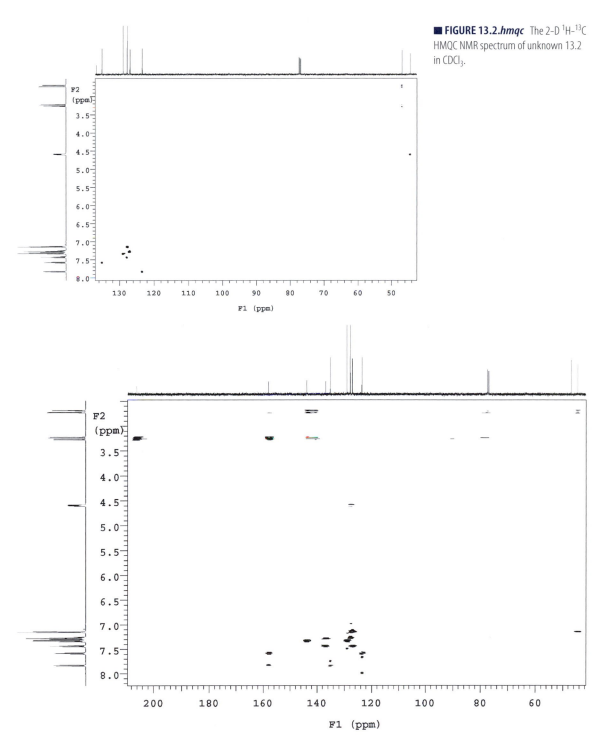

■ **FIGURE 13.2.*hmqc*** The 2-D ¹H-¹³C HMQC NMR spectrum of unknown 13.2 in CDCl₃.

■ **FIGURE 13.2.*hmbc*** The 2-D ¹H-¹³C HMBC NMR spectrum of unknown 13.2 in CDCl₃.

PROBLEM 13.3 **UNKNOWN 13.3 IN CDCl$_3$ (SAMPLE 22)**

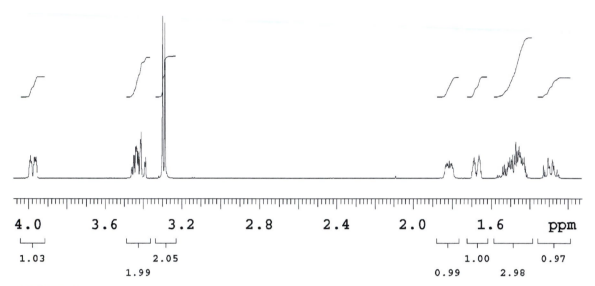

■ **FIGURE 13.3.h** The 1-D ^{1}H NMR spectrum of unknown 13.3 in CDCl$_3$.

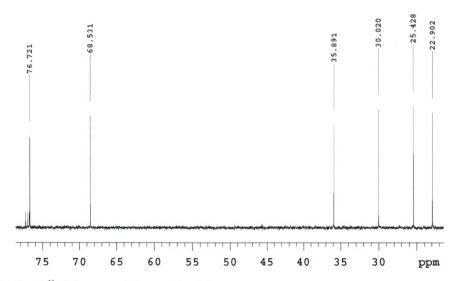

■ **FIGURE 13.3.c** The 1-D ^{13}C NMR spectrum of unknown 13.3 in CDCl$_3$.

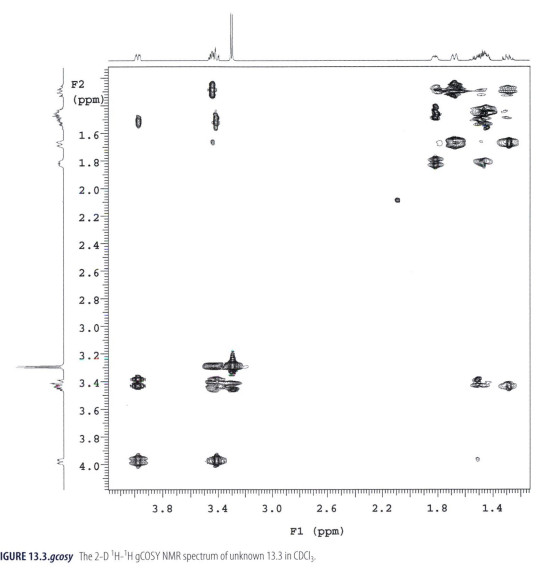

■ **FIGURE 13.3.*gcosy*** The 2-D ¹H-¹H gCOSY NMR spectrum of unknown 13.3 in CDCl₃.

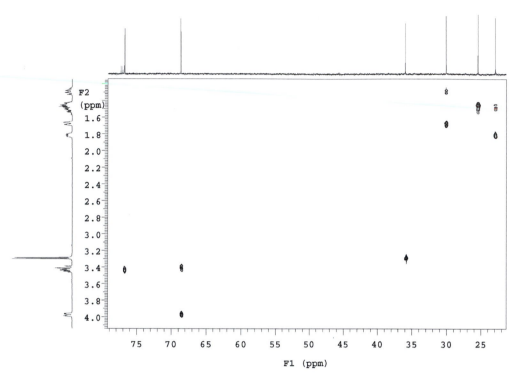

■ **FIGURE 13.3.*hmqc*** The 2-D ^1H-^{13}C HMQC NMR spectrum of unknown 13.3 in $CDCl_3$.

PROBLEM 13.4 **UNKNOWN 13.4 IN CDCl$_3$ (SAMPLE 24)**

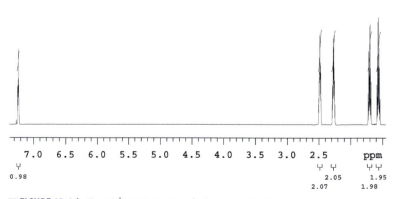

■ **FIGURE 13.4.*h*** The 1-D ^1H NMR spectrum of unknown 13.4 in $CDCl_3$.

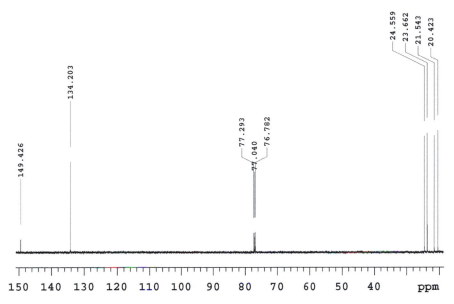

■ **FIGURE 13.4.c** The 1-D ^{13}C NMR spectrum of unknown 13.4 in CDCl$_3$.

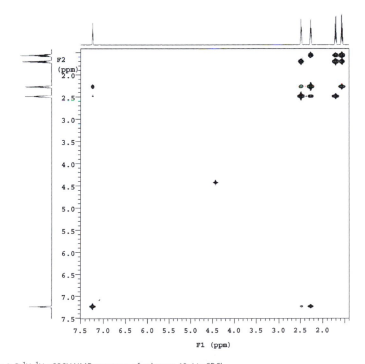

■ **FIGURE 13.4.gcosy** The 2-D ^{1}H-^{1}H gCOSY NMR spectrum of unknown 13.4 in CDCl$_3$.

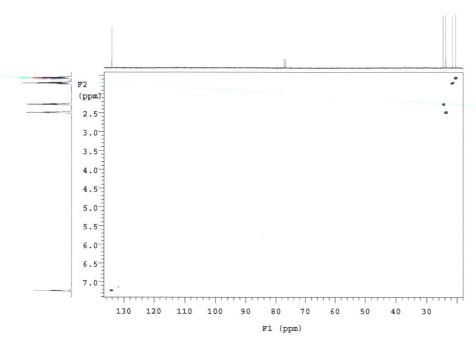

■ **FIGURE 13.4.***hmqc* The 2-D ^{1}H-^{13}C HMQC NMR spectrum of unknown 13.4 in CDCl$_3$.

PROBLEM 13.5 **UNKNOWN 13.5 IN CDCl$_3$ (SAMPLE 34)**

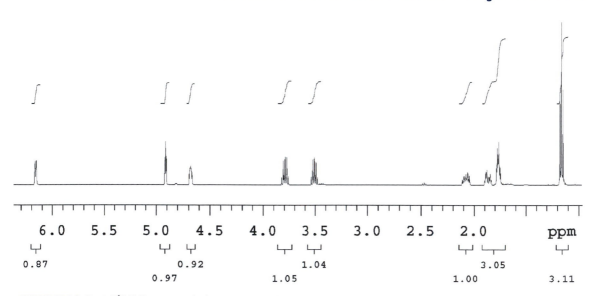

■ **FIGURE 13.5.***h* The 1-D ^{1}H NMR spectrum of unknown 13.5 in CDCl$_3$.

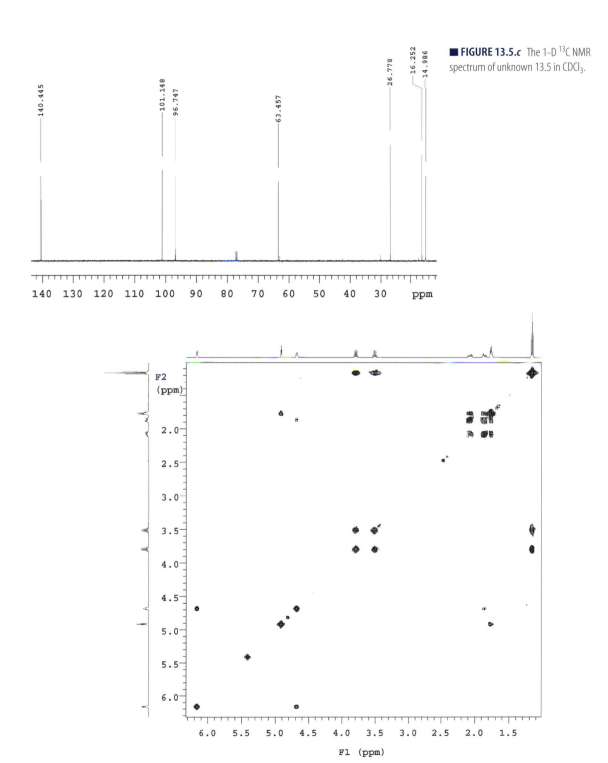

■ **FIGURE 13.5.c** The 1-D ^{13}C NMR spectrum of unknown 13.5 in CDCl$_3$.

■ **FIGU RE 13.5.gcosy** The 2-D ^{1}H–^{1}H gCOSY NMR spectrum of unknown 13.5 in CDCl$_3$.

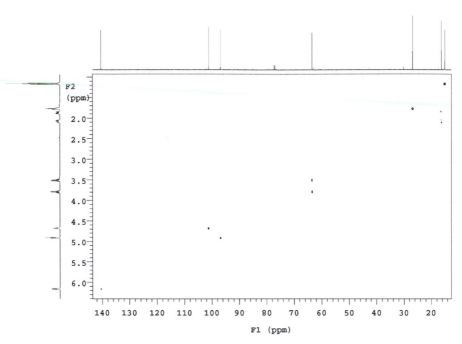

■ **FIGURE 13.5.***hmqc* The 2-D ^{1}H-^{13}C HMQC NMR spectrum of unknown 13.5 in CDCl$_3$.

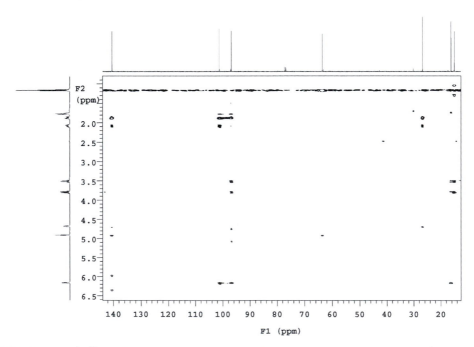

■ **FIGURE 13.5.***hmbc* The 2-D ^{1}H-^{13}C HMBC NMR spectrum of unknown 13.5 in CDCl$_3$.

PROBLEM 13.6 **UNKNOWN 13.6 IN CDCl₃ (SAMPLE 36)**

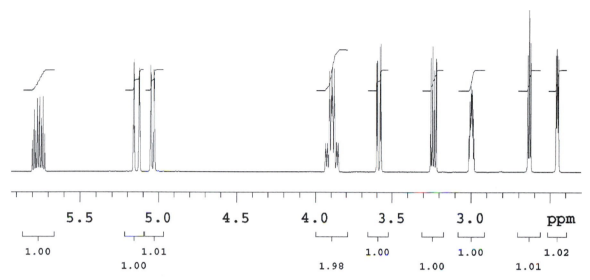

■ **FIGURE 13.6.h** The 1-D ^{1}H NMR spectrum of unknown 13.6 in CDCl₃.

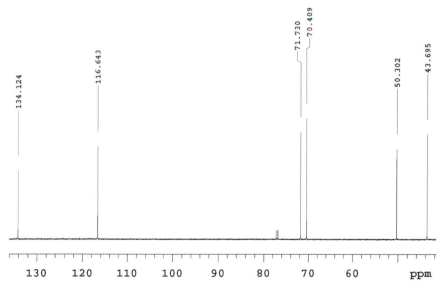

■ **FIGURE 13.6.c** The 1-D ^{13}C NMR spectrum of unknown 13.6 in CDCl₃.

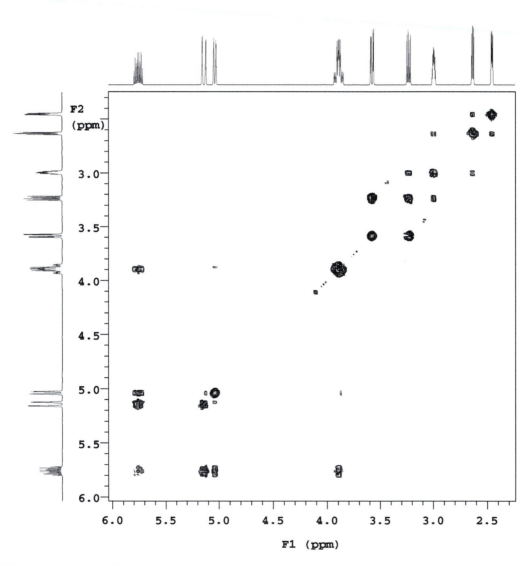

■ **FIGURE 13.6.***gcosy* The 2-D ^1H-^1H gCOSY NMR spectrum of unknown 13.6 in $CDCl_3$.

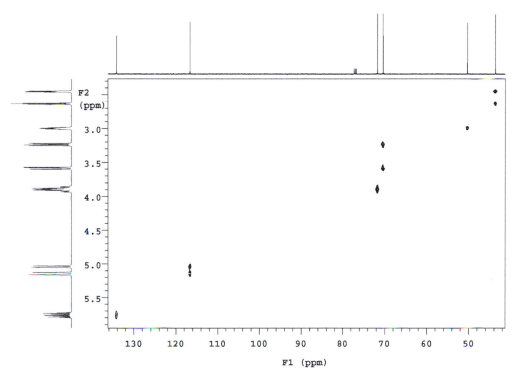

■ **FIGURE 13.6.*hmqc*** The 2-D ¹H-¹³C HMQC NMR spectrum of unknown 13.6 in CDCl₃.

PROBLEM 13.7 **UNKNOWN 13.7 IN CDCl₃ (SAMPLE 50)**

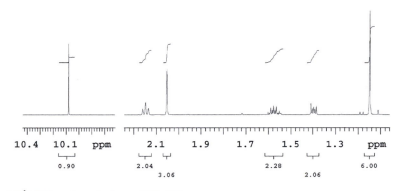

■ **FIGURE 13.7.*h*** The 1-D ¹H NMR spectrum of unknown 13.7 in CDCl₃.

■ **FIGURE 13.7.c** The 1-D ^{13}C NMR spectrum of unknown 13.7 in CDCl$_3$.

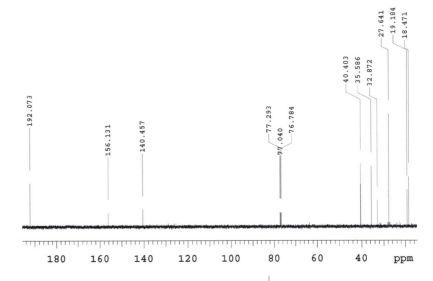

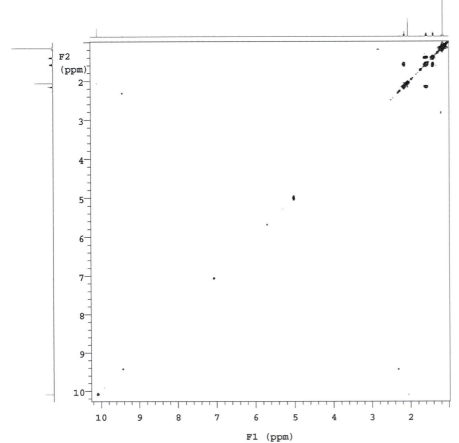

■ **FIGURE 13.7.gcosy** The 2-D ^{1}H-^{1}H gCOSY NMR spectrum of unknown 13.7 in CDCl$_3$.

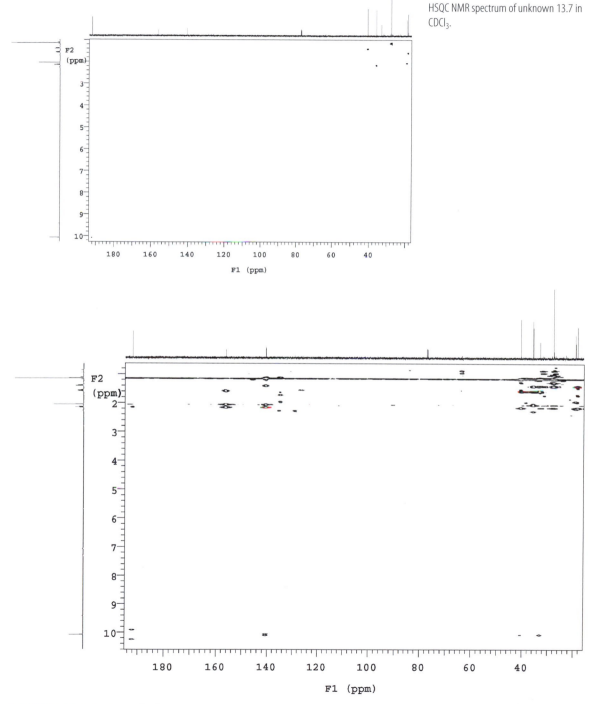

■ **FIGURE 13.7.*hsqc*** The 2-D ¹H-¹³C HSQC NMR spectrum of unknown 13.7 in CDCl₃.

■ **FIGURE 13.7.*ghmbc*** The 2-D ¹H-¹³C gHMBC NMR spectrum of unknown 13.7 in CDCl₃.

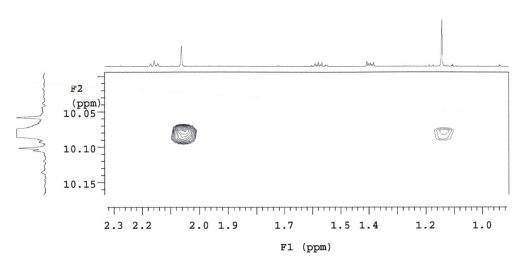

■ **FIGURE 13.7.*roesy*** The 2-D ^1H-^{13}C gHMBC NMR spectrum of unknown 13.7 in CDCl$_3$.

PROBLEM 13.8 **UNKNOWN 13.8 IN CDCl$_3$**
(SAMPLE 83)

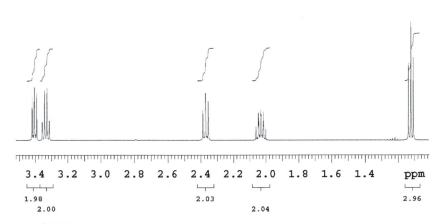

■ **FIGURE 13.8.*h*** The 1-D ^1H NMR spectrum of unknown 13.8 in CDCl$_3$.

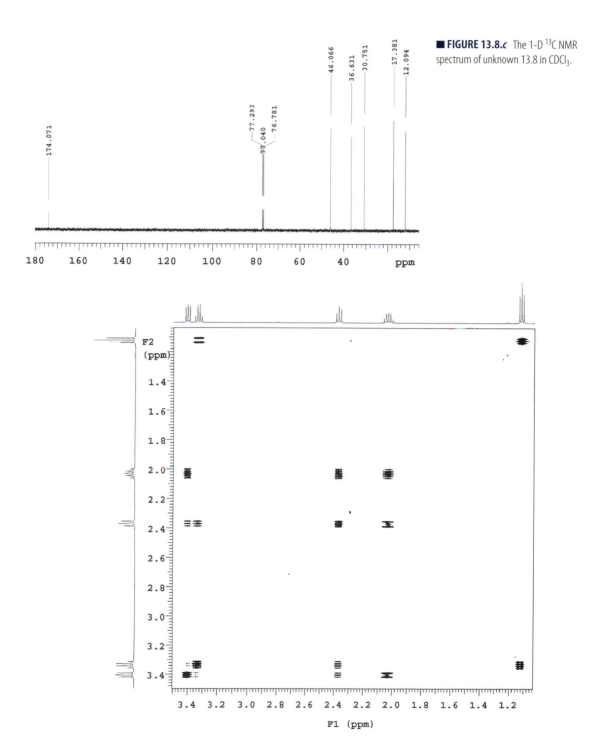

■ **FIGURE 13.8.c** The 1-D ^{13}C NMR spectrum of unknown 13.8 in CDCl₃.

■ **FIGURE 13.8.gcosy** The 2-D ^{1}H–^{1}H gCOSY NMR spectrum of unknown 13.8 in CDCl₃.

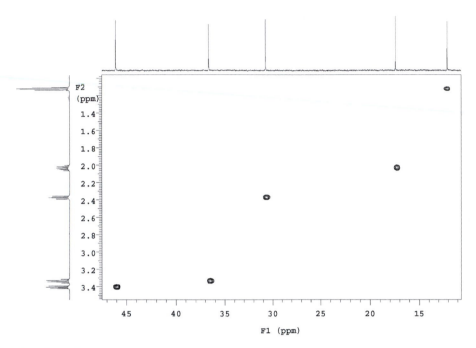

■ **FIGURE 13.8.*ghsqc*** The 2-D ^{1}H–^{13}C gHSQC NMR spectrum of unknown 13.8 in CDCl$_3$.

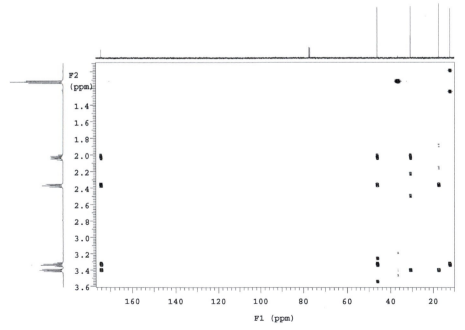

■ **FIGURE 13.8.*ghmbc*** The 2-D ^{1}H–^{13}C gHMBC NMR spectrum of unknown 13.8 in CDCl$_3$.

PROBLEM 13.9 **UNKNOWN 13.9 IN CDCl₃ (SAMPLE 82)**

Although this problem does not include the HMQC spectrum, the position of the HMQC cross peaks can still be inferred from the symmetric spacing of the vertical pairs of HMBC cross peaks.

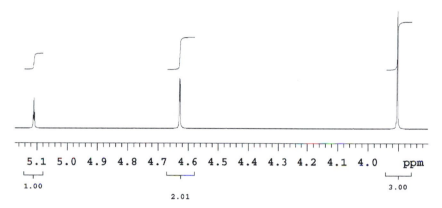

■ **FIGURE 13.9.h** The 1-D ¹H NMR spectrum of unknown 13.9 in CDCl₃.

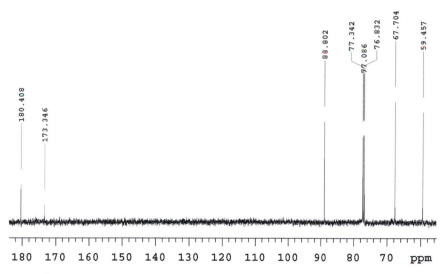

■ **FIGURE 13.9.c** The 1-D ¹³C NMR spectrum of unknown 13.9 in CDCl₃.

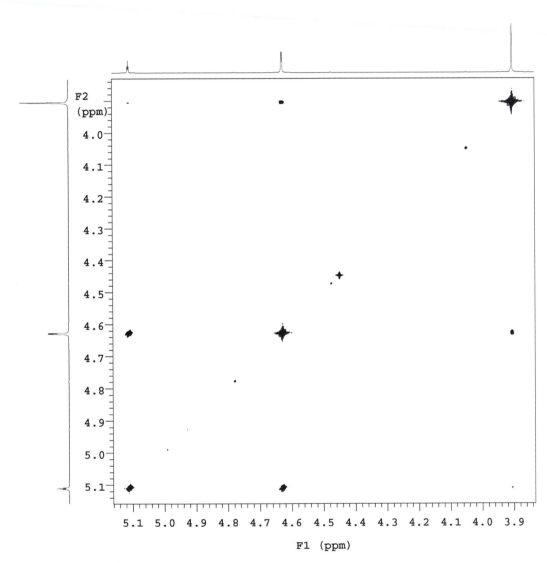

■ **FIGURE 13.9.gcosy** The 2-D ^{1}H–^{1}H gCOSY NMR spectrum of unknown 13.9 in CDCl$_3$.

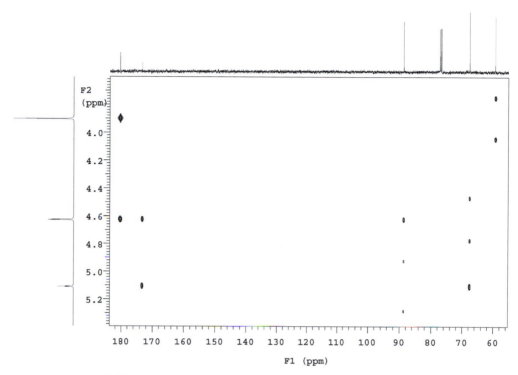

■ **FIGURE 13.9.*ghmbc*** The 2-D ¹H–¹³C gHMBC NMR spectrum of unknown 13.9 in CDCl₃.

PROBLEM 13.10 **UNKNOWN 13.10 IN CDCl₃ (SAMPLE 84)**

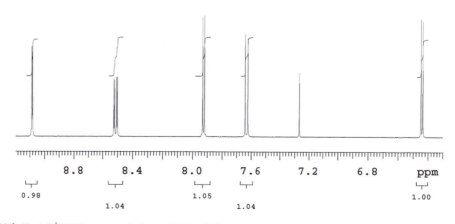

■ **FIGURE 13.10.*h*** The 1-D ¹H NMR spectrum of unknown 13.10 in CDCl₃.

■ **FIGURE 13.10.c** The 1-D ^{13}C NMR spectrum of unknown 13.10 in CDCl$_3$.

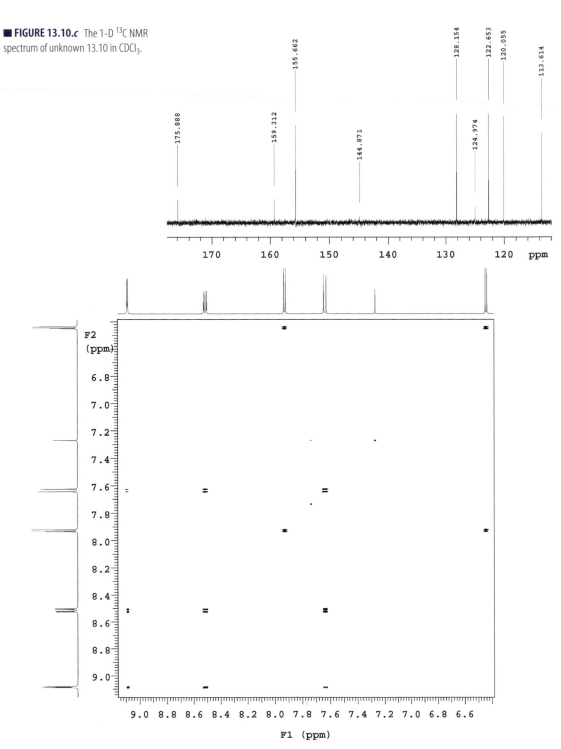

■ **FIGURE 13.10.gcosy** The 2-D ^{1}H-^{1}H gCOSY NMR spectrum of unknown 13.10 in CDCl$_3$.

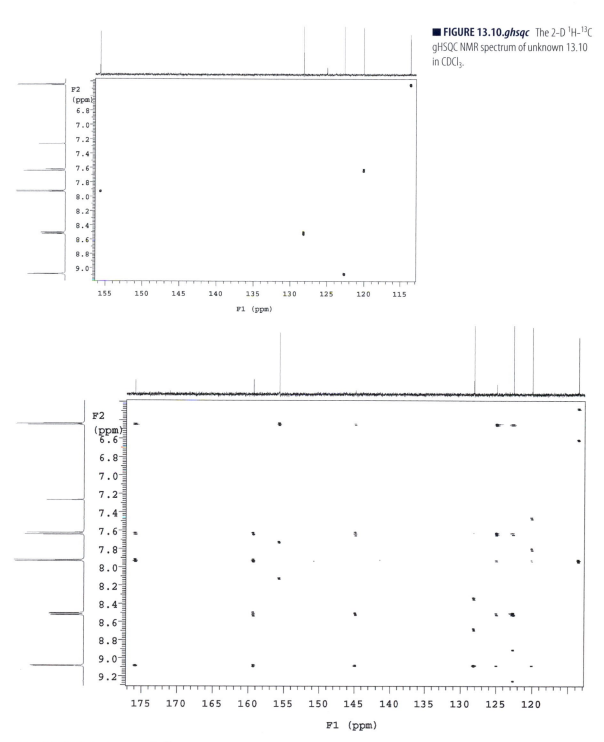

■ **FIGURE 13.10.*ghsqc*** The 2-D ^{1}H–^{13}C gHSQC NMR spectrum of unknown 13.10 in CDCl$_3$.

■ **FIGURE 13.10.*ghmbc*** The 2-D ^{1}H–^{13}C gHMBC NMR spectrum of unknown 13.10 in CDCl$_3$.

14

Chapter

Complex Unknown Problems

This chapter contains more difficult NMR unknown problems. In some cases the molecular weight or empirical formula of the molecule is given, and it is up to the reader to elucidate the structure of the unknown.

PROBLEM 14.1 **UNKNOWN 14.1 IN CDCl$_3$ (SAMPLE 32)**

MW = 192.29

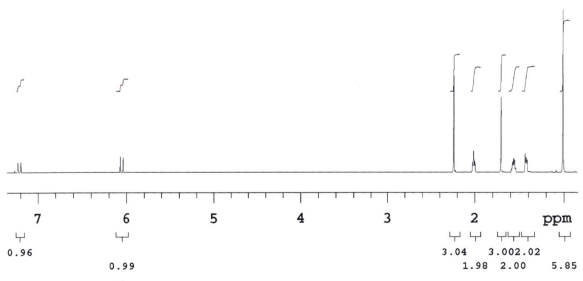

■ FIGURE 14.1.*h*   The 1-D ^{1}H NMR spectrum of unknown 14.1 in CDCl$_3$.

■ **FIGURE 14.1.*c*** The 1-D ^{13}C NMR
spectrum of unknown 14.1 in CDCl$_3$.

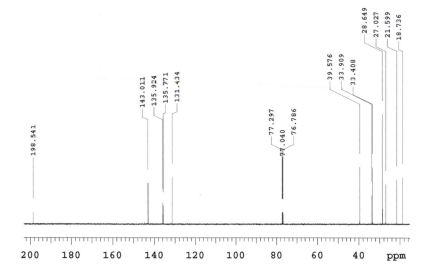

■ **FIGURE 14.1.*gcosy*** The
2-D ^{1}H-^{1}H gCOSY NMR spectrum
of unknown 14.1 in CDCl$_3$.

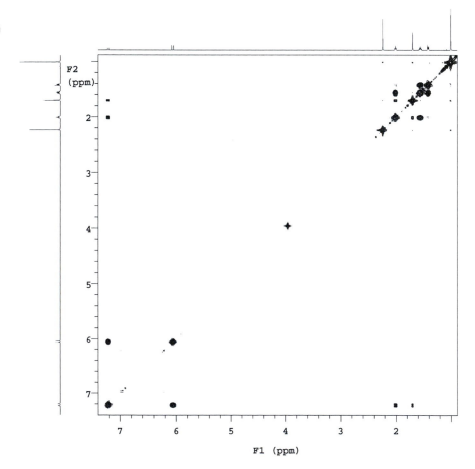

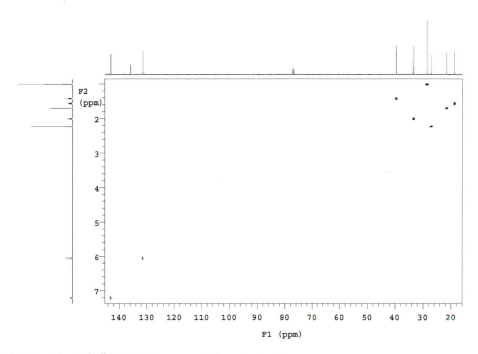

■ **FIGURE 14.1.*hmqc*** The 2-D ¹H-¹³C HMQC NMR spectrum of unknown 14.1 in CDCl₃.

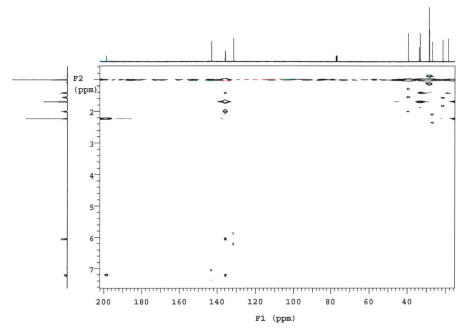

■ **FIGURE 14.1.*hmbc*** The 2-D ¹H-¹³C HMBC NMR spectrum of unknown 14.1 in CDCl₃.

PROBLEM 14.2 **UNKNOWN 14.2 IN CDCl$_3$**
(SAMPLE 33)

MW = 192.29

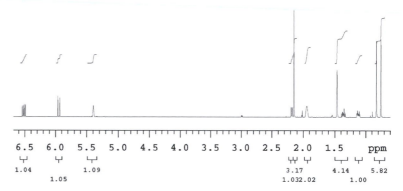

■ FIGURE 14.2.*h* The 1-D ^{1}H NMR spectrum of unknown 14.2 in CDCl$_3$.

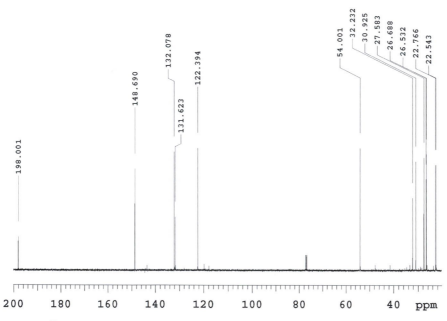

■ FIGURE 14.2.*c* The 1-D ^{13}C NMR spectrum of unknown 14.2 in CDCl$_3$.

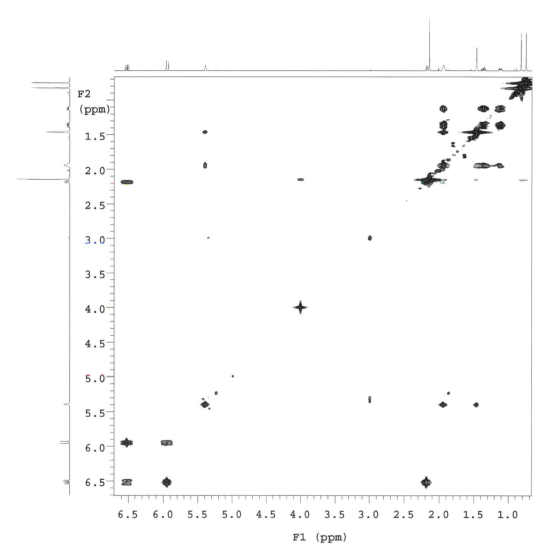

■ **FIGURE 14.2.*gcosy*** The 2-D ¹H-¹H gCOSY NMR spectrum of unknown 14.2 in CDCl₃.

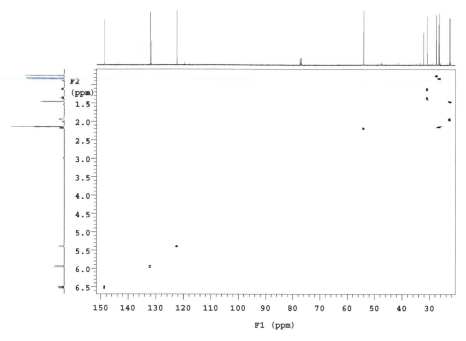

■ **FIGURE 14.2.***hmqc*   The 2-D ^{1}H-^{13}C HMQC NMR spectrum of unknown 14.2 in CDCl$_3$.

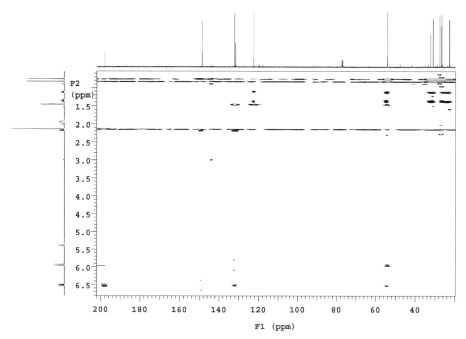

■ **FIGURE 14.2.***hmbc*   The 2-D ^{1}H-^{13}C HMBC NMR spectrum of unknown 14.2 in CDCl$_3$.

PROBLEM 14.3 **UNKNOWN 14.3 IN CDCl₃ (SAMPLE 51)**

MW = 218.33

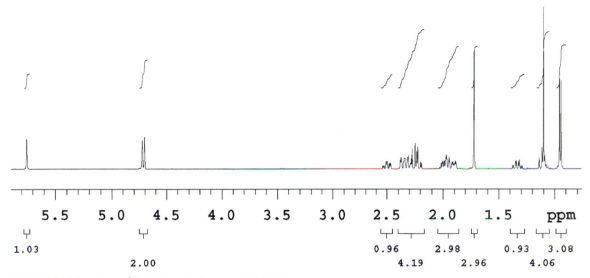

■ **FIGURE 14.3.h** The 1-D ¹H NMR spectrum of unknown 14.3 in CDCl₃.

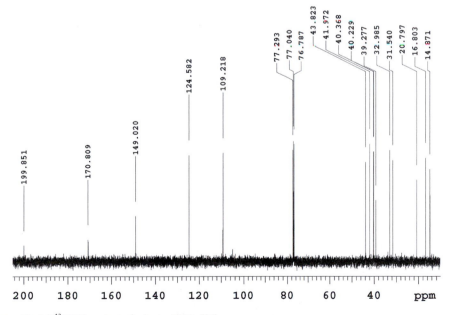

■ **FIGURE 14.3.c** The 1-D ¹³C NMR spectrum of unknown 14.3 in CDCl₃.

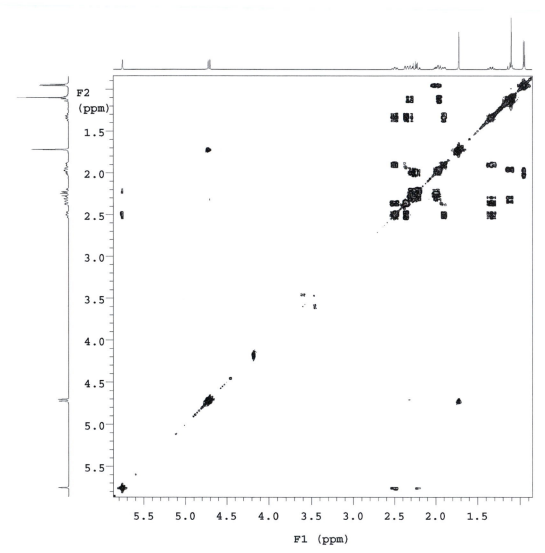

■ **FIGURE 14.3.***gcosy*   The 2-D ¹H–¹H gCOSY NMR spectrum of unknown 14.3 in CDCl₃.

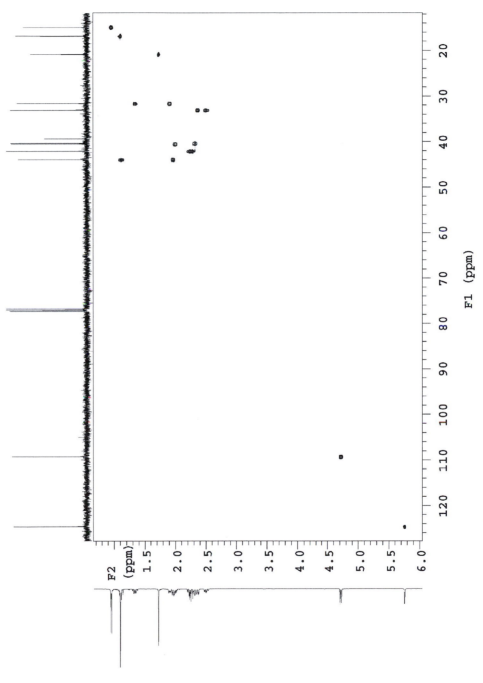

■ **FIGURE 14.3.*hsqc*** The 2-D ¹H–¹³C HSQC NMR spectrum of unknown 14.3 in CDCl₃.

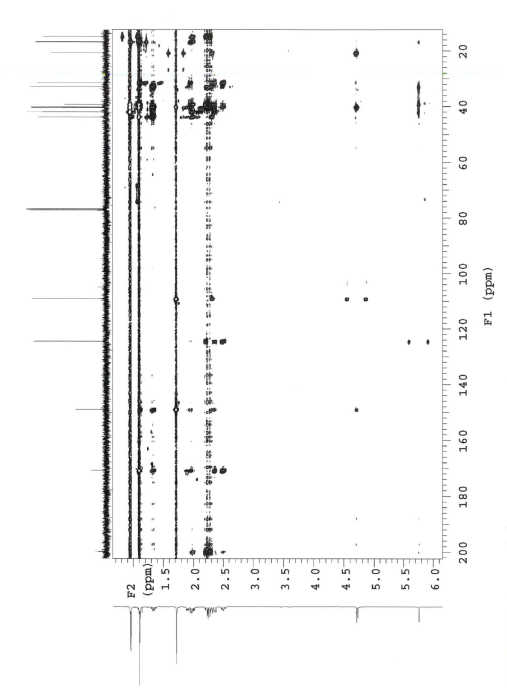

■ **FIGURE 14.3.*ghmbc*** The 2–D ^{1}H–^{13}C gHMBC NMR spectrum of unknown 14.3 in CDCl$_3$.

PROBLEM 14.4 **UNKNOWN 14.4 IN CDCl₃
(SAMPLE 74)**

This sample is a hydrocarbon; its MW can be readily determined
from examination of just the ¹H and ¹³C 1-D spectra.

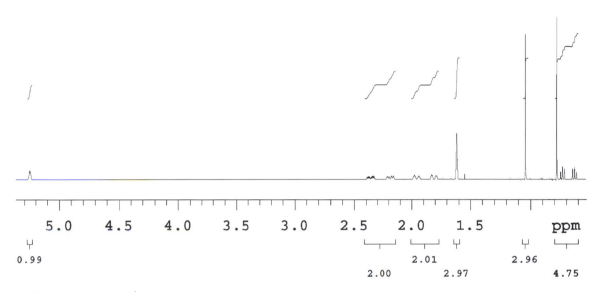

■ **FIGURE 14.4.h**   The 1-D ¹H NMR spectrum of unknown 14.4 in CDCl₃.

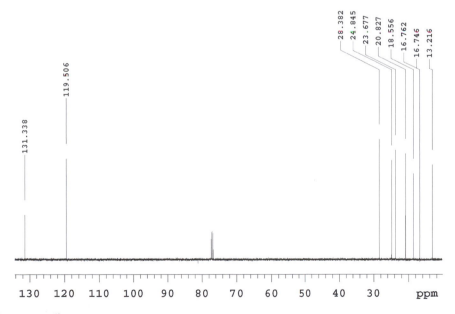

■ **FIGURE 14.4.c**   The 1-D ¹³C NMR spectrum of unknown 14.4 in CDCl₃.

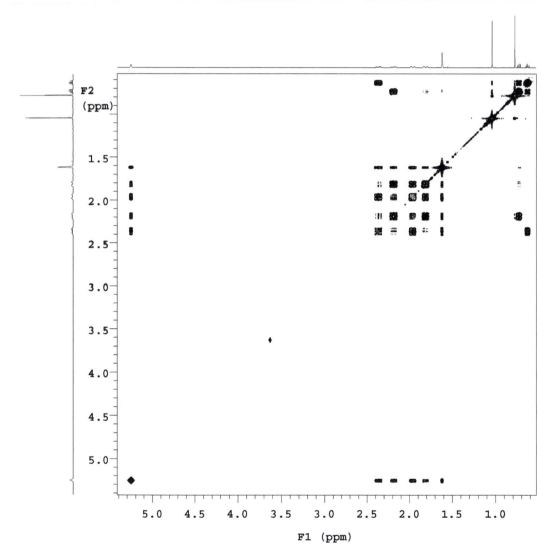

■ **FIGURE 14.4.***gcosy*   The 2-D ¹H-¹H gCOSY NMR spectrum of unknown 14.4 in CDCl₃.

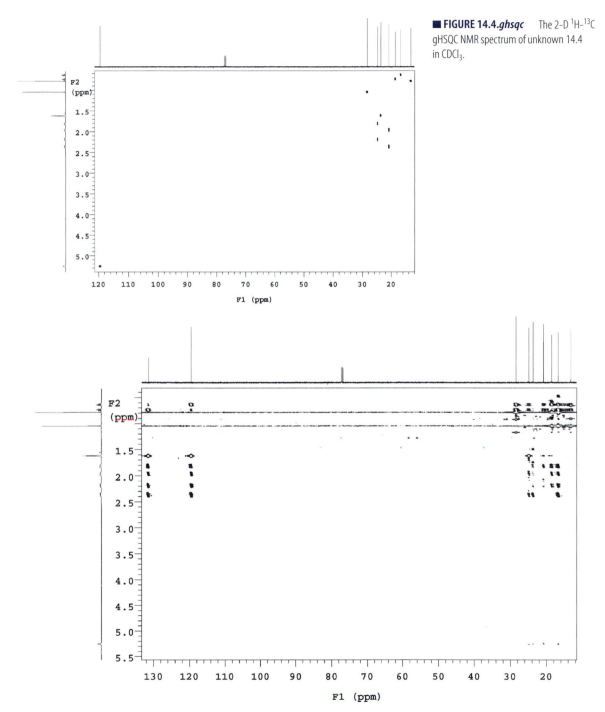

■ **FIGURE 14.4.*ghsqc*** The 2-D ¹H-¹³C gHSQC NMR spectrum of unknown 14.4 in CDCl₃.

■ **FIGURE 14.4.*ghmbc*** The 2-D ¹H-¹³C gHMBC NMR spectrum of unknown 14.4 in CDCl₃.

PROBLEM 14.5 **UNKNOWN 14.5 IN CDCl₃ (SAMPLE 75)**

This unknown is a hydrocarbon whose MW can be readily determined from the ¹H and ¹³C 1-D spectra.

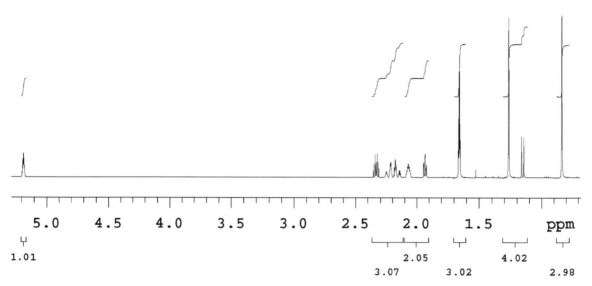

■ FIGURE 14.5.h The 1-D ¹H NMR spectrum of unknown 14.5 in CDCl₃.

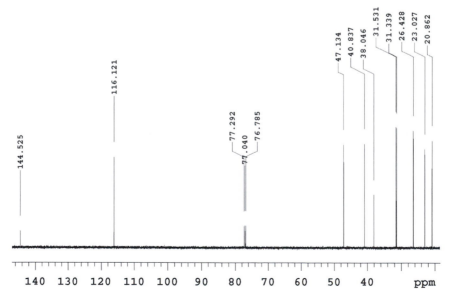

■ FIGURE 14.5.c The 1-D ¹³C NMR spectrum of unknown 14.5 in CDCl₃.

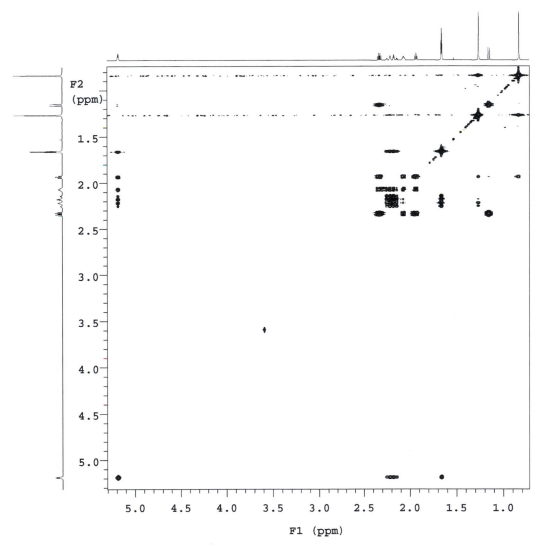

■ **FIGURE 14.5.gcosy** The 2-D ¹H-¹H gCOSY NMR spectrum of unknown 14.5 in CDCl₃.

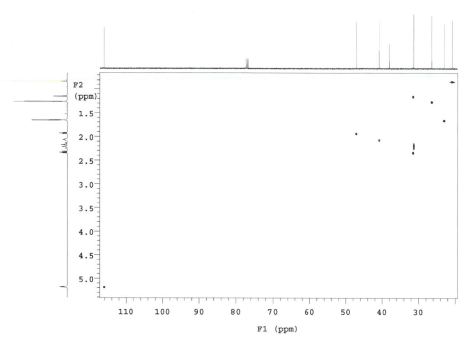

■ **FIGURE 14.5.*ghsqc*** The 2-D ^{1}H-^{13}C gHSQC NMR spectrum of unknown 14.5 in CDCl$_3$.

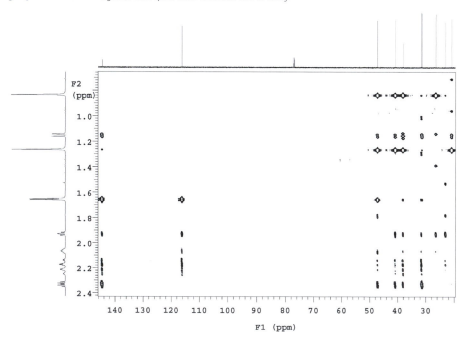

■ **FIGURE 14.5.*ghmbc_upfield*** The upfield ^{1}H portion of the 2-D ^{1}H-^{13}C HMBC NMR spectrum of unknown 14.5 in CDCl$_3$.

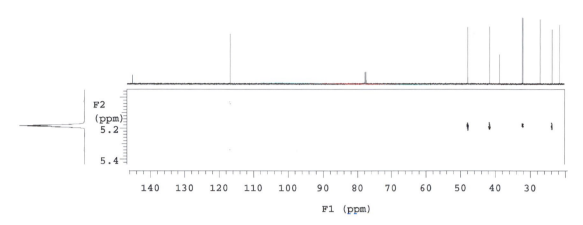

■ **FIGURE 14.5.ghmbc_downfield** The downfield ¹H portion of the 2-D ¹H-¹³C HMBC NMR spectrum of unknown 14.5 in CDCl₃.

PROBLEM 14.6 **UNKNOWN 14.6 IN CDCl$_3$ (SAMPLE 80)**

This unknown has a MW of 251.24 (recall the meaning of an odd MW).

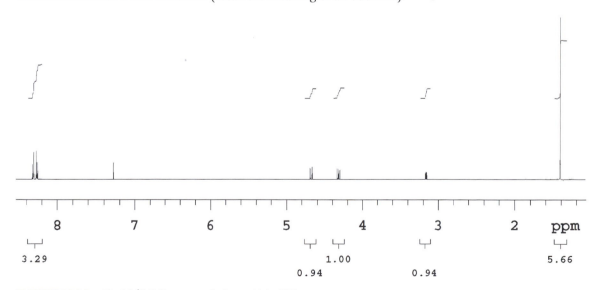

■ **FIGURE 14.6.h** The 1-D ¹H NMR spectrum of unknown 14.6 in CDCl₃.

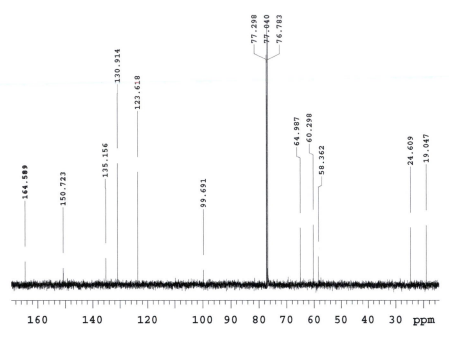

■ **FIGURE 14.6.c** The 1-D ^{13}C NMR spectrum of unknown 14.6 in CDCl$_3$.

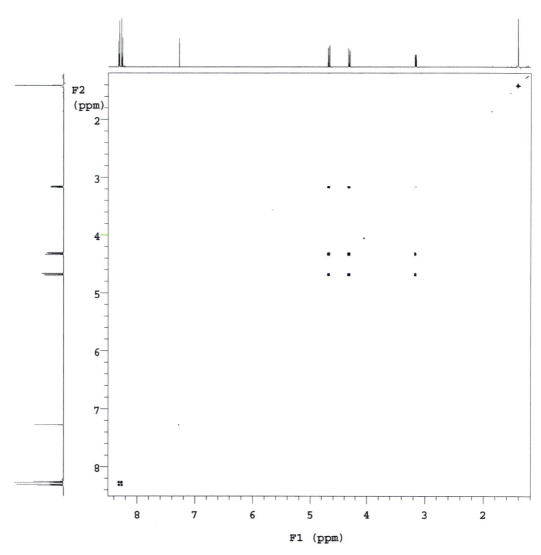

■ **FIGURE 14.6.***gcosy*   The 2-D ¹H–¹H gCOSY NMR spectrum of unknown 14.6 in CDCl₃.

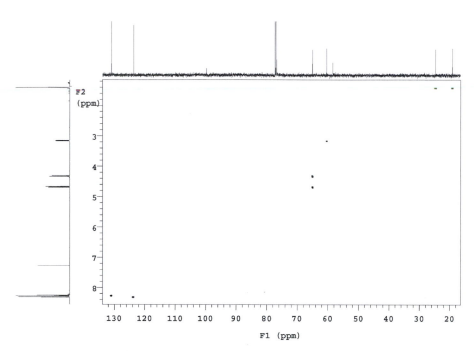

■ **FIGURE 14.6.*ghsqc*** The 2-D ^1H-^{13}C gHSQC NMR spectrum of unknown 14.6 in CDCl$_3$.

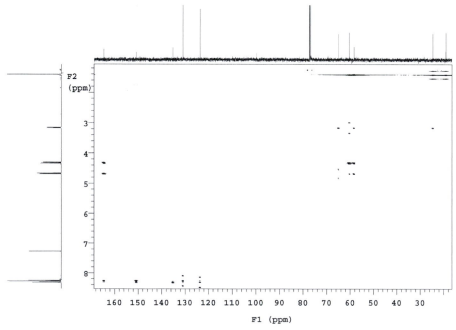

■ **FIGURE 14.6.*ghmbc*** The 2-D ^1H-^{13}C gHMBC NMR spectrum of unknown 14.6 in CDCl$_3$.

PROBLEM 14.7 **UNKNOWN 14.7 IN ACETONE-d_6
(SAMPLE 86)**

This unknown has a MW of 163.10. Other nuclei are NMR-active
besides ^{1}H and ^{13}C.

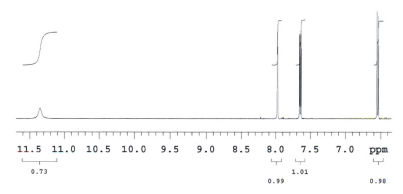

■ **FIGURE 14.7.*h***   The 1-D ^{1}H NMR spectrum of unknown 14.7 in acetone-d_6.

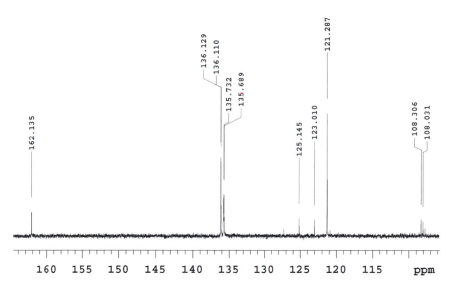

■ **FIGURE 14.7.*c***   The 1-D ^{13}C NMR spectrum of unknown 14.7 in acetone-d_6.

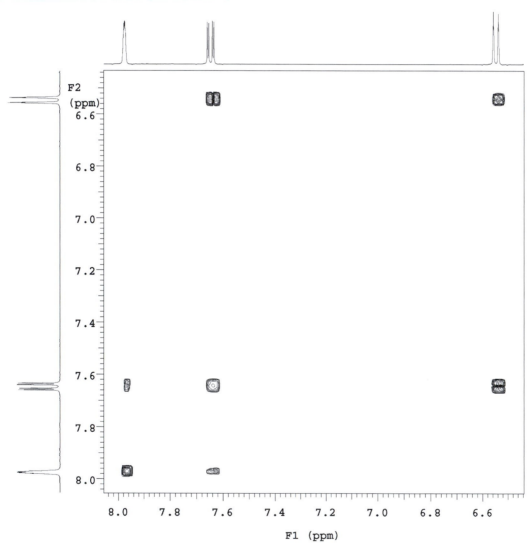

■ **FIGURE 14.7.*gcosy*** The 2-D ¹H–¹H gCOSY NMR spectrum of unknown 14.7 in acetone-d_6.

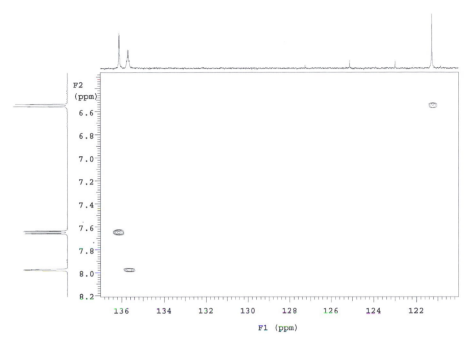

■ **FIGURE 14.7.*ghsqc*** The 2-D ¹H-¹³C gHSQC NMR spectrum of unknown 14.7 in acetone-d_6.

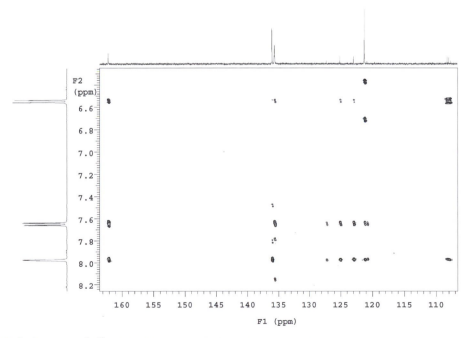

■ **FIGURE 14.7.*ghmbc*** The 2-D ¹H-¹³C gHMBC NMR spectrum of unknown 14.7 in acetone-d_6.

PROBLEM 14.8 **UNKNOWN 14.8 IN CDCl$_3$**
(SAMPLE 87)

This unknown has an empirical formula of $C_{15}H_{18}O_3$. One of the
gHSQC cross peaks is very weak, so the plot threshold had to be lowered, thus allowing some t_1 noise to appear in the plot.

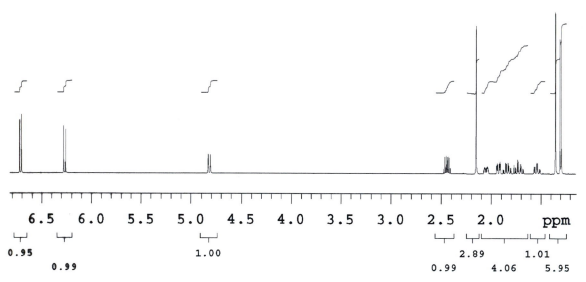

■ **FIGURE 14.8.h** The 1-D ^{1}H NMR spectrum of unknown 14.8 in CDCl$_3$.

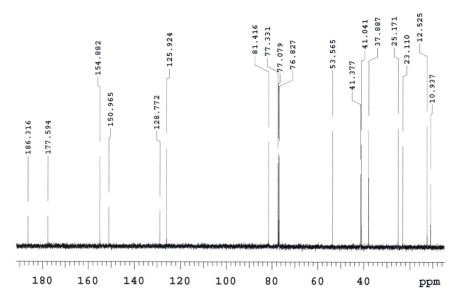

■ **FIGURE 14.8.c** The 1-D ^{13}C NMR spectrum of unknown 14.8 in CDCl$_3$.

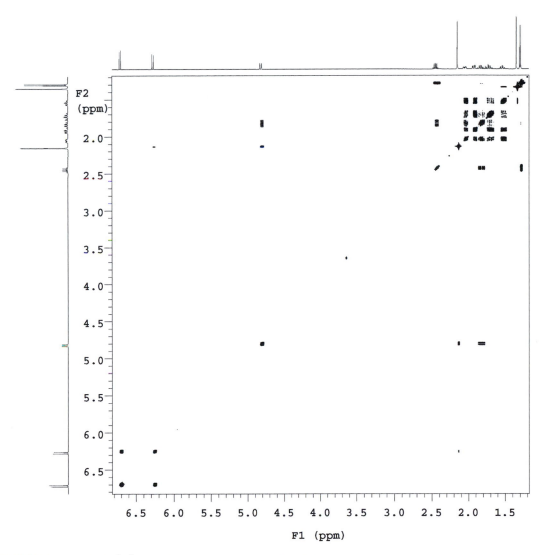

■ **FIGURE 14.8.gcosy** The 2-D ¹H-¹H gCOSY NMR spectrum of unknown 14.8 in CDCl₃.

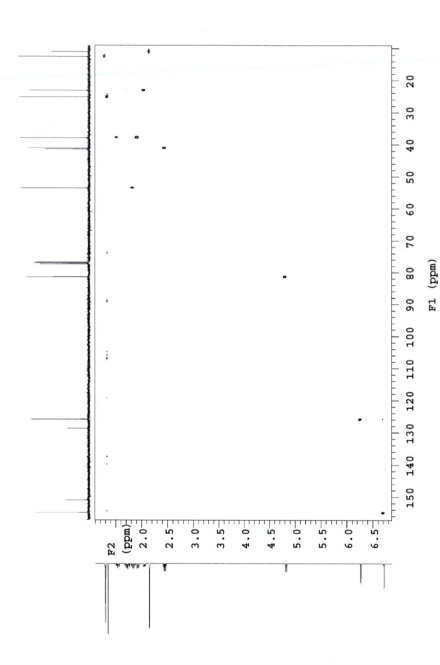

■ **FIGURE 14.8.ghsqc** The 2-D ^{1}H-^{13}C gHSQC NMR spectrum of unknown 14.8 in CDCl$_3$.

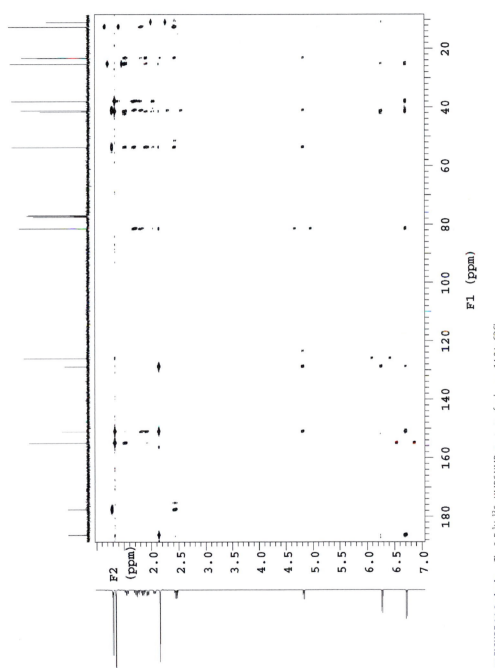

■ **FIGURE 14.8.ghmbc** The 2-D ¹H–¹³C gHMBC NMR spectrum of unknown 14.8 in CDCl₃.

PROBLEM 14.9 **UNKNOWN 14.9 IN CDCl₃ (SAMPLE 88)**

This unknown has an empirical formula of $C_{13}H_{10}O_4$.

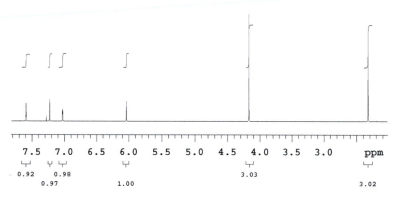

■ **FIGURE 14.9.h**   The 1-D ¹H NMR spectrum of unknown 14.9 in CDCl₃.

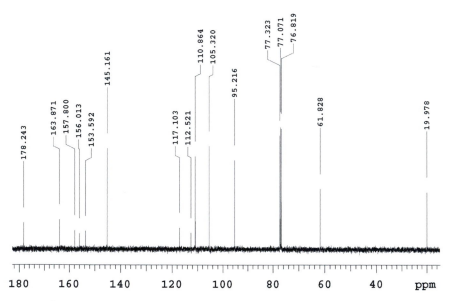

■ **FIGURE 14.9.c**   The 1-D ¹³C NMR spectrum of unknown 14.9 in CDCl₃.

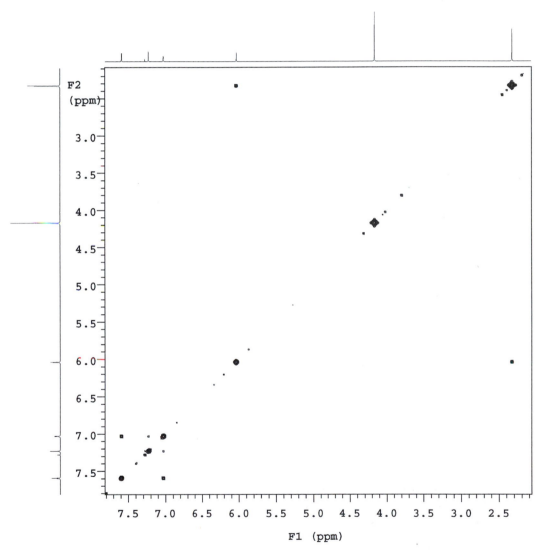

■ **FIGURE 14.9.***gcosy*   The 2-D ¹H–¹H gCOSY NMR spectrum of unknown 14.9 in CDCl₃.

■ **FIGURE 14.9**.*ghsqc* The 2-D ^{1}H-^{13}C gHSQC NMR spectrum of unknown 14.9 in CDCl$_3$.

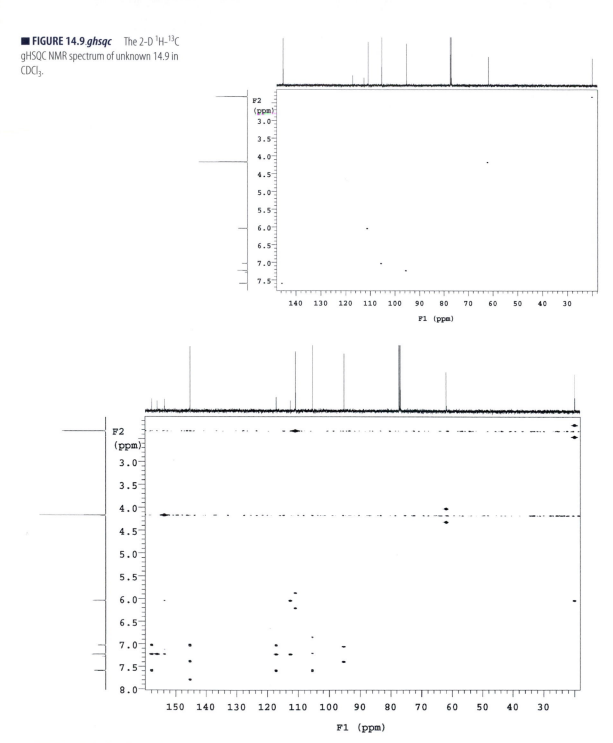

■ **FIGURE 14.9**.*ghmbc* The 2-D ^{1}H-^{13}C gHMBC NMR spectrum of unknown 14.9 in CDCl$_3$.

PROBLEM 14.10 **UNKNOWN 14.10 IN CDCl$_3$ (SAMPLE 72)**

Unknown 14.10 has an empirical formula of C$_9$H$_{15}$NO$_3$.

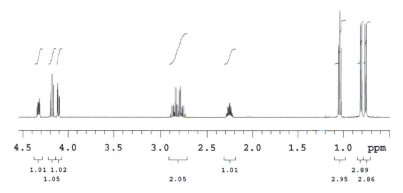

■ **FIGURE 14.10.*h*** The 1-D ^{1}H NMR spectrum of unknown 14.10 in CDCl$_3$.

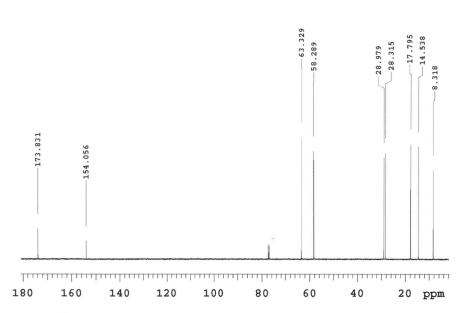

■ **FIGURE 14.10.*c*** The 1-D ^{13}C NMR spectrum of unknown 14.10 in CDCl$_3$.

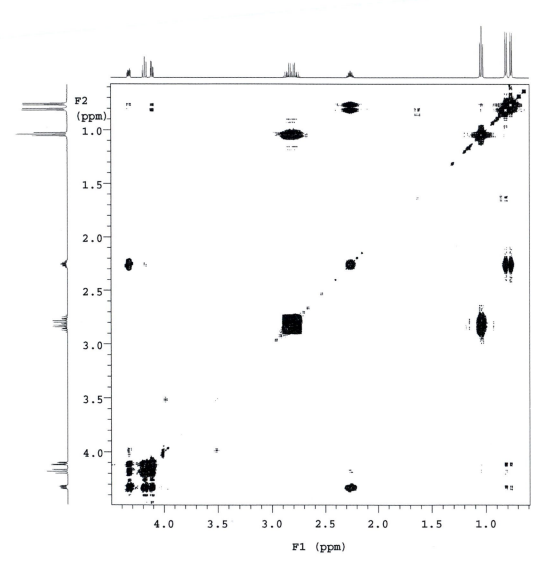

■ **FIGURE 14.10.***gcosy*   The 2-D ^{1}H-^{1}H gCOSY NMR spectrum of unknown 14.10 in CDCl$_3$.

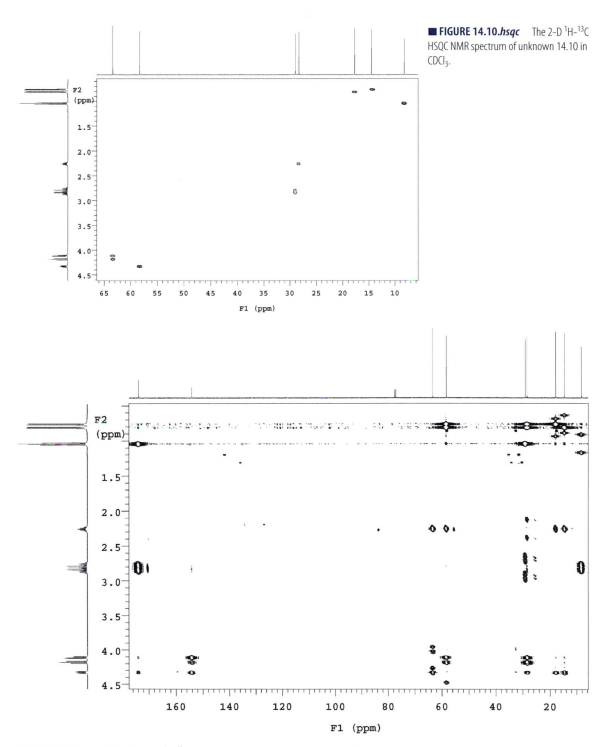

■ **FIGURE 14.10.***hsqc* The 2-D ¹H-¹³C HSQC NMR spectrum of unknown 14.10 in CDCl₃.

■ **FIGURE 14.10.***ghmbc* The 2-D ¹H-¹³C HMBC NMR spectrum of unknown 14.10 in CDCl₃.

Glossary of Terms

accuracy. How close an experimentally determined value is to the true or actual value.

acquisition time, at (Varian), or AQ (Bruker). Syn. detection period. The amount of time that the free induction decay (FID) is digitized to generate an array of numbers; denoted "at" on a Varian instrument, or "AQ" on a Bruker instrument.

activation energy, E_a. The energy barrier that must be overcome to initiate a chemical process.

alpha position. One bond removed from the chemical substituent (atom or functional group) in question.

analog-to-digital converter, A/D. An electronic device that converts an analog voltage into a binary number composed of discreet digits (a series of 1's and 0's).

anisochronous. Two spins are anisochronous when they undergo transitions between their allowed spin states at different frequencies. That is, two spins are anisochronous if their resonances appear at different points in the frequency spectrum.

anomeric methine group. The carbon-hydrogen group in a cyclic sugar that is bound to two oxygen atoms.

apodization. The application (multiplication) of a mathematical function to an array of numbers that represent the time domain signal. An apodization function is often used to force the right edge of a digitized FID to zero to eliminate truncation error, but apodization functions can also be used to enhance resolution or to modulate the time domain signal in other ways. Common apodization functions are Lorentzian line broadening (lb), Gaussian line broadening (gf), (phase-)shifted sine bell, and (phase-)shifted squared sine bell.

apodization function. Syn. window function. The mathematical function multiplied by the time domain signal.

applied field, B_0. Syn. applied magnetic field. The area of nearly constant magnetic flux in which the sample resides when it is inside the probe which is in turn inside the bore tube of the magnet.

attached proton test, APT. An experiment that determines whether or not a given ^{13}C resonance corresponds to a protonated carbon site in a molecule.

backward linear prediction. The use of a mathematical algorithm to predict how the time domain signal would appear during early portions of the digitized FID. Backward linear prediction is normally used to compensate for the corruption of the first few points of a digitized FID by pulse ringdown.

baseline correction. The flattening or zeroing of the regions of a frequency spectrum that are devoid of resonances.

beat. The maximum of one wavelength of a sinusoidal wave.

beta position. Two bonds distant from the chemical substituents (atom or functional group) in question.

bond angle. If one atom is bonded to two other atoms, the bond angle is the geometrical angle between the vectors connecting the common atom to the two other atoms.

bore tube. See magnet bore tube.

broadband channel. The portion of a spectrometer capable of generating RF to excite nuclei with Larmor frequencies less than or equal to that of ^{31}P (at 40% of the ^1H Larmor frequency). In some cases, the broadband channel may be capable of exciting ^{205}Tl (at 57% of the ^1H Larmor frequency). The preamplifier in the broadband channel's receiver will normally feature a low-pass filter and a ^2H bandstop (notch reject) filter.

CAMELSPIN experiment. See rotational Overhauser effect spectroscopy (ROESY).

carrier frequency. Syn. NMR frequency, Larmor frequency, on-resonance frequency, transmitter frequency. The frequency of the RF being generated by a particular channel of the spectrometer. The carrier frequency is located at the center of the observed spectral window for the observed (detected) nuclide.

Carr-Purcell-Meiboom-Gill (CPMG) experiment. An experiment wherein the net magnetization is tipped into the xy plane, and subjected to a series (or train) of RF pulses and delays to refocus the net magnetization. Maintaining the net magnetization in the xy plane allows the measurement of the T_2 relaxation time.

chemically equivalent. Atoms or functional groups in the same chemical environment.

chemical shift (δ). The alteration of the resonant frequency of chemically distinct NMR-active nuclei due to the resistance of the electron cloud to the applied magnetic field. The point at which the integral line of a resonance rises to 50% of its total value.

chemical shift anisotropy, CSA. The variation of the chemical shift as a function of molecular orientation with respect to the direction of the applied magnetic field.

chemical shift axis. The scale used to calibrate the abscissa (x-axis) of an NMR spectrum. In a one-dimensional spectrum, the chemical shift axis typically appears underneath an NMR frequency spectrum when the units are given in parts-per-millions (as opposed to Hz, in which case the axis would be termed the frequency axis).

chiral molecule. A group of atoms bonded together that cannot be superimposed upon its mirror image.

coalescence point. The moment in time or the temperature at which two resonances merge to become one resonance. Mathematically, coalescence occurs when the curvature of the middle of the observed spectral feature changes sign from positive to negative.

coherence selection. The isolation of a particular component of the total magnetization, often accomplished with the application of pulsed field gradients (PFGs).

coil configuration. The relative orientation of the highband and broadband coils with respect to their placement around the sample-containing NMR tube. A normal coil configuration locates the broadband coil closer to the sample, while an inverse configuration locates the highband (normally tuned to ^{1}H, but possibly ^{19}F) closer to the sample.

complex data point. A data point consisting of both a real and an imaginary component. The real and imaginary components allow an NMR data set to be phase sensitive, insofar as there can occur a partitioning of the data depicted between the two components. An array of complex data points therefore consists of two arrays of ordinates as a function of the abscissa.

complex impedance. Electrical resistance as a function of frequency.

console computer. The computer built into the NMR console (usually one or two 19″ wide electronics racks). The console computer normally communicates via an Ethernet connection with the host computer, which is the computer the NMR operator uses to initiate experiments, etc.

continuous wave decoupling, CW decoupling. The application of a single frequency of RF to a sample for the purpose of selectively irradiating (saturating) a particular resonance, thus perturbing other resonances from spins that interact with the spins corresponding to the irradiated resonance.

continuous wave RF. Syn. CW RF. A sinusoidally varying electrical current or photons with a frequency in the radio-frequency region of the electromagnetic spectrum, usually with a low amplitude, applied for a relatively long period of time (typically one or more seconds) to the sample.

correlation time, τ_c. The amount of time required for a molecule to diffuse one molecular diameter or to rotate one radian (roughly $^1/_6$ of a complete rotation).

cross peak. The spectral feature in a multidimensional NMR spectrum that indicates a correlation between a frequency position on one axis with a frequency position on another axis. Most frequently, the presence of a cross peak in a 2-D spectrum shows that a resonance on one chemical shift axis somehow interacts with a different resonance on the other chemical shift axis. In a homonuclear 2-D spectrum, a cross peak is a peak that occurs off of the diagonal. In a heteronuclear 2-D spectrum, any observed peak is, by definition, a cross peak.

cross product. A geometrical operation wherein two vectors will generate a third vector orthogonal (perpendicular) to both vectors. The cross product also has a particular handedness (we use the right-hand rule), so the order of how the vectors are introduced into the operation is often important.

^{13}C satellite peaks. Syn. ^{13}C satellites. Small peaks observed in the NMR frequency spectrum found on either side of the resonance of an NMR-active nuclide (most often ^{1}H) bound to carbon. Because ^{13}C is 1.1% abundant relative to ^{12}C, each ^{13}C satellite peak will be 0.55% of the central peak's intensity.

Dach effect. Syn. roof effect. The skewing of the intensities of the individual peaks (legs) of a multiplet caused by the close proximity (in the spectrum) of another resonance to which the resonance in question is coupled. The Dach effect is due to nonfirst-order coupling behavior.

data point. Two numerical values (i.e., an x,y pair) corresponding to intensity as a function of time, or to intensity as a function of frequency.

decoupling. The practice of irradiating one set of spins to simplify or otherwise perturb the appearance of other sets of spins through the suppression of one or more spin-spin couplings. Decoupling can be homonuclear or heteronuclear. Decoupling can be applied either continuously or in discreet bursts.

degenerate. Two spin states are said to be degenerate when their energies are the same.

dephasing. The spreading out of the individual components that comprise a net magnetization vector so the summation is zero.

depolarization. The equalization of the populations of two or more spin states.

deshielded group. A chemical functional group deprived of its normal complement of electron density.

detection period. The time period in the pulse sequence during which the FID is digitized. For a 1-D pulse sequence, this time period is denoted t_1. For a 2-D pulse sequence, this time period is denoted t_2.

deuterium lock. See field lock.

deuterium lock channel. Syn. lock channel. The RF channel in the NMR console devoted to maintaining a constant applied magnetic field strength through the monitoring of the Larmor frequency of the ^2H's in the solvent and adjusting the field with the z0 (Varian) or FIELD (Bruker) shim to keep the ^2H Larmor frequency constant.

diastereotopic group. Two chemical moieties that appear to be related by symmetry but are not. Isotopic labeling of one of two diastereopic groups produces a molecule that cannot be superimposed on the molecule produced when the other group is labeled (even following energetically plausible rotations about single bonds).

digital resolution. The number of Hz per data point in a spectrum. The digital resolution is the sweep width divided by the number of data points in one channel of the spectrum.

digitization. The conversion of an analog voltage to a digital, binary number amenable to subsequent computational manipulation.

dihedral angle (Φ). The angle between two planes defined by four atoms connected by three bonds, the middle two of which share a common bond.

dipolar relaxation rate constant, W. For a two spin system, W_0 is the rate constant for the zero quantum spin flip, W_1 is the rate constant for the single quantum spin flip, and W_2 is the rate constant for the double quantum spin flip. For short correlation times, the ratio of $W_2:W_1:W_0$ is $1:\frac{1}{4}:\frac{1}{6}$.

distortionless enhancement through polarization transfer (DEPT). A ^{13}C-detected one-dimensional experiment that can determine the number of ^1H's bound to a ^{13}C.

double-precision word. A computer memory allocation used to store a single number that contains twice the number of bytes as a normal (single-precision) word.

double quantum spin flip rate constant, W_2. The kinetic rate constant controlling the simultaneous change

in both spin states for a two-spin system where both spin-$\frac{1}{2}$ spins go from the α to the β state or from the β to the α state.

doublet. A resonance splitting pattern wherein the resonance appears as two peaks, lines, or legs.

doublet of doublets. A single resonance split into a multiplet containing four (roughly) equal intensity peaks. Two coupling constants can be extracted from this multiplet by numbering the peaks from left to right and obtaining the shift differences (in Hz) between some of the peak pairs. The 1–2 and 3–4 differences give the smaller coupling, and the 1–3 and 2–4 differences give the larger coupling.

downfield. The left side of the chemical shift scale. This corresponds to higher frequency, and the higher resonant frequency in turn indicates a lack of electron density.

A resonance is downfield if it is located on the left side of the spectrum or if it is observed to appear to the left of its expected value. By convention, lower frequencies (also lower chemical shifts values in ppm) appear on the right side of the spectrum and higher frequencies appear on the left side of the spectrum. Because the first generation of NMR instruments (CW instruments) operated with a constant RF frequency and a variable applied magnetic field strength, lower fields (downfield) were required to make higher frequency resonances come into resonance. Therefore, when an NMR spectrum is plotted as a function of field strength, the lower values appear on the left and higher values appear on the right.

dummy scans. Syn. steady state scans. Executions of the pulse sequence carried out prior to the saving of any of the data for the purpose of ensuring that the net magnetization has reached a constant magnitude following the relaxation delay at the start of each execution of the pulse sequence. Steady-state scans also help ensure that the sample reaches a constant temperature, which is especially important in carrying out any 2-D experiment involving X-nucleus decoupling (HMQC and HSQC) or a spin-lock (TOCSY and ROESY).

dwell time. The time interval between sampling events for the digitization of the analog signal arising from the FID; equal to the reciprocal of the sampling rate.

electron-donating group, EDG. A functional group that donates electron density through chemical bonds to other groups or atoms nearby. Electron density donation may occur either directly to groups that are alpha to the EDG, or to groups that are more distant through the phenomenon of (chemical) resonance.

Effective electron donors include atoms or functional groups with lone pairs on their attachment points and functional groups containing atoms with low electronegativities.

electronegativity. A number reflecting the affinity of an atom for electron density.

electron withdrawing group, EWG. A functional group that withdraws electron density through chemical bonds from other groups or atoms nearby. Electron density withdrawal may occur either directly from groups that are alpha to the EWG, or from groups that are more distant through the phenomenon of (chemical) resonance. Effective electron withdrawers include atoms with large electronegativities, and functional groups that can, through (chemical) resonance, assume a negative formal charge (and still comply with the octet rule).

enantiotopic. Two atoms or functional groups in a molecule are said to be enantiotopic if marking one of the atoms or groups renders a molecule that is the mirror image of the molecule obtained when the other atom or group is marked. The resonances of NMR-active enantiotopic groups or atoms will be isochronous in the absence of a chiral compound.

enhancement, η. Syn. NOE enhancement. The numerical factor by which the integrated intensity of a resonance increases as the result of irradiation of a spin that is nearby in space. For the irradiation of nuclear spins, the upper limit for the observation of a nearby spin is on the order of five angstroms.

ensemble. A large number of NMR-active spins.

entry point. The initial pairing of a readily recognizable spectral feature to the portion of the molecule responsible for the feature.

Ernst angle. The optimal tip angle for repeated application of a 1-D pulse sequence based on the relaxation time of the spin being observed and the time required to execute the pulse sequence a single time.

Ethernet card. A printed circuit board that resides in a digital device (a host computer or NMR console) that allows communication between the digital device and one or more other digital devices.

Ethernet connection. The link between two or more digital devices through their respective Ethernet cards.

evolution time, t_1. The time period(s) in a 2-D pulse sequence during which a net magnetization is allowed to precess in the xy plane prior to (separate mixing and) detection. In the case of the COSY experiment, the evolution and mixing times occur simultaneously. Variation of the t_1 delay in a 2-D pulse sequence generates the t_1 time domain.

excitation. The perturbation of spins from their equilibrium distribution of spin state populations.

exponentially damped sinusoid. A sine wave whose amplitude decays exponentially with time. The analog signal induced in the receiver coil of the NMR instrument will consist of one or more exponentially damped sinusoids.

F_1 axis, f_1 axis. Syn. f_1 frequency axis. The reference scale applied to the f_1 frequency domain. The f_1 axis may be labeled with either ppm or Hz.

F_2 axis, f_2 axis. Syn. f_2 frequency axis. The reference scale applied to the f_2 frequency domain. The f_2 axis may be labeled with either ppm or Hz.

f_1 frequency domain. The frequency domain generated following the Fourier transformation of the t_1 time domain. The f_1 frequency domain most often used for ^1H or ^{13}C chemical shifts.

f_2 frequency domain. The frequency domain generated following the Fourier transformation of the t_2 time domain. The f_2 frequency domain is almost exclusively used for ^1H chemical shifts.

f_1 projection. The summation or maxima picking of a 2-D data matrix parallel to the f_2 axis. If the f_1 axis is vertical, then the f_1 projection will normally be shown on the left side of the data matrix. The projection is obtained by summing all points or finding the maximum value of each row of the matrix.

f_2 projection. The summation or maxima picking of a 2-D data matrix parallel to the f_1 axis. If the f_2 axis is horizontal, then the f_2 projection will normally be shown on the top of the data matrix. The projection is obtained by summing all points or finding the maximum value of each column of the matrix.

fast-exchange limit. The fast-exchange limit is said to be reached when no further increase in the rate at which a dynamic process occurs will alter observed spectral features. Normally, we speak of resonance coalescence as occurring when the fast-exchange limit is reached.

field gradient pulse. Syn. gradient pulse. The short application (10 ms is typical) of an electrical current in a coil of wire in the probe that surrounds the detected region of the sample and causes the strength of the applied magnetic field to vary as a function of displacement along one or more axes.

field heterogeneity. The variation in the strength of the applied magnetic field within the detected or scanned region of the sample. The more heterogeneous the field, the broader the observed NMR resonances. Field heterogeneity is reduced through adjustment of shims and, in some cases, through sample spinning.

field homogeneity. The evenness of the strength of the applied magnetic field over the volume of the sample from which signal is detected. The more homogeneous the field, the narrower the observed NMR resonances. Field homogeneity is achieved through adjustment of shims and, in some cases, through sample spinning.

field lock. Syn. deuterium field lock, ^{2}H lock, lock. The holding constant of the strength of the applied magnetic field through the monitoring of the Larmor frequency of one nuclide (normally ^{2}H, but possibly ^{19}F) in the solution and making small field strength adjustments.

filter. Syn. RF filter. An electronic device used to limit the passage of RF of specific frequencies from one side of the filter to the other side. There are four types of filters. A high-pass filter only allows RF with frequencies above a given value to traverse the filter. A low-pass filter only allows RF with frequencies below a given value to traverse the filter. A band-pass filter only allows RF with frequencies that fall within a certain frequency range to traverse the filter. A band-stop filter only prevents RF with a certain range of frequencies from traversing the filter.

filtering. The limiting of the frequencies of RF that may pass from one side of a filter to the other.

first-order phase correction. Syn. first-order phasing. The variation in the proportion of amplitude data taken from two orthogonal arrays (or matrices) wherein the proportion varies linearly as a function of the distance from the pivot point.

flip-flop transition. Syn. zero quantum transition, W_0 transition, zero quantum spin flip. When two spins undergo simultaneous spin flips such that the sum of their spin quantum numbers is the same before and after the transition takes place. For example, if spins A and B undergo a flip-flop transition, then if spin A goes from the α to the β spin state, then spin B must simultaneously goes from the β to the α spin state.

folding. Syn. foldover, aliasing, fold back. The appearance of an NMR resonance at a incorrect position on the frequency axis because the true position lies outside the limits of the spectral window.

forward linear prediction. The addition of data points past the last point in time that was digitized.

Forward linear prediction can be used to add points on to the end of a digitized FID, or it can be used to add to a 2-D data matrix additional digitized FIDs corresponding to those that would be obtained if longer t_1 evolution times were used.

forward power. Power delivered by the NMR instrument to the sample.

Fourier ripples. Constantly spaced bumps in the frequency spectrum found on either side of a peak. In a 1-D spectrum or a half-transform 2-D data matrix, these ripples are found when the apodized intensity has not faded to the level of the background noise by the time the digitization of the FID ceases. In a fully transformed 2-D spectrum, Fourier ripples parallel to the f_1 frequency axis are observed when an inappropriate t_1 apodization function is used prior to conversion of the t_1 time domain to the f_1 frequency domain.

Fourier transform, FT. A mathematical operation that converts the amplitude as a function of time to amplitude as a function of frequency.

free induction decay, FID. The analog signal induced in the receiver coil of an NMR instrument caused by the **xy** component of the net magnetization. Sometimes the FID is also assumed to be the digital array of numbers corresponding to the FID's amplitude as a function of time.

frequency domain. The range of frequencies covered by the spectral window. The frequency domain is located in the continuum of all possible frequencies by the frequency of the instrument transmitter's RF (this frequency is also that of the rotating frame) and by the rate at which the analog signal (the FID) is digitized.

frequency synthesizer. A component of the NMR instrument that generates a sinusoidal signal at a specific frequency.

full cannon homospoil. Application of an RF pulse at or near full transmitter power with a duration 200 times greater than that required tip the net magnetization by 90° for the express purpose of dephasing the net magnetization in **xz** or **yz** plane. Small phase errors accumulate with this absurdly large tip angle because of the slight inhomogeneities of the applied RF's B_1 field across the volume of the sample excited by the pulse.

γH_2 calibration procedure. A decoupler calibration procedure wherein one irradiates the sample with continuous wave RF whose frequency is at a known distance from spin A's frequency and measures the partial collapse of the multiplet from spin B (spins A and B must be coupled, normally $A = {}^1H$ and $B = {}^{13}C$ or vice versa).

gated decoupling. Gated decoupling occurs when the decoupling RF is turned on and off (gated on or off) at particular points in the pulse sequence. The most common use of this method arises when one wishes to acquire a quantitative 1-D spectrum with decoupling. The decoupler is gated off during the relaxation delay and on during the acquisition time (the time during which the FID is digitized). This protocol prevents varying amounts of NOE enhancement (which can vary from site to site within a molecule) from skewing the relative intensities of the components of the net magnetization generated by each site in a molecule, but at the same time preserves the more lucid presentation of resonances as singlets (multiplets collapsed into a single peak), not to mention the better signal-to-noise ratio associated with the placement of all the intensity of a given resonance into a single peak.

Gaussing line broadening function. Syn. Gaussian function, Gaussian, Gaussian apodization function. An apodization function commonly used to weight the signal-rich initial portion of a digitized FID relative to the noise-rich tail portion of a digitized FID. The weighted digitized FID is then Fourier transformed to convert the time domain to the frequency domain. In the frequency spectrum, the Gaussian function is more effective than the Lorentzian function at suppressing truncation error, improving the spectrum's signal-to-noise ratio, and minimizing the peak broadening inherent in the use of most apodization functions.

geminal. Two atoms or functional groups are geminal if they are both bound to the same atom. Geminal atoms or functional groups discussed in the context of NMR spectroscopy are usually but not always the same species.

gradient probe. See pulsed field gradient probe.

gradient pulse. See field gradient pulse.

gyromagnetic ratio, γ. Syn. magnetogyric ratio. A nuclide-specific proportionality constant relating how fast spins will precess (in radians $\cdot$ sec^{-1}) per unit of applied magnetic field (in T).

heteronuclear decoupling. The decoupling of spins of one nuclide to favorably affect the signal observed from a second nuclide, e.g., to collapse the multiplets observed in the resonances of a second nuclide.

homogeneous. Constant throughout.

homonuclear decoupling. The simultaneous decoupling and observation of spins of the same nuclide, normally accomplished with the application of low-power, single frequency RF that is gated off only long enough to digitize each point of the FID. Because the homonuclear decoupling RF is low power, the pulse ringdown delays normally associated with the application of high power (hard) pulses can be foregone. The location of the decoupled resonance may lie within the limits of the observed spectral window, so great care must be taken to avoid overloading the receiver.

homospoil methods. A method of eliminating residual net magnetization achieved through temporary disruption of the applied magnetic field's homogeneity so that the spins precess at different frequencies and become dephased.

homotopic. Two or more atoms or groups in a molecule are homotopic if labeling one generates the same molecule as labeling the other. Homotopic atoms or groups always generate resonances with the same chemical shift.

host computer. The computer attached directly to the NMR instrument.

integral. The numerical value generated by integration.

integration. The measurement of the area of one or more resonances in a 1-D spectrum, or the measurement of the volume of a cross peak in a 2-D spectrum.

interferogram. A 2-D data matrix that has only undergone Fourier transformation along one axis to convert the t_2 time domain to the f_2 frequency domain. An interferogram will therefore show the f_2 frequency domain on one axis and the t_1 time domain on the other axis.

internuclear distance, r. The through-space distance between two nuclei.

isochronous. Two atoms or functional groups are isochronous when they generate NMR resonances with the same chemical shift.

J-coupling. Syn. coupling, scalar coupling, spin-spin coupling. The alteration of the spin-state transition frequency of one spin by the spin state of a second spin. J-coupling is a through-bond effect and is, for a given system (molecule), invariant as a function of the applied magnetic field strength. When a J-coupling is described in writing, a leading superscript denotes how many bonds separate the two spins, and a trailing subscript may denote the identities of the two coupled spins.

Johnson noise. Syn. thermal noise. The electrical noise caused by the Brownian motion of ions in a conducting material, e.g., a wire. This type of electrical noise varies with temperature.

Karplus diagram. A plot showing the Karplus relationship.

Karplus relationship. A mathematical function based on orbital overlap that relates the magnitude of J-coupling as a function of the geminal bond angle or the vicinal dihedral angle.

Larmor frequency. Syn. precession frequency, nuclear precession frequency, NMR frequency, rotating frame frequency. The rate at which the **xy** component of a spin precesses about the axis of the applied magnetic field. The frequency of the photons capable of inducing transitions between allowed spin states for a given NMR-active nucleus.

lattice. The rest of the world. The environment outside the immediate vicinity of a spin.

leg. Syn. line. One individual peak within a resonance split into a multiplet through the action of J-coupling.

linear prediction. A mathematical operation that generates new or replaces existing time domain data points with predicted ones. Linear prediction can add to the end of an array of digitized FID data points, can extend the number of t_1 time domain data points in a 2-D interferogram, or can replace the initial points in a digitized FID that may have been corrupted by pulse ringdown.

line broadening. Syn. Apodization (not strictly correct). Any process that increases the measured width of peaks in a spectrum. This can either be a natural process we observe with our instrument, or the post-acquisition processing technique of selectively weighting different portions of a digitized FID to improve the signal-to-noise ratio of the spectrum obtained following conversion of the time domain to the frequency domain with the Fourier transformation.

line broadening function. A mathematical function multiplied by the time domain data to smooth out noise by emphasizing the signal-rich beginning of the digitized FID and deemphasizing the noise-rich tail of the digitized FID. In a 2-D interferogram, application of a line broadening function emphasizes the signal-rich f_2 spectra (usually those with shorter t_1 evolution times) and deemphasizes the noise-rich f_2 spectra (usually with longer t_1 evolution times).

lock. Syn. field lock. The maintenance of a constant applied field strength through the use of an active feedback mechanism.

lock channel. See deuterium lock channel.

lock frequency. The Larmor frequency of the ^2H's in the solvent.

locking. The act of establishing the condition of a stable deuterium lock.

Lorentzian line broadening function. The apodization function most commonly used to emphasize the signal-rich initial portion of a digitized FID.

magnet bore tube. Syn. bore, bore tube. The hollow, cylindrical tube that runs vertically (for an NMR magnet, horizontally for an MRI magnet) through the interior of a cryomagnet. The magnetic field maximum occurs within the interior of the bore tube. The room temperature (RT) shims are a hollow cylinder that is inserted inside the bore tube, and the probe is inserted inside the RT shims. Samples are lowered pneumatically down the upper bore tube, which is a smaller tube that rests on top of the RT shims and probe assembly.

magnetically equivalent. Two nuclei are magnetically equivalent when they are of the same nuclide and when both have exactly the same geometrical relationship to every other NMR-active nucleus in a molecule.

magnetic moment. A vector quantity expressed in units of angular momentum that relates the torque felt by the particle to the magnitude and direction of an externally applied magnetic field. The magnetic field associated with a circulating charge.

magnetic susceptibility. The ability of a material to accommodate within its physical being magnetic field lines (magnetic flux).

magnetization. See net magnetization vector.

mixing. The time interval in a 2-D NMR pulse sequence wherein t_1-encoded phase information is passed from spin to spin.

mixing down. Syn. mixdown. The reduction of an analog signal from a high frequency (typically tens or hundreds of MHz) to a lower frequency range (typically below 100 kHz).

multiplet. A resonance showing multiple maxima; the amplitude distribution, often showing a high degree of symmetry, in a frequency spectrum arising from a single NMR-active atomic site in a molecule that is divided (split) into multiple peaks, lines, or legs.

multiplicity. Into how many peaks of what relative intensities the resonance from a single NMR-active atomic site in a molecule is divided.

net magnetization vector, M. Syn. magnetization. The vector sum of the magnetic moments of an ensemble of spins.

Newman projection. A graphical image useful for exploring rotational conformation wherein one sights down the bond about which rotation occurs.

90° RF pulse. Syn. 90° pulse. An RF pulse applied to the spins in a sample to tip the net magnetization vector of those spins by 90°.

NMR frequency. See Larmor frequency.

NMR instrument. A host computer, console, preamplifier, probe, cryomagnet, pneumatic plumbing, and cabling that together allow the collection of NMR data.

NMR probe. Syn. probe. A nonferrous metal housing consisting of a cylindrical upper portion that fits inside the lower portion of the magnet bore tube. The probe contains electrical conductors, capacitors, and inductors, as well as a Dewared air channel with a heater coil and a thermocouple. It may also contain one or more coils of wire wound with a geometrical configuration such that passing current through these coils will induce a magnetic field gradient across the volume occupied by the sample when it is in place.

NMR time scale. The time scale of dynamic processes that can be observed with an NMR spectrometer.

no-bond resonance. An extension of traditional resonance structure formulations wherein one violates the octet rule with a two-electron deficit being placed on an alkyl group to explain the electron-donating nature of alkyl groups.

node. A point where a function—e.g., the excitation profile of an RF pulse—has zero amplitude.

nonfirst-order behavior. Syn. Dach effect, intensity skewing. The deviation of the intensity of the individual peaks in a multiplet from those intensities predicted using Pascal's triangle.

nuclear Overhauser effect, NOE, nOe. The perturbation of the populations of one set of spins achieved through saturation of a second set of spins less than five angstroms distant.

nuclear spin. The circular motion of the positive charge of a nucleus.

1-D NMR pulse sequence. A series of delays and RF pulses culminating in the detection, amplification, mixing down, and digitization of the FID.

1-D NMR spectrum. A linear array showing amplitude as a function of frequency, obtained by the Fourier transformation of an array with amplitude as a function of time.

1-D NOE difference experiment. The subtraction of a 1-D spectrum obtained by irradiating a single resonance at low power with CW RF from a 1-D spectrum obtained by irradiating a resonance-free region in or near the same spectral window. The resulting spectrum shows the irradiated resonance phased negatively, and any resonance that has its equilibrium spin population perturbed through cross relaxation with the irradiated resonance shows a positive integral.

off resonance. A spin is off resonance when the spin's resonant frequency is not at the center of the spectral window (the center of the spectral window corresponds to the frequency of the rotation of the rotating frame of reference).

offset. The small amount a coarse frequency value (typically tens or hundreds of MHz) may be adjusted up or down.

one-pulse experiment. The simplest 1-D NMR experiment consisting of only a relaxation delay, a single RF pulse, and detection of the FID.

on resonance. A spin is said to be on resonance if the spin's resonant frequency lies at the center of the spectral window.

oversampling. The collection of data points at a rate faster than that called for by the sweep width being used, thus allowing the subsequent averaging of the extra points to yield more accurate amplitude values spaced at the correct dwell time.

paramagnetic relaxation agent. A nonreactive chemical additive (often containing europium) introduced into a sample containing unpaired electrons that has the effect of reducing the spin-lattice relaxation times for the spins in the solute molecules.

paramagnetic species. A molecule containing unpaired electrons.

peak picking. The determination of the location of peak maxima in a frequency spectrum, often done by software with minimal user guidance.

phase. The point along one wavelength of a sine wave where the waveform starts. The phase of an RF pulse also determines the direction in the rotating frame of reference that the net magnetization vector will tip relative to its initial orientation. The phase of an RF pulse is denoted with a subscript to indicate the axis of the rotating frame axis about which rotation occurs.

phaseable. The ability to effectively control the relative absorptive versus dispersive character of an NMR spectrum through partitioning of the displayed data between orthogonal (real and imaginary) subsets.

phase character. The absorptive or dispersive nature of a spectral peak. The angle by which magnetization precesses in the xy plane over a given time interval.

phase correction. The balancing of the relative emphasis of two orthogonal data arrays (or matrices for a 2-D spectrum) to generate a frequency spectrum with peaks that have a fully absorptive (or, in some cases, fully dispersive) phase character.

phase cycling. The alternation of the phase of applied RF pulses and/or receiver detection on successive passes through a pulse sequence without the variation of pulse lengths or delays. Data collected when only RF pulse phases are varied is often added to data collected with different RF pulse phases to cancel unwanted components of the detected signal, or to cancel artifacts either inherent in single executions of a specific pulse sequence or inherent to unavoidable instrumental limitations.

phase sensitive. The collection of an NMR data set involving the use of a 90° phase shift in the receiver and also possibly in the phase of one of the RF pulses of the pulse sequence, thus allowing the storage of the digitized data points into two separate memory locations to allow phase correction during processing.

phasing. The manipulation of a frequency spectrum through the weighting of points from two orthogonal data arrays (or matrices) to generate spectral features that are most often purely absorptive and positive.

point. See data point.

polarization. The unequal population of two or more spin states.

preamplifier. Syn. preamp. An electronic device housed inside a metal box very close to the magnet containing circuitry to amplify the low-level NMR signal coming from the probe.

precession frequency. Syn. Larmor frequency, NMR frequency. The frequency at which a nuclear magnetic moment rotates about the axis of the applied magnetic field.

precision. The magnitude of the smallest discernable variations in an experimental measurement.

preparation. The placement of magnetization into the xy plane for subsequent detection.

probe. See NMR probe.

probe tuning. The adjustment of the complex impedance of the probe to maximize the delivery of RF power to the sample (forward power), to minimize reflected RF power, and to maximize the sensitivity of the instrument receiver to the NMR signal emanating from the sample following the application of the pulse sequence.

prochiral. An atom or functional group that is part of an enantiotopic pair which, upon (isotopic) labeling, generates a chiral compound.

pro-R. An atom or functional group that is part of an enantiotopic or diastereotopic pair which, upon (isotopic) labeling such that it acquires a higher precedence, generates an R chiral center.

pro-S. An atom or functional group that is part of an enantiotopic or diastereotopic pair which, upon (isotopic) labeling such that it acquires a higher precedence, generates an S chiral center.

proton channel. Syn. highband channel. The RF channel of an NMR instrument devoted to the generation and detection of the highest frequencies of which the instrument is capable. The highband channel can also normally generate the RF suitable for carrying out ^{19}F NMR experiments. Although ^{3}H (tritium) has a higher Larmor frequency than ^{1}H, in practice this frequency is rarely called for.

proton decoupling. The irradiation of ^{1}H's in a molecule for the purpose of collapsing the multiplets one would otherwise observe in a ^{13}C (or other nuclide's) NMR spectrum. Proton decoupling will also likely alter the signal intensities of the observed spins of other nuclides through the NOE. For ^{13}C, proton decoupling enhances the ^{13}C signal intensity.

pseudodiagonal. The line connecting the upper-right corner to the lower-left corner of a heteronuclear 2-D spectrum, especially a ^{1}H-^{13}C HMQC or HSQC 2-D spectrum.

pseudoquartet, Ψq. A quartet-like splitting pattern caused by the identical coupling of the resonance of the observed spin to three other spins not related by symmetry.

pseudotriplet, Ψt. A triplet-like splitting pattern caused by the identical coupling of the resonance of the observed spin to two other spins not related to each other by symmetry.

pulse. Syn. RF pulse. The abrupt turning on of a sinusoidal waveform with a specific phase for a specific duration, followed by the abrupt turning off of the sinusoidal waveform.

pulse calibration. The correlation of RF pulse duration (at a given transmitter power) to net magnetization tip angle.

pulsed field gradient, PFG. The transient application of an electric current through a coil of wire wound to induce a change in magnetic field that varies linearly with position along the x-, y-, or z-axis of the probe.

pulsed field gradient probe, PFG Probe. Syn. gradient probe. An NMR probe equipped with one or more gradient coils capable of altering linearly the strength of the applied field as a function of position.

pulse ringdown. The lingering effects of an RF pulse applied just before digitization of the NMR signal.

pulse roll-off. The diminution of tip angle that results from the accumulated error caused by the difference between the frequency of the applied RF pulse and the frequency of a given resonance.

pulse sequence. A series of timed delays, RF pulses, and gradient pulses that culminates in the detection of the NMR signal.

quartet. Four evenly spaced peaks in the frequency spectrum caused by the splitting of a single resonance by J-coupling to three identical spin-½ nuclei to give a multiplet with four peaks with relative intensity ratios of 1:3:3:1.

quintet. Five evenly spaced peaks in the frequency spectrum caused by the splitting of a single resonance by J-coupling to four identical spin-½ nuclei to give a multiplet with five peaks with relative intensity ratios of 1:4:6:4:1.

radio frequency, RF. Electromagnetic radiation with a frequency range from 3 kHz to 300 GHz.

rapid chemical exchange. A chemical exchange process that occurs so rapidly that two or more resonances coalesce into a single resonance.

read pulse. The RF pulse applied just before the digitization of the FID.

receiver coil. An inductor in a resistor-inductor-capacitor (RLC) circuit that is tuned to the Larmor frequency of the observed nuclide and is positioned in the probe so that it surrounds a portion of the sample.

reflected power. The portion of the power of an applied RF pulse that fails to be dissipated in the sample and instead returns through the cable connecting the probe to the rest of the instrument.

relaxation. The return of an ensemble of spins to the equilibrium distribution of spin state populations.

relaxation delay. The initial period of time in a pulse sequence devoted to allowing spins to return to equilibrium.

resolution enhancement. The application of an apodization function that emphasizes the later portions of the digitized FID (at the expense of the signal-to-noise ratio) that, upon Fourier transformation, will generate a spectrum with peaks whose widths are narrower than their natural line widths.

resonance. An NMR signal consisting of one or more relatively closely spaced peaks in the frequency spectrum that are all attributable to a unique atomic species in a molecule.

resonance broadening. The spreading out, in the frequency spectrum, of one or more peaks. Resonance broadening can either be homogeneous or inhomogeneous. An example of homogeneous resonance broadening is the broadening caused by a short T_2^*. An example of inhomogeneous resonance broadening is the broadening caused by the experiencing of a ensemble of molecular environments (that are not averaged on the NMR time scale).

RF. See radio frequency.

RF channel. The portion of the instrument devoted to generating RF with a specific frequency. There are four types of RF channels that may be found in an NMR instrument: a highband channel (for ^1H and ^{19}F, and maybe ^3H), a broadband channel, for all nuclides with Larmor frequencies at that of ^{31}P and lower, a lock channel (devoted exclusively ^2H), and a fullband channel (any nucleus). Most instruments have one of the first three channels listed above.

ring current. The circulation of charge in cyclic, conjugated aromatic systems that is induced by the applied magnetic field. The ring current will vary as a function of the orientation of the π-electron system to the axis of the applied magnetic field, with the maximum ring current occurring when the vector normal to the plane of the ring is parallel to the applied field axis.

roll-off. See pulse roll-off.

rotamer. Syn. rotational isomer. An isomer generated by rotation (usually 120°) about a chemical bond.

rotating frame. An alternate Cartesian coordinate system (x′, y′, z′) sharing its z-axis with that of the laboratory (stationary) frame of reference. The rotating frame of reference rotates at the Larmor frequency of the nuclide being observed.

rotational isomerism. Interconversion of rotational isomers or rotamers.

rotational Overhauser effect spectroscopy, ROESY. Syn. CAMELSPIN experiment. A 2-D NMR experiment similar to the 2-D NOESY experiment, except that the ROESY experiment employs a spin-lock using the B_1 field of the applied RF, thus skirting the problem of the cancellation of the NOE cross peak when correlation times become long enough to reduce the rate constant for the dipolar double-quantum spin flip.

Ruben-States-Haberkorn method. Syn. States method. A phase-cycling method for making the indirectly detected dimension (f_1) in a 2-D spectrum phase sensitive. Phase sensitivity is realized by varying the phase of one of the RF pulses in the pulse sequence by 90° for pairs of digitized FIDs obtained using the same t_1 time increment.

sample spinning. The rotation, using an air bearing, of the NMR tube/spinner assembly, used to average, on the NMR time scale, the strength of the applied magnetic field experienced by molecules in the sample solution. Sample spinning narrows the line widths of the peaks we observe and is almost exclusively employed in the collection of 1-D spectra.

saturation. The application of RF tuned to a specific transition between spin states for the purpose of equalizing the populations of the affected spin states.

scalar coupling. See J-coupling.

scan. A single execution of a pulse sequence ending in the digitization of a FID.

sensitivity. The ability to generate meaningful data per unit time.

septet. Seven evenly spaced peaks in the frequency spectrum caused by the splitting of a single resonance by J-coupling to six identical spin-½ nuclei to give a multiplet with seven peaks with relative intensity ratios of 1:6:15:20:15:6:1.

sextet. Six evenly spaced peaks in the frequency spectrum caused by the splitting of a single resonance by J-coupling to five identical spin-½ nuclei to give a multiplet with six peaks with relative intensity ratios of 1:5:10:10:5:1.

shielded group. A functional group with extra electron density.

shifted sine bell function. An apodization function with the amplitude of the sinusoidal pattern starting at a maximum and dropping to zero. The first quarter of a cosine waveform.

shifted squared sine bell function. An apodization function with the amplitude of a squared sinusoidal pattern starting at a maximum and dropping to zero. The first quarter of a squared cosine waveform.

shim (v). The variation of current in a number of coils of wire, each wrapped in such a way as to produce a different geometrical variation in the strength of the applied magnetic field, in order to make the magnetic field experienced by the portion of the sample residing in the detected region of the NMR probe as homogeneous as possible.

shim (n). One of a number of coils of wire surrounding the sample and probe wrapped so that a current passing through this coil induces a change in the strength of the applied magnetic field with a prescribed geometry.

shimming. The act of varying the currents in the shims to achieve a more homogeneous applied magnetic field. Shimming most often entails maximizing the level of the signal of the lock channel, as rendering the field more homogeneous reduces the solvent's ^2H line width, which, given that the area of the ^2H peak is constant, must necessarily increase the height of the ^2H solvent peak, i.e., the lock level.

shim set. A group of shims.

shot noise. Electrical noise resulting from the movement of individual charge quanta, like raindrops on a tin roof. With a low flux, individual drops are heard in a random pattern, but as the flux increases, the impact of the individual drop is lost in the continuum of many drop impacts per unit time.

signal. An electrical current containing information.

signal-to-noise ratio, S/N. The height of a real peak (measure from the top of the peak to the middle of the range of baseline noise) divided by the amplitude of the baseline noise over a statistically reasonable range.

single quantum spin flip rate constant, W_1. The kinetic rate constant controlling the change in the spin state of a single spin-½ spin from either the α to the β state or from the β to the α state.

singlet. A resonance that appears in the frequency spectrum as a single peak.

sinusoid. A sine wave.

soft atom. An atom with a low electronegativity, and often with a higher atomic number (Z). These atoms have electron clouds that are more easily distorted by external forces and, as such, they are often the source of electron density that shields unexpectedly (alters the chemical shifts to values that are anomalously upfield) nearby atoms.

spectral density function, $J(v)$, $J(\omega)$. A mathematical function that describes how energy is spread about as a function of frequency.

spectral window, SW. The range of frequencies spanned by a spectrum, whose location in the frequency spectrum is determined by both the dwell time and the frequency subtracted from the time domain analog signal prior to digitization.

spectrometer frequency, sfrq. The frequency of the RF applied to the sample for the observe channel, also the frequency of the rotating frame for the observe nuclide.

spin echo. Magnetization allowed to dephase in the **xy** plane will, following an appropriately phased 180° RF pulse, refocus to give a sharp spin echo. Sampling of the spin echo maxima for a series of refocusing events allows the determination of T_2, the spin-spin relaxation rate constant. Interestingly, the signal we collect in this experiment (see CPMG) is not technically a FID, because we perturb the magnetization periodically in between the collection of data points for a each single passage through the pulse sequence.

spin-lattice relaxation. Syn. T_1 relaxation. Relaxation involving the interaction of spins with the rest of the world (the lattice).

spin lock. The placement of magnetization into the **xy** plane, followed by the application of CW RF with a 90° phase shift. The spin lock can also be accomplished with a series of 180° pulses to generate repeated spin echoes. If the frequency of the spin lock RF is in the center of the spectral window and the 90° pulse width is around 30 μs, the spin lock gives optimal TOCSY mixing. If the frequency of the spin lock RF is well outside the spectral window and the 90° pulse width is around 90 μs, then the spin lock gives optimal ROESY mixing. Using a longer 90° pulse width helps minimize the development of TOCSY cross peaks in the ROESY spectrum.

spin-spin coupling. See J-coupling.

spin-spin relaxation. Syn. T_2 relaxation. Relaxation involving the interaction of two spins.

spin state. Syn. spin angular momentum quantum number. The projection of the magnetic moment of a spin onto the **z**-axis. The orientation of a component of the magnetic moment of a spin relative to the applied field axis (for a spin-½ nucleus, this can be +½ or −½).

spin system. A group of spins within a molecule that all couple to one or more other members of the same group.

spin-tickling experiment. The low power CW RF irradiation of one resonance to observe the effect this irradiation has on the resonances generated by other spins, which may be partial or complete multiplet collapse or a change in integrated intensity.

splat-90-splat. A method for destroying residual magnetization involving the applying a z-gradient pulse, applying a 90° pulse, and then another z-gradient pulse.

splitting. The coupling-induced division of the resonance from an NMR-active atom in a single atomic site in a molecule into two or more peaks, legs, or lines.

splitting pattern. The division by J-coupling of a resonance into a multiplet with a recognizable ratio of peak intensities and spacings.

static frame. Syn. laboratory frame. The frame of reference corresponding to the physical world in which the experiment is carried out.

steady-state scans. See dummy scans.

stray field. The magnetic field lines that extend beyond the physical dimensions of the NMR magnet's cryostat.

sweep width, SW (note that spectral window is not the same thing but is also denoted SW—a great source of confusion). The amount of the frequency spectrum spanned, which is controlled by the dwell time.

symmetry operation. A geometrical manipulation involving rotation, inversion, reflection, or some combination thereof.

T_1 relaxation. See spin-lattice relaxation.

t_1 ridge. Syn. t_1 noise. A stripe of noise often observed in a 2-D spectrum for a given chemical shift value of the f_2 chemical shift axis which runs parallel to the f_1 chemical shift axis. A t_1 ridge occurs with an f_2 chemical shift corresponding to one or more of the most intense peaks in the spectrum of the 1-D spectrum of the f_2 nuclide.

t_1 time. The first time delay in a pulse sequence used to establish a time domain that will subsequently be converted to the frequency domain f_1.

t_1 time increment. A discreet variation (normally an increase by a fixed amount) in the t_1 time delay in the NMR pulse sequence.

T_2 relaxation. See spin-spin relaxation.

t_2 time increment. The second time delay in a pulse sequence used to establish a time domain that will subsequently be converted to the frequency domain f_2.

$T_1\rho$ relaxation. The diminution of the net magnetization vector in the rotating frame of reference as the net magnetization vector is subjected to a **B_1** spin lock.

thermal energy, kT. The random energy present in all systems that varies in proportion to temperature.

time domain. The range of time delays spanned by a variable delay (t_1 or t_2) in a pulse sequence.

time-proportional phase incrementation method, TPPI method. A method for imparting phase sensitivity into either the indirectly (f_1) detected dimension of a 2-D experiment or the directly detected dimension of a 1-D experiment (or the f_2 dimension of a 2-D experiment). The TPPI method involves sampling points at half the dwell time prescribed for a given sweep width (for the directly detected dimension), or using a t_1 time increment that is half of that prescribed for the f_1 sweep width.

transition. The change in the spin state of one or more NMR-active nuclei.

transmitter frequency. Syn. carrier frequency, NMR frequency, on-resonance frequency. The rate at which the maxima of the sinusoidal wave of the RF generated by the observe nuclide's RF channel occur.

transmitter glitch. A small spectral artifact often observed in the very center of the spectral window which is caused by a small amount of the RF generated in the console getting through to the receiver.

triplet. Three evenly spaced peaks in the frequency spectrum caused by the splitting of a single resonance by J-coupling to two identical spin-½ nuclei to give a multiplet with three peaks with relative intensity ratios of 1:2:1, or to one spin-1 nucleus to give a multiplet with three peaks with relative intensity ratios of 1:1:1.

truncation error. Regularly spaced bumps or ripples observed on either side of the narrow peaks in a frequency spectrum which are caused by the failure of the digitized FID's amplitude to fall to the level of the background noise before the end of the digitization of the FID.

tuning. See probe tuning.

upfield. The right side of the chemical shift scale, corresponding to lower frequency, and the lower resonant frequency in turn indicates additional electron density. A resonance is upfield if it is located on the right side of the spectrum or if it is observed to appear to the right of its expected value. By convention, lower frequencies (also lower chemical shifts values in ppm) appear on the right side of the spectrum and higher frequencies appear on the left side of the spectrum. Because the first generation of NMR instruments (CW instruments) operated with a constant RF frequency and a variable applied magnetic field strength, higher fields (upfield) were required to make lower frequency resonances come into resonance. Therefore, when an NMR spectrum is plotted as a function of field strength, the lower values appear on the left and higher values appear on the right.

upper bore tube. Syn. upper magnet bore assembly. A second metal tube (plus air lines, and possibly spin sensing components and PFG wiring), residing inside the upper portion of the magnet bore tube, through which the spinner/tube assembly passes via pneumatics en route between the top of the magnet and its operating position just above the probe inside the magnet.

valence shell electron pair repulsion, VSEPR. A mental tool used to predict deviations from standard hybridized bond angles based on how much of the surface area of an atom a given electron pair in its outer shell occupies.

vicinal. Two atoms or functional groups are vicinal if they are separated by three bonds, meaning that they are bound to two atoms sharing a common bond. Vicinal atoms or functional groups discussed in the context of NMR spectroscopy are usually but not always the same species.

viscosity-induced resonance broadening. Syn. viscosity broadening. The increase in the line width of peaks in a spectrum caused by the decrease in the T_2 relaxation time that results from a slowing of the molecular tumbling rate. Saturated solutions and solutions at a temperature just above their freezing point often show this broadening behavior.

VSEPR. See valence shell electron pair repulsion.

W-coupling. A four-bond J-coupling occurring when two NMR-active spins are separated by four bonds held in a static, planar conformation that forms the letter "W."

window function. See apodization function.

word. A portion of computer memory devoted to the storage of one number. A word will normally consist of four or eight bytes (one byte is eight bits, or eight 1's or 0's).

Zeeman effect. The linear divergence of the energies of the allowed spin states of an NMR-active nucleus as a function of applied magnetic field strength.

zero filling. Addition of 0's at the tail end of time domain data for the purpose of improving the digital resolution in the frequency spectrum following Fourier transformation. Zero filling increases the number of points per unit frequency (a good thing).

zero quantum spin flip rate constant, W_0. The kinetic rate constant controlling the simultaneous change in both spin states for a two-spin system where one of the spin-½ spins goes from the α to the β state while the other spin-½ spin goes from the β to the α state.

zero-order phase correction. Syn. zero-order phasing. The variation in the proportion of amplitude data taken from two orthogonal arrays (or matrices) that is applied evenly to each point in the spectrum.

z^4 hump. An asymmetric peak shape, usually consisting of a low-intensity peak offset but overlapping with the main peak (a shoulder), that results from a poorly set z^4 RT shim current. The presence of z^4 humps are particularly troublesome when one wishes to observe a weak peak whose chemical shift is near that of an intense peak. ^1H Shifts near 4.65 ppm when the solvent is 10% D_2O/90% H_2O suffer particularly from the presence of a z^4 hump.

Glossary of Acronyms and Abbreviations

AC. Alternating current.

A/D. Analog-to-digital converter.

APT. Attached proton test.

AQ. Acquisition time (Bruker).

at. Acquisition time (Varian).

COSY. Correlation spectroscopy.

CPMG. Carr-Purcell-Meiboom-Gill.

CSA. Chemical shift anisotropy.

CW. Continuous wave.

d1. Relaxation delay (Varian).

DC. Direct current.

DEPT. Distortionless enhancement through polarization transfer.

DQFCOSY. Double quantum filtered correlation spectroscopy.

EDG. Electron donating group.

EWG. Electron-withdrawing group.

EXSY. Exchange spectroscopy experiment.

FID. Free induction decay.

FT. Fourier transform.

gCOSY. Gradient-selected absolute value correlation spectroscopy.

gHMBC. Gradient-selected heteronuclear multiple bond correlation.

gHSQC. Gradient-selected heteronuclear single-quantum correlation.

HETCOR. Heteronuclear correlation.

HMBC. Heteronuclear multiple bond correlation.

HMQC. Heteronuclear multiple quantum correlation.

HSQC. Heteronuclear single quantum correlation.

INADEQUATE. Incredible natural abundance double quantum transfer.

NMR. Nuclear magnetic resonance.

NOE. Nuclear Overhauser effect.

NOESY. Nuclear Overhauser effect spectroscopy.

o1. Frequency offset, channel 1 (Bruker).

o2. Frequency offset, channel 2 (Bruker).

PFG. Pulsed field gradient.

ppm. Parts per million.

RD. Relaxation delay.

RF.. Radio frequency.

RLC. Resistor-inductor-capacitor.

ROESY. Rotational Overhauser effect spectroscopy.

RT. Room temperature.

sfrq. Spectrometer frequency.

S/N. Signal-to-noise ratio.

SW. Spectral window or sweep width.

TFA. Trifluoroacetic acid.

TFE. Trifluoroethanol.

TMS. Tetramethylsilane.

TOCSY. Total correlation spectroscopy.

tof. Transmitter frequency offset (Varian).

TPPI. Time-proportional phase incrementation.

VT. Variable temperature.

Glossary of Symbols

B_0. A vector denoting the applied magnetic field strength and direction.

B_1. A vector denoting the strength and direction of the magnetic field component of an RF pulse applied at the same frequency of the observe nucleus.

c_i. Sample concentration.

δ. Chemical shift, a unitless quantity given in parts-per-million.

dB. Decibel.

$\Delta\delta$. The difference between two chemical shifts.

ΔE. The energy gap (difference) between two spin states.

η. NOE enhancement.

E_a. Activation energy.

f_1, F_1. The frequency domain resulting from the Fourier transformation of the t_1 time domain.

f_2, F_2. The frequency domain resulting from the Fourier transformation of the t_2 time domain.

γ. Gyromagnetic ratio.

h. Planck's constant.

$\hbar$ Planck's constant divided by 2π.

Hz. Hertz.

I. Electrical current.

$J(v)$, $J(\omega)$. Spectral density function.

k. Boltzmann constant.

k. 1024.

kT. Thermal energy.

K. Exchange rate.

μ. nuclear magnetic moment.

M. Net magnetization vector.

v. Frequency, in events per second.

n. Number of scans.

N. Number of excess spins in one spin state compared to another.

N. Noise.

ω. Angular frequency, in 2π radians per second.

Φ. Dihedral angle.

P. Power.

r. Internuclear distance.

R. Resistance.

S_i. Signal from component i.

τ_c. Correlation time.

Θ. Bond angle.

t. Time allowed for relaxation.

T. Tesla.

t_1. Evolution time in a 2-D pulse sequence.

T_1. Spin-lattice relaxation time constant.

t_2. Detection period in a 2-D pulse sequence.

T_2. Spin-spin relaxation time constant.

V. Voltage.

W. Dipolar relaxation rate constant.

Index

DATE DUE

Demco, Inc. 38-293